By-catch Reduction in the World's Fisheries

Reviews: Methods and Technologies in Fish Biology and Fisheries

VOLUME 7

Series editor:

Jennifer L. Nielsen

U.S. Geological Survey,
Biological Resources Division,
Anchorage, Alaska

By-catch Reduction in the World's Fisheries

edited by

Steven J. Kennelly

Cronulla Fisheries Research Centre of Excellence,
Cronulla, Australia

Springer

A C.I.P. Catalogue record for this book is available from the Library of Congress.

ISBN 978-1-4020-6077-9 (HB)
ISBN 978-1-4020-6078-6 (e-book)

Published by Springer,
P.O. Box 17, 3300 AA Dordrecht, The Netherlands.

www.springer.com

Printed on acid-free paper

Preface

Humans have been harvesting fish for at least 90,000 years using technologies that have developed from simple harpoons through to huge factory trawlers. For most of this history, the driving force behind developments in fishing technology has been to develop methods that catch ever-greater quantities of fish of an ever-increasing diversity. This theme changed dramatically during the last few decades in the light of one of the world's most serious and controversial fishing issues – the waste associated with the incidental capture, mortality and discarding of unwanted by-catch. In response to these by-catch issues, the field of fishing technology altered its focus to one where fishing techniques are developed to be selective in what they catch, so that targeted species (and targeted sizes of species) are caught whilst unwanted by-catches are not. In more recent times, this field has expanded even further, to address problems associated with fishing gears (especially dredges and trawls) impacting on the benthos and seabed ecosystems.

This focus on by-catch reduction and ecosystem-effects of fishing has resulted in many successful changes in fishing practices which are estimated to be conserving millions of fish and other organisms throughout the world. These successes have occurred in many types of fisheries and have improved many of the world's most non-selective and problematic fishing techniques. This book provides, in one volume, a timely aggregation of many of these developments in this relatively new field. Incorporating a plethora of case-studies, this book summarises, analyses and provides future directions for most aspects of this field including: the methodologies used; the key locations where the work has been done; the various fishing methods examined (particularly the most problematic methods); and the all-important methods used to ensure the uptake of newly developed techniques by fishers.

The publication of this book marks a very successful period of achievement by the world's by-catch reduction specialists and gear technologists in ameliorating some of the most critical problems facing the world's fisheries. It also provides templates for how to continue this work and how to broaden the lessons learned to address other emerging fisheries issues.

STEVEN J KENNELLY
December 2006

The publication of this book would not have been possible without the professionalism and considerable skills of my Executive Officer, Ms Tracey McVea

– SJK

Contents

Contributors

Editor

Steven J. Kennelly

NSW Department of Primary Industries, Cronulla Fisheries Research Centre of Excellence, P.O. Box 21, Cronulla, NSW, 2230, Australia, email: steve.kennelly@dpi.nsw.gov.au

Chapter authors

Matt K. Broadhurst

NSW Department of Primary Industries, Conservation Technology Unit, c/- National Marine Science Centre, PO Box J321, Coffs Harbour, NSW, 2450, Australia, email: matt.broadhurst@dpi.nsw.gov.au

Charles Gray

NSW Department of Primary Industries, Cronulla Fisheries Research Centre of Excellence, P.O. Box 21, Cronulla, NSW, 2230, Australia, email: charles.gray@dpi.nsw.gov.au

John W. Watson

NOAA Fisheries Service, Southeast Fisheries Science Center, Mississippi Laboratories, P.0. Drawer 1207, Pascagoula, MS, 39563, USA, email: john.watson@noaa.gov

Petri Suuronen

Finnish Game and Fisheries Research Institute, Viikinkaari 4, P.O. Box 2, FIN-00791 Helsinki, Finland, email: petri.suuronen@rktl.fi

Francesc Sarda

Instituto de Ciencias del Mar (CMIMA-CSIC), Department de Recursos Marins Renovables, Passeig Maritim de la Barceloneta 37-49, 08003 Barcelona, Spain, email: siscu@icm.csic.es

Miguel B. Gaspar

Instituton Nacional de Investigação agrária e das Pescas (INIAP/IPIMAR), Centro Regional de Investigação Pesqueira do Sul (CRIPSul), Av. 5 de Outubro s/n, 8700-305 Olhão, Portugal, email: mbgaspar@cripsul.ipimar.pt

Luis M. Chícharo

Universidade do Algarve, Faculdade de Ciências do Mar e Ambiente (FCMA), Centro de Ciências do Mar do Algarve (CCMAR), Campus de Gambelas, 8005-139 Faro, Portugal, email: lchichar@ualg.pt

John Stewart

NSW Department of Primary Industries, Cronulla Fisheries Research Centre of Excellence, P.O. Box 21, Cronulla, NSW, 2230, Australia, email: john.stewart@dpi.nsw.gov.au

Steven J. Cooke

Fish Ecology and Conservation Physiology Laboratory, Department of Biology, Carleton University, 1125 Colonel By Drive, Ottawa, Ontario, K1S 5B6, Canada, email: steven_cooke@carleton.ca

Gene R. Wilde

Wildlife and Fisheries Management Institute, Department of Natural Resources Management, Texas Tech University, Box 42125, Lubbock, Texas, 79409, USA, email: gene.wilde@ttu.edu

Pingguo He

Institute for the Study of Earth, Oceans and Space Ocean Process Analysis Laboratory, New Hampshire Sea Grant, University of New Hampshire, 142 Morse Hall, Durham, NH, 03824. USA, email: Pingguo.He@unh.edu

Martin A. Hall

Inter-American Tropical Tuna Commission, 8604 La Jolla Shores Dr., La Jolla, California, 92037, USA, email: mhall@iattc.org

Hideki Nakano

Japan Fisheries Agency, Research and Technology Guidance Division, 1-2-1, Kasumigaseki, Chiyoda-ku, Tokyo, 100-8907, Japan, email: hnakano@affrc.go.jp

Shelley Clarke

Imperial College London, Division of Biology, Silwood Park Campus, Buckhurst Road, Ascot, Berkshire SL5 7PY, United Kingdom

Simon Thomas

Southern Seabird Solutions Trust, 194 Reikorangi Road, Waikanae, New Zealand, email: southernseabirds@doc.govt.nz

Janice Molloy

Southern Seabird Solutions Trust, 194 Reikorangi Road, Waikanae, New Zealand, email: southernseabirds@doc.govt.nz

S. Hoyt Peckham

Department of Ecology and Evolutionary Biology, University of California at Santa Cruz, 100 Shaffer Road, Santa Cruz, California, 95060, USA. email: hoyt@biology.ucsc.edu

Johath Laudino Santillán

Grupo Tortuguero, Cuauhtemoc 155, Pueblo Nuevo, La Paz, Baja California Sur, CP 23060, México, email: johath@grupotortuguero.org

Wallace J. Nichols

The Ocean Conservancy, POB 324, Davenport, California, 95017, USA, email: jnichols@oceanconservancy.org

Eric Gilman

Fisheries By-catch Program, Blue Ocean Institute, 2718 Napuaa Place, Honolulu, Hawaii, 96822, USA, [also affiliated with the University of Tasmania School of Geography & Environmental Studies, Australia], email: egilman@blueocean.org

Sean Martin

Hawaii Longline Association, 45-519 Mokulele Dr., Kaneohe, Hawaii 96744, USA, email: sean@pop-hawaii.com

Jim Cook

Hawaii Longline Association, 45-519 Mokulele Dr., Kaneohe, Hawaii 96744, USA, email: jim@pop-hawaii.com

Stephen J. Hall

The Worldfish Centre, P.O Box 500, G.P.O. Box 10670, Penang, Malaysia, email: s.hall@cgiar.org

John P. Croxall

British Antarctic Survey, Natural Environment Research Council, Madingley Road, Cambridge CB3 0ET, United Kingdom; Present Address: BirdLife International, Wellbrook Court, Girton, Cambridge CB3 0NA, United Kingdom, email: john.croxall@birdlife.org

Kim Rivera

NOAA Fisheries Service, Alaska Region, Juneau, 99802, USA, email: kim.rivera@noaa.gov

Carlos A. Moreno

Instituto de Ecologia y Evolucion, Campus Isla Teja, Universidad Austral de Chile, Valdivia, Chile, email: cmoreno@uach.cl

1 Strategies for Improving the Selectivity of Fishing Gears

MATT K. BROADHURST, STEVEN J. KENNELLY AND CHARLES GRAY

1.1 The Issue of By-catch

From the earliest evidence of fishing more than 90 000 years ago (Yellen et al. 1995) to the present day, humans have exponentially advanced their harvesting methods. The clear focus of these developments has been to maximise the catches of an ever-increasing diversity of targeted species, with little or no regard for the incidental catches (termed 'by-catch', sensu Saila 1983). A progression from simple harpoons, hooks and traps deployed from the shore, through nets set from boats, to the industrial factory trawlers of developed countries has culminated in technology which, in many cases, far exceeds the sustainability of local resources. This excess was evident at the end of the 20th century by the collapse of many commercially-important stocks, a plateau in the world's total landed wild catch (at less than 100 million t) and the volumes of by-catch discarded in pursuit of targeted catches (Alverson et al. 1994; Kelleher 2005).

While recognition of the potentially negative impacts of unchecked fishing technology date back to the 14th century (Dyson 1977), it is only during the last few decades that coordinated attempts have been directed towards improving the selectivity of fishing gears and reducing unwanted fishing mortalities. Relevant reviews of the published literature suggest that nearly all fishing gears and methods have received at least some attention (e.g., gillnets – Hamley 1975; longlines – Løkkeborg and Bjordal 1992; traps – Mahon and Hunte 2001), although the majority of effort has been directed towards benthic trawl fisheries and especially those targeting shrimp (Andrew and Pepperell 1992; Broadhurst 2000; Broadhurst et al. 2006). This has occurred in response to the disproportional ratio of retained-to-discarded catches and the amount of unwanted catch discarded each year by shrimp trawlers; recently estimated at more than 1.8 million tonnes per year (Kelleher 2005). While the absolute volume of by-catch associated with shrimp trawling clearly makes it one of the most the most problematic

S. J. Kennelly (ed.), By-catch Reduction in the World's Fisheries, 1–21.

fishing methods, many other gears including fish trawls, seines, gillnets, traps and longlines have, in recent times, been identified as having significant selectivity issues and have consequently been associated with prolonged calls for improvements coming from a variety of environmental groups, recreational fishers, interacting commercial fisheries and the general public.

1.2 Solving By-catch Problems

During the past two decades, problems surrounding the issue of by-catch have shifted the focus of fishing gear technology from catching as much of the target species as possible (with little regard for collateral impacts) to improving selectivity, both in terms of the species targeted and their desired sizes (Valdemarsen and Suuronen 2003). In many cases, the successful development and adoption of solutions to improve selection in problematic gears can be summarised in a simple framework (see also Kennelly and Broadhurst 1996) which involves industry and researchers each applying their respective areas of expertise to the particular problem. This framework comprises five key steps: (i) quantifying by-catches (mostly via observer programs), (ii) identifying the main by-catch species and their sizes of concern, (iii) developing alterations to existing fishing gears and practices that minimise the mortality of these species, (iv) testing these alternatives in appropriately-designed field experiments and (v) gaining acceptance of the new technology throughout the particular fishery and interested stakeholders.

The protocol for completing the framework is quite straightforward and has been described with numerous examples by Kennelly and Broadhurst (1996) and Kennelly (1997). The crucial and most difficult step (i.e., step (iii) above) is the actual development of appropriate solutions that improve the selectivity of existing fishing gears for the targeted catch and reduce the mortality of unwanted by-catch (Broadhurst et al. 2006). Depending on the type of gear and its particular problems, solutions may involve simple adjustments to operational procedures and/or existing components of the gear, like changing the size and/or shape of meshes or hooks. Alternatively, for many towed gears, more complicated modifications that include physical by-catch reduction devices (BRDs) may need to be invented or modified from other fisheries (Valdemarsen and Suuronen 2003). Owing to their relative complexity, these types of modifications frequently require detailed adjustment and reassessment to exclude specific sizes of individuals or species, yet maintain targeted catches.

While the above-mentioned framework summarises several successful attempts at addressing the problems of by-catch in different fisheries throughout the world, in many cases the established protocols for improving inherently problematic gears has restricted fishing technologists in terms of working towards the ultimate goal of perfect selectivity. A reason for this is that to ensure the industry adoption and acceptance of modified designs that reduce by-catch (i.e., step (v) above), nearly all researchers have aimed to achieve 100% retention of the targeted species (during step (iii)). Theoretically, it should be possible to improve the selectivity of most fishing gears dramatically, provided some concomitant sacrifice in their overall efficiency is permitted. The issue would then become what is an acceptable loss of the targeted catch in order to improve selectivity and reduce by-catch. An extreme solution for achieving 'perfect selectivity' may be to re-order the above logic and, using traditional gears and established by-catch reduction methods, approach a 100% exclusion rate of unwanted catch at any cost to the desired catch. This approach could be appropriate in tightly-regulated fisheries where there is imminent threat of closure due to discarding. Reductions in gear efficiency could also be offset via some compensatory increases in the value of the targeted catch through 'eco-labelling'. These sorts of strategies would not be feasible, however, in the vast majority of countries and especially those where artisanal fisheries represent the main source of income for communities. For these fisheries, by-catch reduction clearly needs to be maximised with minimal impact on the efficiency of the gear in catching the targets.

1.3 Maximising Gear Development within Existing By-catch Reduction Frameworks

To approach maximum by-catch reduction with no loss of the targeted catch during step (iii) of the framework described above, there needs to be a general estimate of what is achievable for particular gears. As a starting point, this requires an assessment of the limits of established modifications for improving selectivity. For many conventional towed gears, different sizes and/or shapes of mesh are among the simplest alterations and their utility is often (or at least should be) determined first. Under the framework proposed by Broadhurst (2000), this involves testing beyond what might intuitively be appropriate, so that the limits of a particular range of mesh sizes or shapes can be quantified and defined. If the solution to reducing particular by-catch species of concern is not apparent within the boundaries of the simple alterations tested, then more complex modifications (including physical BRDs) will warrant examination. Specific designs of BRDs should also be tested to define their limits. For example, if mechanical-sorting grids are required to exclude organisms larger than the targeted species, then a range of configurations that include very narrow and wide bar spacings, and small and larger profiles or angles of orientation should be examined (e.g., Broadhurst et al. 2004b). Similarly, because factors like relative water flow can strongly influence the performance of BRDs that operate by exploiting differences in the behaviour of species, these sorts of modifications need to be tested at different positions throughout the gear (e.g., Graham and Kynoch 2001; Broadhurst et al. 2002). Coherent hypotheses encompassing the full range of key factors influencing the performance of modifications will facilitate the accurate assessment of the extent to which selectivity might be improved. Quantifying basic, gear-related parameters and their boundaries in terms of reducing by-catch while still maintaining the target catches can save considerable time and effort towards the longer-term development of more selective gears.

In addition to identifying what might be achievable using established technologies, we propose that it is necessary to also consider alternative methods which are not part of the existing conventional gear. That is, one should examine completely different methods for catching the target species and determine if these have particular attributes that might be used to modify the gear of interest. In the best case, a consideration of alternative methods might provide a completely different fishing gear that could be simply substituted for the problematic gear. But, even in the worst case scenario, comparing alternative methods could highlight specific selection mechanisms that provide new directions for modifying a problematic gear.

This latter, lateral approach is not commonly adopted in studies to improve gear selectivity, although there are numerous examples where key mechanisms associated with one gear have been used to improve the efficiencies of others. One high-profile example involves the use of fire as a visual stimulus (i.e., light) throughout prehistory to augment the catches of simple hooks, spears and clubs (Yami 1976). The benefits of fishing with light were subsequently realised in a plethora of artisanal and industrial fisheries using both static (e.g., gillnets and traps) and active gears (e.g., purse seines) (Yami 1976; Sainsbury 1996). A more recent example is the use of baits to stimulate chemoreception in fish towards traps and longlines being extended to other static gears like gillnets (Engås et al. 2000).

Although not aimed at improving the selectivity or efficiency of particular gears, many other studies have compared alternative and/or competing fishing methods, including longlines versus gillnets (e.g., Santos et al. 2002; Stergiou et al. 2002; Erzini et al. 2003), longlines versus trawls (e.g., Hovgård and Riget 1992; Otway et al. 1996; Halliday 2002), longlines versus trawls versus gillnets (e.g., Huse et al. 1999; 2000), gillnets versus trammelnets (e.g., Matsuoka et al. 1990; Acosta and Appeldoorn 1995), gillnets versus electro fishing (e.g., Colvin 2002), gillnets versus trap nets (e.g., Hanchin et al. 2002) and gillnets versus trammelnets versus seines

(Stergiou et al. 1996). In most cases, these comparisons were done to reduce sampling bias and improve resource estimates. An indirect benefit, however, is some information on the relative selection between different gears and methods, and the general conclusion that static gears typically are more size- and species-selective than towed gears (e.g., Hovgård and Riget 1992; Løkkeborg and Bjordal 1992; Huse et al. 1999; Stergiou et al. 2002). Despite these, and other known differences among fishing methods and gears, few studies have attempted to use this sort of information to resolve by-catch issues. The potential benefits of such a lateral approach towards improving the selectivity of problematic gears are explored in the following experimental case study comparing active and static artisanal fishing gears for penaeid prawns in New South Wales (NSW) Australia.

1.4 A Case Study of Gear-Specific Selection for Penaeid Prawns in Estuaries in NSW, Australia

1.4.1 Introduction

Commercial fisheries for penaeid prawns occur throughout rivers and coastal lagoons in NSW, Australia and catches include three species, although school prawns, *Metapenaeus macleayi*, account for more than 90% of the annual production (approximately 925 t). These penaeids are targeted using several types of small-scale fishing gears that include trawls (Fig. 1.1A), seines (Fig. 1.2A) and trap nets (Fig. 1.3A).

Trawls and seines are active gears, designed to direct organisms along their wings, through a main body and into a collection bag (termed the 'codend'), where most of the size selection is believed to occur (Fig. 1.1 and 1.2). Prawn trawls used in NSW estuaries have a head line length less than 11 m and are dragged along in single or multi-rigs behind small vessels (Fig. 1.1C) at approximately 1.2 ms^{-1}. Seines have comparatively longer wings (head line lengths of at least 20 m) than trawls and are set using anchors, buoys and 100-m ropes in a semi-circular configuration around the area to be fished (Fig. 1.2). Immediately after setting, the wings are hauled together and the seine retrieved at a stationary vessel (Fig. 1.2B). In contrast to trawls and seines, trap nets are static gears and catch prawns by exploiting their migratory behaviour within estuaries, mostly at night and between the last and first quarter phases of the moon. Trap nets comprise a long (130 m) wall of mesh (like the wing of a trawl or seine) secured between a vertical stanchion located near the shore and the horizontal gunwale of a dory anchored on a lake (Fig. 1.3). The prevailing current

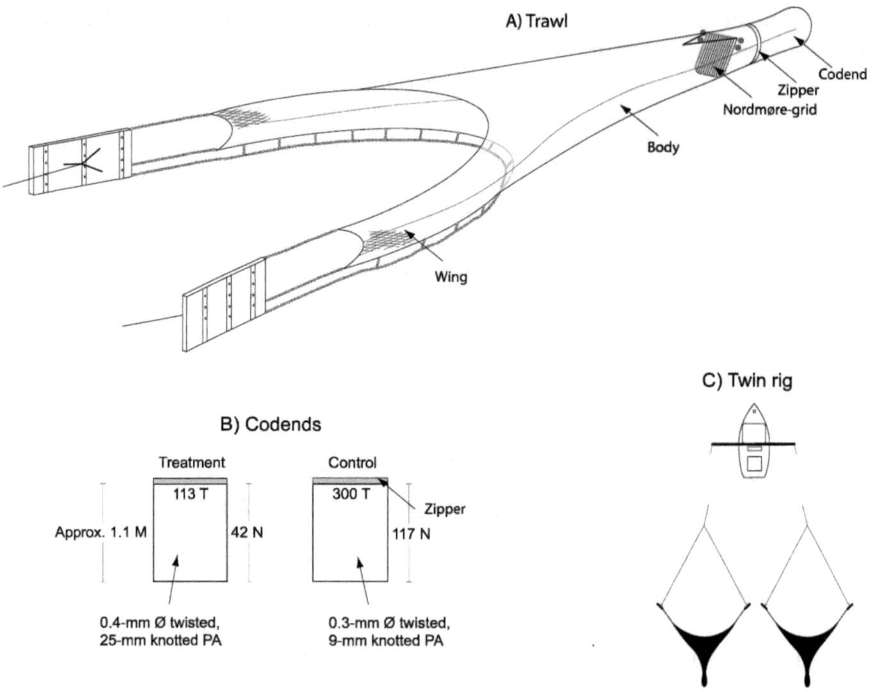

Fig. 1.1. The A) trawl, B) treatment and control codends and C) towing arrangement used in the study. T transversals, N normals, M meters, ø diameter, PA polyamide

causes the netting to assume a parabolic shape, effectively trapping moving prawns and directing them along the wall of netting towards the horizontally-orientated bag (termed 'bunt') at the dory. Fishers facilitate this movement of catch by regularly lifting and hauling sections of the trap net over a dory, so that it passes underneath the trap net and the catch is rolled towards the bunt (Fig. 3B).

All trawls, seines and trap nets used to catch prawns in NSW are managed by a range of gear-specific regulations that include minimum legal-mesh openings throughout of 40, 30 and 25 mm, respectively. Fishers generally target prawns larger than 15 mm carapace length (CL), but most of the gears retain by-catch that comprises at least some proportion of small, unwanted conspecifics. The selection of these individuals appears to be at least partially gear dependant, with previous studies indicating that, despite having the smallest size of mesh (i.e., 25 mm), trap nets have low by-catches and select relatively large prawns across a narrow range of sizes (Broadhurst et al. 2004a; 2004c). Our aims in this case study were: (i) to test the hypothesis of gear-specific selection by comparing the selectivity

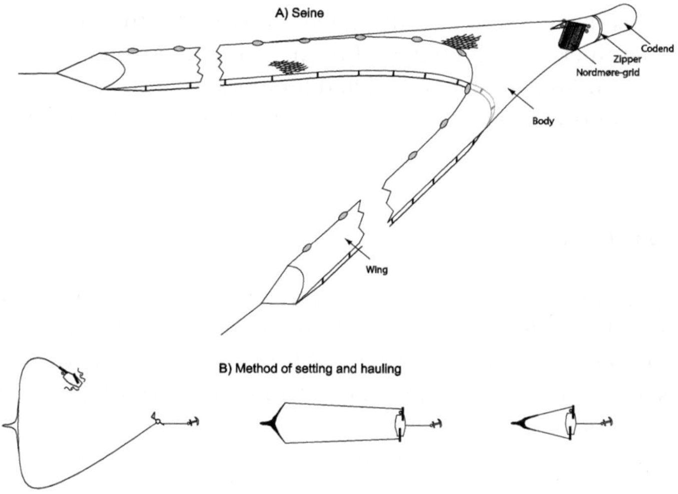

Fig. 1.2. The A) seine and B) method of setting and hauling used in the study

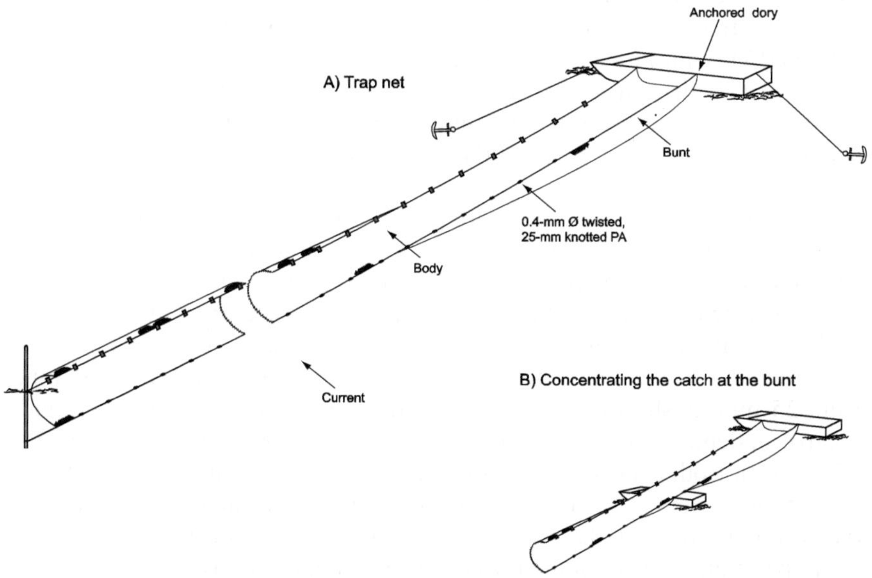

Fig. 1.3. The A) trap net and B) method of concentrating the catch at the bunt. ø, diameter; PA, polyamide

of a trawl, seine and trap net, all rigged with exactly the same size, hanging ratio, twine thickness and material of mesh in the codend or bunt (i.e., collection bag) across similar spatial and temporal scales; and (ii) determine if this sort of information can be used to further the development of selective gears.

1.4.2 Methods

This experiment was done on commercial prawn-trawl grounds in Lake Wooloweyah (29°26'S, 153°22'E) during two weeks in August 2003 using chartered commercial prawn fishers. All fishing was done over a combination of sandy and mud bottoms in depths ranging from 1 to 3 m and within an area of approximately 5 hectares.

Three commercial fishing gears were used in the study: (i) a Florida Flyer trawl (7.32-m headline) made from 40-mm knotted mesh (1.2-mm diameter (ø), 3-strand twisted polyethylene (PE) twine) throughout the wings and body (1N3B taper) and rigged as part of a twin gear configuration (the 'twin' trawl was not used in the study) (Fig. 1.1); (ii) a seine (headline length of 20 m) made from 30-mm knotted mesh (the same twine as above) through the wings and body (same taper as above) (Fig. 1.2); and (iii) a trap net (headline length and stretched depth of 130 and 6 m, respectively) made from 25-mm knotted mesh (0.4 mm ø, 3-strand twisted polyamide (PA) twine) hung at a ratio (*E*) of 0.5 throughout. An additional trap net (130 x 6 m) was constructed from 9.5-mm PA netting (0.7-mm ø, braided twine) and used as the control for the commercial trap net design above (see Broadhurst et al. 2004c, for a detailed description of trap net designs).

Two identical plastic Nordmøre-grids (600 x 400 mm) were installed into the aft bodies of the trawl and seine without guiding panels (see Broadhurst and Kennelly 1996, for details on construction). Zippers (1.45 m in length) were attached immediately posterior to the Nordmøre-grids to facilitate changing codends (Fig. 1A and 2A). Two codends were constructed for use with the trawl and seine: a treatment codend made from 25-mm mesh (identical to that used in the commercial trap net) and a control codend made from 9-mm mesh (0.3 mm ø, 3 strand twisted PA) (Fig. 1.1B). Both codends had a stretched length of 1.05 m and were attached to 1.45-m zippers at a hanging ratio of 0.5 (i.e., the same hanging ratio as that used throughout the trap nets). Using the zippers, the two codends could be interchanged on the trawl and seine net bodies (Fig. 1A and 2A).

On separate days or nights, each of the three treatment gears described above (i.e., the trawl and seine with the 25-mm codend attached and the 25-mm trap net) were alternately fished with their respective control gears.

The trawl and seine were hauled at commercial speeds of 1.2 and 0.13 ms^{-1}, respectively. We attempted four replicate, alternate 14-minute randomly-located hauls or deployments of each treatment gear and their respective control per day or night. A total of 8 balanced replicates were completed over two days for each of the trawl and seine (between 08:00 and 15:00) and over three nights (between 18:30 and 24:00) for the trap net. With the exception of the common soak time (14 minutes), no attempt was made to standardise effort between the gears.

The following categories of data were collected from each replicate deployment: the weight of total school prawns and a subsample (at least 250 prawns from each codend or bunt) of their lengths (to the nearest 1 mm CL); the number of total school prawns and the number and weight of retained school prawns (> 15 mm CL – estimated from the measured subsample); the weights of total by-catch (comprising discarded school prawns and fish) and fish by-catch; and the numbers of all fish.

Non-metric multivariate analyses were used to test the hypothesis of there being no differences in the structures of catches between the three treatment gears. Counts for all species were $log(x+1)$ transformed (to enhance the contributions of species caught in low abundances) and used to develop similarity matrices based on the Bray-Curtis similarity measure. Multidimensional relationships among ranks of the similarities from individual deployments of each of the three treatment gears were displayed graphically in a multidimensional scaling (MDS) ordination (Shepard 1962; Clarke 1993). One-way analyses of similarity (ANOSIM) were used to test for differences in catch assemblages between the three gears over their 8 replicate deployments. Similarity percentage (SIMPER) analyses were used to identify those species responsible for discrimination between the treatment gears.

One-factor analyses of variance (ANOVA) was used to test the hypothesis of no differences in the ratios of catches of retained school prawns to key by-catches. Prior to analyses, data were $log(x+1)$ transformed (to account for multiplicatively) and tested for heterocedasticity using Cochran's test. Significant F-ratios were investigated using Student-Newman-Keuls (SNK) multiple comparisons.

Size frequencies of school prawns were combined across all tows for each of the three treatment gears and their controls. Because previous studies have demonstrated sigmoid selection for all three gears (Broadhurst et al. 2004a, 2004c), parametric curves (logistic and Richards) were fitted to these data using maximum likelihood and REP corrected for overdispersion arising from between-haul variation (Millar et al. 2004). These fits used the estimated-split SELECT model for trouser trawls (Millar and Walsh 1992) and were implemented using a free R function available from *www.stat.auckland.ac.nz/~millar/selectware/code.html.* Model fits were assessed by likelihood ratio tests and comparing deviance residuals. Pairwise bivariate Wald statistics (Kotz et al. 1982) were calculated using the estimated parameter vectors of appropriate models to test for differences between the selectivities of the treatment gears.

1.4.3 Results

More than 23 species were recorded in the treatment gears, although 98% of catches comprised 6 species, all smaller than approximately 150 mm total length: school prawns (86.9%), southern herring, *Herklotsichthys castelnaui* (6.4%), Ramsey's perchlet, *Ambassis marianus*, (2.4%) silver biddy, *Gerres subfaciatus* (1.0%), pink breasted siphon fish, *Siphamia roseigaster* (0.8%) and yellowfin bream, *Acanthopagrus australis* (0.7%). MDS had a stress of 0.12 and 0.20 for the best 3- and 2- dimensional ordinations, respectively (Fig. 1.4). Catch structures were significantly different between gears (ANOSIM Global $R = 0.54$, $P < 0.01$, pairwise comparisons $P < 0.01$ in all cases; Fig. 1.4). SIMPER analyses showed that of the species comprising by-catch, Ramsey's perchlet, southern herring and pink-breasted siphonfish contributed the most towards distinguishing catches between the gears (Table 1.1).

The mean ratios of weight of retained school prawns to weights of total by-catch and fish by-catch ranged from 1:2.9 to 1:4 kg and were not significantly different between the treatment gears (Table 1.2). Similarly,

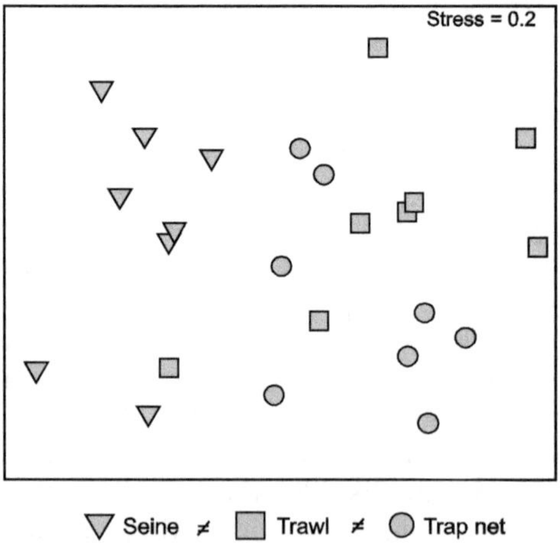

Fig. 1.4. Two-dimensional ordination for the numbers of all species captured in the treatment gears during the experiment

Table 1.1 Contribution of 99% of species to the similarity matrix of catches in the treatment seine, trawl and trap net

Species	% Contribution	Cumulative %
Seine		
School prawn	93.18	93.18
Ramsey's perchlet	4.12	97.31
Silver biddy	1.25	98.56
Yellowfin bream	0.54	99.10
Trawl		
School prawn	81.04	81.04
Southern herring	7.01	88.05
Pink breasted siphon fish	3.15	91.21
Ramsey's perchlet	2.91	94.12
Yellowfin bream	2.52	96.64
Silver biddy	2.14	98.77
Bottle squid	0.63	99.40
Trap Net		
School prawn	77.29	77.29
Southern herring	14.99	92.28
Ramsey's perchlet	5.39	97.67
Fantail mullet	0.61	98.28
Yellowfin bream	0.54	98.81
Toadfish	0.36	99.17

Table 1.2 Mean (± se) ratios of school prawns-to-catch variables, F-ratios from the 1-factor ANOVA to determine the effects on these variables due to fishing with the different treatment gears (i.e., the 25 mm trap net and the trawl and seine with the 25 mm codend) and where required, Student-Newman-Keuls tests of means. All data were *log(x+1)* transformed

Ratio	Seine	Trawl	Trap net	F ratio	SNK test
Weight (1 kg prawns):					
Total by-catch	4.00 (2.22)	3.15 (1.09)	3.04 (1.05)	0.06 ns	na
Number (1 prawn):					
Fish by-catch	3.56 (2.13)	2.73 (1.08)	2.86 (1.01)	0.09 ns	na
Yellowfin bream	0.05 (0.04)	0.09 (0.03)	0.02 (0.01)	1.10 ns	na
Southern herring	0.001 (0.001)	0.21 (0.09)	0.62 (0.25)	4.75*	seine<trawl=trap net
Ramsey's perchlet	0.54 (0.45)	0.11 (0.06)	0.14 (0.05)	0.56 ns	na

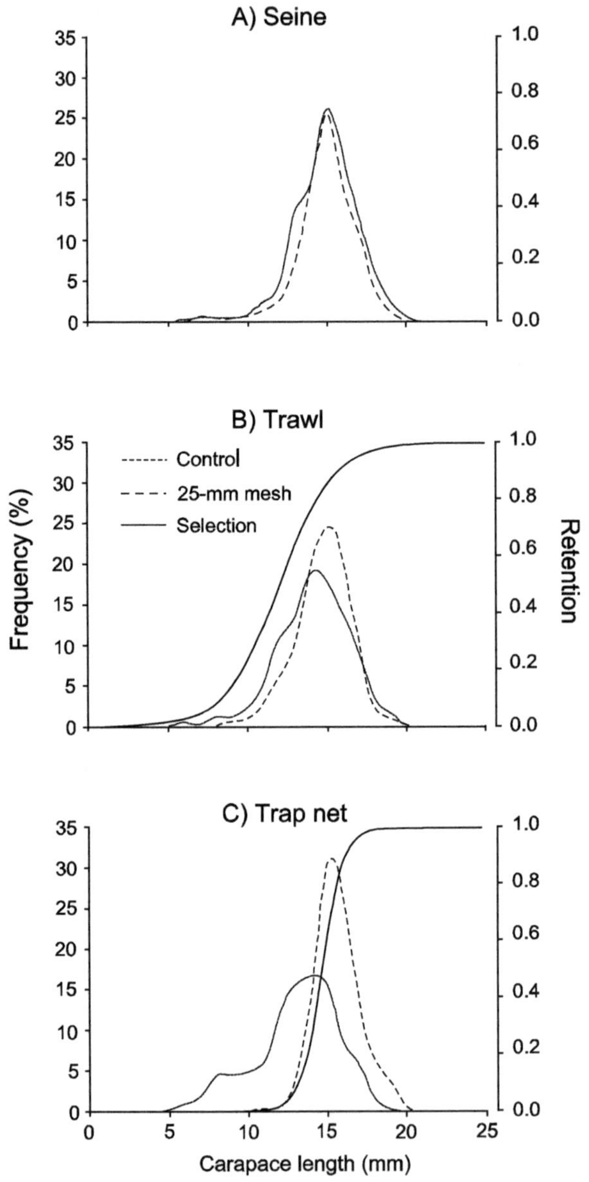

Fig. 1.5. Size-frequency distributions and, where appropriate, logistic selection curves for the A) seine, B) trawl and C) trap net all rigged with the same size and type mesh (25 mm) in the codend or bunt

Table 1.3. Carapace lengths at 25, 50 and 75% probability of retention (L_{25}, L_{50}, and L_{75}, respectively), selection ranges (SR) and relative fishing efficiencies (p) for school prawns caught by the treatment trawl and trap net (the seine was non-selective at $P > 0.05$).

Parameter	Trawl		Trap net	
L_{25}	10.16	(0.70)	13.88	(0.28)
L_{50}	12.02	(0.89)	14.63	(0.37)
L_{75}	13.89	(1.46)	15.38	(0.49)
SR	3.73	(1.12)	1.49	(0.28)
p	0.44	(0.04)	0.75	(1.49)

there were no significant differences for the ratios of number of school prawns to numbers of yellowfin bream and Ramsey's perchlet (Table 1.2). Significant differences were detected for southern herring, with a lower ratio recorded in the seine (1:0.001) compared to the trawl (1:0.21) and trap net (1:0.62) (Table 1.2).

Similar cohorts of school prawns were retained by the three control gears, particularly in the seine and trawl (Fig. 1.5). Using these data, appropriate parametric selection curves were converged for all three treatment gears, although for the seine, these models were not significantly different from the null model (i.e., no selectivity at $P > 0.05$) and so were not presented (Fig. 1.5A). For the trawl and trap net, there was no significant reduction in deviance associated with using a Richard's curve ($P > 0.05$) and so the simpler logistic model was applied (Fig. 1.5B and C; Table 1.3). Pairwise bivariate Wald tests detected significant differences in parameter estimates of these curves (χ^2 test, $P < 0.01$) with the trap net selecting school prawns at a significantly greater L_{50} and across a considerably lower SR (Table 1.3). Figure 4C shows that the parameter estimates for the trap net corresponded to an almost vertical logistic curve (i.e., almost 'knife-edged' selection).

1.4.4 Discussion

The results demonstrate considerable gear-specific differences in selectivity, partly evident by the significant dissimilarity of catches between the treatment gears, but mostly by the size distributions of the targeted school prawns retained. These results clearly delineate the three gears and illustrate the utility of trap netting as a method for selectively harvesting prawns. Prior to a discussion of the consequences of this sort of information in terms of a lateral approach towards improving the relatively poorly-selective active gears, some explanation of the mechanisms that contributed towards the observed results is required.

All three gears had significantly different assemblages of catches (Fig. 1.4). Considering the temporal scales involved in the experiment, some of the variations in assemblages between the active gears and the trap net could be attributed to diurnal fluctuations in the abundances and distributions of the key species across the area fished. But for the seine and trawl, which were used at similar times, these variations are more likely to be related to the gears used, reflecting their different net geometries (e.g., mesh sizes through anterior sections) and operational characteristics (e.g., towing speeds). Despite the different assemblages of catches, apart from a significant reduction in the ratio of numbers of retained school prawns to southern herring in the seine (compared to the other two gears), the individual proportions of key species or total by-catches retained remained similar among the three gears. For the seine and trawl, these similar ratios were probably influenced by the presence of the Nordmøre-grid. This is the most effective BRD available for towed prawn gears in NSW, mechanically separating all organisms larger than 20 mm in width and previously demonstrated to exclude up to 90% of the total by-catch from trawls used in Lake Woolooweyah, with no significant loss of prawns (Broadhurst and Kennelly 1996). Varying quantities of by-catch probably entered both the trawl and seine but, because of the Nordmøre-grid, only some proportion of those very small individuals were retained by both gears. While the numbers of these individuals and/or species varied (contributing to the differences in the assemblages of catches observed above), the total weight of by-catch consistently remained quite low and comparable between the active gears. In contrast to the seine and trawl, the trap net (which had no BRD) retained all individuals that encountered the gear. The similar ratios of retained school prawns to by-catches between the 'modified' active gears and the conventional static trap net therefore highlights the inherent selectivity of the latter fishing gear.

The selective characteristics of the trap net are best demonstrated by the sizes of school prawns retained (Fig. 1.5). This gear selected individuals at an L_{50} of 14.63 mm and across a SR of 1.49 mm. Despite having the same size and hanging ratio of mesh in the codend, the trawl selected prawns at an L_{50} 1.2 times lower and across a SR more than 2.5 times larger, while the seine was essentially non-selective (i.e., the 25-mm codend retained the same sizes of individuals as the 9-mm control codend). These considerable differences between gears rigged with the same size of mesh in their codend/bunt can be attributed to their different geometries and methods of operation.

For the trap net, hauling the headline and footrope of the entire gear (i.e., 130 m) over the second dory effectively spread the entire transverse section of the netting (i.e., > 3 m) and maintained maximum mesh openings at an area where the catch was dispersed and being progressively rolled towards the bunt (Fig. 1.3C). By facilitating multiple contacts between all school prawns and the open meshes, this process provided numerous opportunities for selection to occur along the entire gear. In contrast, the probability of prawns encountering open meshes in the seine and trawl was considerably reduced. Unlike the trap net, most of the meshes in the bodies of active gears were orientated at a shallow angle to the movement of catch, effectively reducing the likelihood of prawns encountering meshes as they moved towards the codend. More importantly, owing to variables such as towing speed, drag, twine diameter and taper of the net body, the meshes throughout these active gears would have opened at only a small fraction of their overall size (Reeves et al. 1992; Lowery and Robertson 1996). Therefore, those prawns that did encounter meshes probably had relatively little chance of escaping. This is particularly evident in the seine where we observed that, owing to the slow hauling speed (0.13 ms^{-1}), most prawns remained in the wings and net body and only passed into the codend during the final stages of hauling when the gear was lifted onboard. This explains the apparent lack of selection in the codend of this gear (i.e., no difference in selectivity between the 25-mm treatment and 9-mm control codends) and also, given the relatively small mesh size in the body of the seine (compared to the trawl), why few individuals would have been able to escape through the meshes in the body and wings.

This experiment illustrated the potential for static trap nets to harvest prawns considerably more selectively than either of the active gears, and especially the seine. This information, along with the identification of gear-specific selection mechanisms, is considered below in discussing a lateral approach towards improving selectivity in problematic fishing gears.

1.5 Using the Lateral Approach

The above case-study demonstrated the utility of testing very different fishing gears for providing comparative information on relative selection and, more importantly, for identifying key mechanisms that may provide direction for improving selection in problematic gears. Although we did not consider the fishing effort involved in the three different methods, the natural temporal restrictions on prawn trap netting (i.e., typically done at night and usually between the last and first quarter moon phases) combined with the relatively labour-intensive operation means that far fewer prawns would probably be caught by this gear compared to the trawl or seine. However, the small-scale trap-net operation described here could easily be enlarged (e.g., by using much longer and/or wider walls of netting) and mechanised (e.g., replacing the hauling crew with net drums) without compromising the mechanisms that contributed towards the observed results. Assuming the catchability of the targeted prawns is maintained, this type of fishing gear (or a modified version) could replace problematic trawls or seines in some artisanal fisheries.

A less extreme option would be to adapt some of the key processes that contribute towards selection in the trap netting to the active gears. Specifically, the method of hauling the trap net and maintaining contact between prawns and areas of netting where the transversal mesh openings are maximised is a vital selective attribute in this gear. It should be possible to emulate this mechanism in towed gears by: (i) providing and maintaining sufficient openings in key areas; and (ii) increasing the probability that prawns encounter these openings to the level that occurs in trap nets.

The starting point for such modifications in trawls is the codend, since this is where most size selection is believed to occur (e.g., Wileman et al. 1996). Selection in conventional diamond-mesh codends is highly variable and influenced by numerous factors including the hanging ratio and length of netting, the size, shape and twine thickness of mesh, the towing speed of the trawl and the weight of the catch (Reeves et al. 1992; Lowry and Robertson 1996; Lök et al. 1997; Dahm et al. 2002). Catches in conventional diamond-mesh codends tend to spread horizontally (Fig. 1.6A), which effectively masks large areas of mesh in the posterior section and, due to the associated drag, often closes meshes throughout the anterior extension. One of the simplest ways to reduce variability in the size selection of trawls for prawns and other crustaceans is to open the meshes in the codend by orientating them on the bar so that they are square-shaped (Fig. 1.6B) (e.g., Thorsteinsson 1992). By maintaining consistent openings throughout the codend, square-mesh specifically addresses the first key selective attribute of the trap net examined in the case study (i.e., providing and maintaining sufficient openings in key areas). The second attribute (i.e., increasing encounter probability) might be achieved by considerably reducing the diameter of square-mesh codends (Fig. 1.6C). Individuals in a very narrow square-mesh codend would have a shorter distance to travel towards the netting than in conventional codends and therefore a greater probability of randomly encountering openings. It might also be possible to increase the frequency of encounters between individuals and open meshes by generating turbulent flow in the codend using strategically-positioned panels.

For the seine used in our case-study, improved selection may be achieved by reducing the net taper and increasing mesh openings in the net body (where most selection apparently occurs) and/or increasing the hauling speed. A steeper body taper could increase the probability of prawns

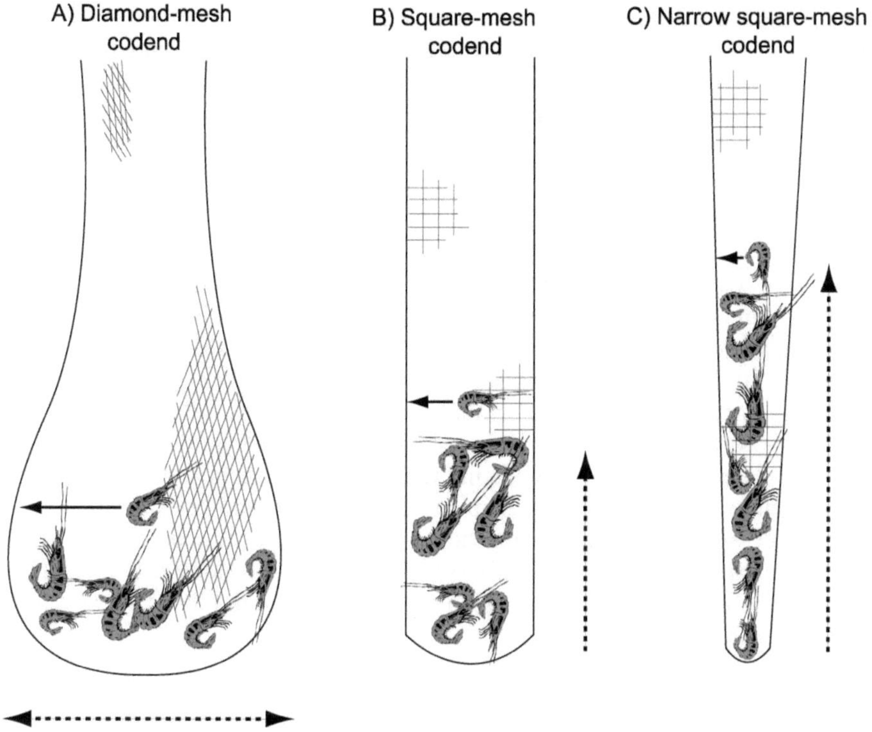

Fig. 1.6. Catch distribution (dashed arrows) and the relative distance (black arrows) an individual has to travel to encounter meshes in A) diamond-mesh, B) square-mesh and C) narrow square-mesh codend

encountering the sides of the seine and so, providing mesh openings are maintained, improve size selection (Broadhurst et al. 2000). A faster hauling speed would augment these modifications and also direct more of the catch into the codend where, like the trawl, selection could be further improved by the changes to codend geometry suggested above.

Improvements to gears like the trawls and seines examined above, should not only be limited to attempts at mimicking the key attributes of one particular type of inherently selective gear. Instead, it should be possible to identify the attributes of a range of different gears and their methods of operation and examine their utility for increasing selection in problematic gears. For example, another modification that could reduce the by-catch of fish from crustacean trawls and seines involves attaching hydrophones at the mouth of the net. Many artisanal fisheries throughout the world have traditionally used noise, generated during physical disturbances, to herd fish into static gears like gillnets (e.g., Gray et al. 2005). Using the same logic, an appropriate volume and frequency emitted from hydrophones positioned at the otter boards or wings could initiate a response in some fish species that causes them to avoid the mouth of trawls and seines and so improve species selection. These sorts of simple modifications could provide the key to improving selection in many problematic gears.

1.6 Conclusions

Like the selection mechanisms identified in the case-study, solutions to problematic selectivity issues are gear- and fishery-specific. Obviously no single solution will be appropriate for all gears and fishing methods. But fishers and fishing technologists should consider other gears and other fisheries because sometimes, critical solutions to selectivity problems will reside there and not in the particular fishery and gear under examination. We believe that only by fully testing the limits of what is achievable within the confines of the fishing gear under examination and considering selection processes in other gears and how they may be used, can one continue to strive towards 'perfect selectivity' in fishing technology. Such a lateral approach should ensure progression towards incrementally more selective fishing gears.

Acknowledgements

Thanks are extended to Will Macbeth, Michael Wooden, Don Johnson, Barry Johnson, Steve Everson and Kevin Crofton for assistance in the field. The case study was funded by NSW Fisheries and the Fisheries Research and Development Corporation as part of Project 2001/031.

References

Acosta AR, Appeldoorn RS (1995) Catching efficiency and selectivity of gillnets and trammel nets in coral reefs from southwestern Puerto Rico. Fish Res 22:175–196

Alverson DL, Freeberg MH, Murawski SA, Pope JG (1994) A global assessment of fisheries bycatch and discards. FAO Fish Tech Pap No. 339. Rome, Italy, 233 pp

Andrew NL, Pepperell JG (1992) The by-catch of shrimp trawl fisheries. Oceanogr Mar Biol Annu Rev 30:527–565

Broadhurst MK (2000) Modifications to reduce bycatch in prawn trawls: A review and framework for development. Rev Fish Biol Fish 10:27–60

Broadhurst MK, Kennelly SJ (1996) Rigid and flexible separator-panels in trawls that reduce the by-catch of small fish in the Clarence River prawn-trawl fishery, Australia. Mar Freshw Res 47:991–998

Broadhurst MK, Kennelly SJ, Gray CA (2002) Optimal positioning and design of behavioural-type bycatch reduction devices (BRDs) involving square-mesh panels in penaeid prawn-trawl codends. Mar Freshw Res 53:813–823

Broadhurst MK, McShane PE, Larsen RB (2000) Effects of twine diameter and mesh size in the body of prawns trawls on bycatch reduction in Gulf St. Vincent, Australia. Fish Bull 98:463–473

Broadhurst MK, Young DJ, Damiano C (2004b) Effects of Nordmøre-grid angles, profiles and other industry-developed modifications on catches in an Australian penaeid-trawl fishery. Cienc Mar 30:155–168

Broadhurst MK, Wooden MEL, Young DJ, Macbeth WG (2004c) Selectivity of penaeid trap nets in south-eastern Australia. Sci Mar 68(3):445–455

Broadhurst MK, Suuronen P, Hulme A (2006) Estimating collateral mortality from towed fishing gear. Fish Fish 7:180–218

Broadhurst MK, Millar RB, Kennelly SJ, Macbeth WG, Young DJ, Gray CA (2004a) Selectivity of conventional diamond- and novel square-mesh codends in an Australian estuarine penaeid-trawl fishery. Fish Res 67:183–194

Clarke KR (1993) Non-parametric multivariate analyses of changes in community structure. Aust J Ecol 18:117–143

Colvin MA (2002) A comparison of gill netting and electrofishing as sampling techniques for white bass in Missouri's large reservoirs. N Amer J Fish Manag 22:690–702

Dahm E, Wienbeck H, West CW, Valdemarsen JW, O'Neill FG (2002) On the influence of towing speed and gear size on the selective properties of bottom trawls. Fish Res 55:103–119

Dyson J (1977) Business in great waters: the story of British fishermen. Angus and Robertson, United Kingdom, pp 37–39

Engås A, Jørgensen T, Angelsen KK (2000) Effects on catch rates of baiting gillnets. Fish Res 45:265–270

Erzini K, Gonçalves JMS, Bentes L, Lino PG, Ribeiro J, Stergiou KI (2003) Quantifying the roles of competing static gears: comparing selectivity of longlines and monofilament gill nets in a multi–species fishery of the Algarve (southern Portugal). Sci Mar 67(3):341–352

Graham N, Kynoch RJ (2001) Square mesh panels in demersal trawls: some data on haddock selectivity in relation to mesh size and position. Fish Res 49:207–218

Gray CA, Johnson DD, Broadhurst MK, Young DJ (2005) Seasonal, spatial and gear-related influences on relationships between retained and discarded catches in a multi-species gillnet fishery. Fish Res 75:56–72

Halliday RG (2002) A comparison of size selection of Atlantic cod (*Gadus morhua*) and haddock (*Melanogrammus aeglefinus*) by bottom longlines and otter trawls. Fish Res 57:63–73

Hamley JM (1975) Review of gillnet selectivity. Journal of the Fish Res Board of Canada 32:1943–1969

Hanchin PA, Willis DW, St Sauver TR (2002) Comparison of concurrent trap-net and gill-net samples for black bullheads. J Freshwater Ecol 17(2):233–237

Hovgård H, Riget FF (1992) Comparison of longline and trawl selectivity in cod surveys off West Greenland. Fish Res 13:323–333

Huse I, Gundersen AC, Nedreaas KH (1999) Relative selectivity of Greenland halibut (*Reinhardtius hippoglossoides*, Walbaum) by trawls, longlines and gillnets. Fish Res 44:75–93

Huse I, Løkkeborg S, Soldal AV (2000) Relative selectivity in trawl, longline and gillnet fisheries for cod and haddock. ICES J Mar Sci 57:1271–1282

Kelleher K (2005) Discards in the world's marine fisheries. An update. FAO Fish Tech Pap 470. Rome, Italy, 131 pp

Kennelly SJ (1997) A framework for solving by-catch problems: examples from New South Wales, Australia, the eastern Pacific and the northwest Atlantic. In: Hancock DA, Smith DC, Grant A, Beumer JP (eds) Developing and sustaining world fisheries resources: the state of science and management (2nd World Fisheries Congress proceedings). ISBN 0 643 05985 7, CSIRO Press, Australia, pp 544–550

Kennelly SJ, Broadhurst MK (1996) Fishermen and scientists solving by-catch problems: examples from Australia and possibilities for New England. In: Solving Bycatch: Considerations for Today and Tomorrow. Alaska Sea Grant College Program Report No. 96-03, University of Alaska Fairbanks, pp 121–128

Kennelly SJ, Broadhurst MK (2002) Bycatch begone: changes in the philosophy of fishing technology. Fish Fish 3:340–355

Kotz S, Johnson NL, Read CB (1982) Encyclopaedia of statistical sciences, vol 9. Wiley, New York, 762 pp

Lök A, Tokaç A, Tosunoğlu Z, Metin C, Ferro RST (1997) The effects of different cod–end design on bottom trawl selectivity in Turkish fisheries of the Aegean Sea. Fish Res 32:149–156

Løkkeborg S, Bjordal Å (1992) Species and size selectivity in longline fishing: a review. Fish Res 13:311–322

Lowery N, Robertson JHB (1996) The effect of twine thickness on cod-end selectivity of trawls for haddock in the North Sea. Fish Res 26:353–363

Mahon R, Hunte W (2001) Trap mesh selectivity and the management of reef fishes. Fish Fish 2:356–375

Matsuoka T, Sibaganei T, Kasu J (1990) A comparison of selectivity of gillnet and trammelnet. Sci New Guinea 16(3):100–108

Millar RB, Walsh SJ (1992) Analysis of trawl selectivity studies with an application to trouser trawls. Fish Res 13:205–220

Millar RB, Broadhurst MK, Macbeth WG (2004) Modelling between-haul variability in the size selectivity of trawls. Fish Res 67:171–181

Otway NM, Craig JR, Upston JM (1996) Gear-dependent size selection of snapper, *Pagrus auratus*. Fish Res 28:119–132

Reeves SA, Armstrong DW, Fryer RJ, Coull KA (1992) The effects of mesh size, cod-end extension length and cod-end diameter on the selectivity of Scottish trawls and seines. ICES J Mar Sci 49:279–288

Saila SB (1983) Importance and assessment of discards in commercial fisheries. FAO Fisheries Circular No. 765. Rome, Italy, 62 pp

Sainsbury JC (1996) Commercial fishing methods: an introduction to vessels and gears, 3rd edn. Fishing News Books, Oxford, England, 359 pp

Santos MN, Gasper MB, Monteiro CC, Vasconcelos P (2002) Gill net and longline catch comparisons in a hake fishery: the case of southern Portugal. Sci Mar 66(4):433–441
Shepard RN (1962) The analysis of proximities: multidimensional scaling with an unknown distance function. Psychometrika 27:125–140
Stergiou KI, Petrakis G, Politou CY (1996) Small-scale fisheries in the South Euboikos Gulf (Greece): species composition and gear competition. Fish Res 26:325–336
Stergiou KI, Moutopoulos DK, Erzini K (2002) Gill net and longlines fisheries in Cyclades waters (Aegean Sea): species composition and gear competition. Fish Res 57:25–37
Thorsteinsson G (1992) The use of square mesh codends in the Icelandic shrimp (*Pandalus borealis*) fishery. Fish Res 13:255–266
Valdemarsen JW, Suuronen P (2003) Modifying fishing gear to achieve ecosystem objectives. In: Sinclair M, Valdimarsson G (eds) Responsible fisheries in the Marine Ecosystem. FAO, Rome, pp 321–341
Wileman DA, Ferro RST, Fonteyne R, Millar RB (1996) Manual of methods of measuring the selectivity of towed fishing gears. ICES Cooperative Research Report No. 215. Denmark, Copenhagen, 126 pp
Yami B (1976) Fishing with light. Fishing News Books, Farnham, Surry, England, 121 pp
Yellen JE, Brooks AS, Cornelissen E, Mehlman MJ, Stewart K (1995) A middle stone age worked bone industry from Katanda, Upper Semliki Valley, Zaire. Science 268:553–555

2 Reconciling Fisheries with Conservation Through Programs to Develop Improved Fishing Technologies in the United States

JOHN WATSON

2.1 Introduction

Technological advancements have revolutionised fishing vessels and fishing gear allowing effective harvesting of target species, but these have also placed harvest pressure on many fish stocks including by-catch. Discarded by-catch can adversely affect the population size and structure of impacted stocks, and reduce the availability of by-catch species that are targeted by other users. The discarding of unwanted and regulated by-catch and the inadvertent capture and mortality of protected species by commercial and recreational fisheries has become an increasingly significant problem in the worldwide effort to conserve and manage marine fisheries resources. By-catch in fisheries worldwide was estimated to be approximately 27 million mt by Alverson et al. (1994). As better estimates of the magnitude of by-catch in fisheries have been made available through fishery observer programs, concern over the impact of by-catch has increased.

Since the mid-1970's, concern over by-catch in the United States has intensified among state and federal fisheries managers, conservationists, fishers and the general public. The incidental capture of endangered and threatened species was the first problem to be addressed and substantial progress has been made to reduce the impacts of fishing gears on their populations (NMFS 2006). More recently, as in other countries, concern over by-catch in the United States has broadened to include the incidental capture of finfish and other living marine resources. Increased world demand for protein has focused attention on the need to minimise waste in all fisheries. In 1996, the United States Congress passed legislation amending the

S. J. Kennelly (ed.), By-catch Reduction in the World's Fisheries, 23–36.

Magnuson-Stevens Act of 1976 which required fishery managers to improve limits on by-catch in fishery management plans. The above history has resulted in United States programs to develop more selective harvesting gears, and examples of these programs are presented in this chapter. The common feature of these programs is that they are all designed to reconcile fisheries with conservation through improved fishing technologies.

2.2 Development of the Turtle Excluder Device (TED)

The inadvertent capture of sea turtles and marine mammals (most of which are protected species) is problematic for many fisheries in the United States. In 1978, all species of marine turtles which occur in United States waters were listed as either threatened or endangered under the Endangered Species Act of 1976. A report by the United States National Academy of Sciences, National Research Council (1990) determined that the penaeid shrimp fishery was the single largest cause of sea turtle mortality. The implications of the listing of sea turtle species under the Endangered Species Act and these findings of the National Academy of Sciences represented severe implications for the valuable penaeid shrimp fishery of the United States – especially in the Gulf of Mexico (the world's largest shrimp fishery) unless mitigation measures could be developed to allow continued fishing without jeopardising the recovery of sea turtles.

In response to the mandates of the United States Endangered Species Act, the United States National Marine Fisheries Service initiated a research program to investigate methods to reduce the incidental capture and mortality of sea turtles in the shrimp trawl fishery. Alternatives considered included: spatial and seasonal closures, restricted tow times, and modifications to fishing gear. It was felt that closures would be only minimally effective due to the widespread distribution of sea turtles and also they would be economically detrimental to the industry. Further, it was considered that restrictions on tow times would not be enforceable. An intensive gear development program was therefore conducted by the National Marine Fisheries Service between 1978 and 1980 which resulted in the development of a metal grid device (turtle excluder device or TED, Fig. 2.1) that was placed in the codend extension of the trawl (Watson and Seidel 1980). The grid mechanically separates and excludes sea turtles and other large objects and organisms while shrimp pass through the grid bars into the codend. The prototype design was developed based on behavioural observations of sea turtles encountering shrimp trawls (Ogren et al. 1977) and was similar to a device developed by shrimp fishers to exclude jellyfish (Seidel and McVea 1982). The original TED demonstrated a 97% reduction in sea turtle captures with less than a 3% reduction in shrimp

Catch Per Unit of Effort (CPUE) (Watson and Seidel 1980). In 1981, a program was initiated to encourage the voluntary use of TEDs by United States shrimp fishers. The United States government sponsored industry workshops and technical demonstrations of the device to encourage voluntary adoption. However, fishers who tested the device were concerned that it was too large, cumbersome and complicated to use.

The government continued research and development of the TED design between 1980 and 1984 in an effort to improve its handling characteristics. Additional modifications were made to reduce fish by-catch as an incentive to encourage adoption (Watson et al. 1986) and in 1985 and 1986, additional workshops and demonstrations were conducted to encourage voluntary acceptance of the gear. After an intensive technology transfer and demonstration effort, it became clear the industry would not voluntarily use the technology. In 1986, several non-governmental organisations (NGOs) filed intents to sue the federal government if it did not enforce the protection of sea turtles as required under the Endangered Species Act. In response, the federal government called for mediation between representatives of the shrimp industry and NGOs which resulted in regulations requiring mandatory use of the grid device by the industry.

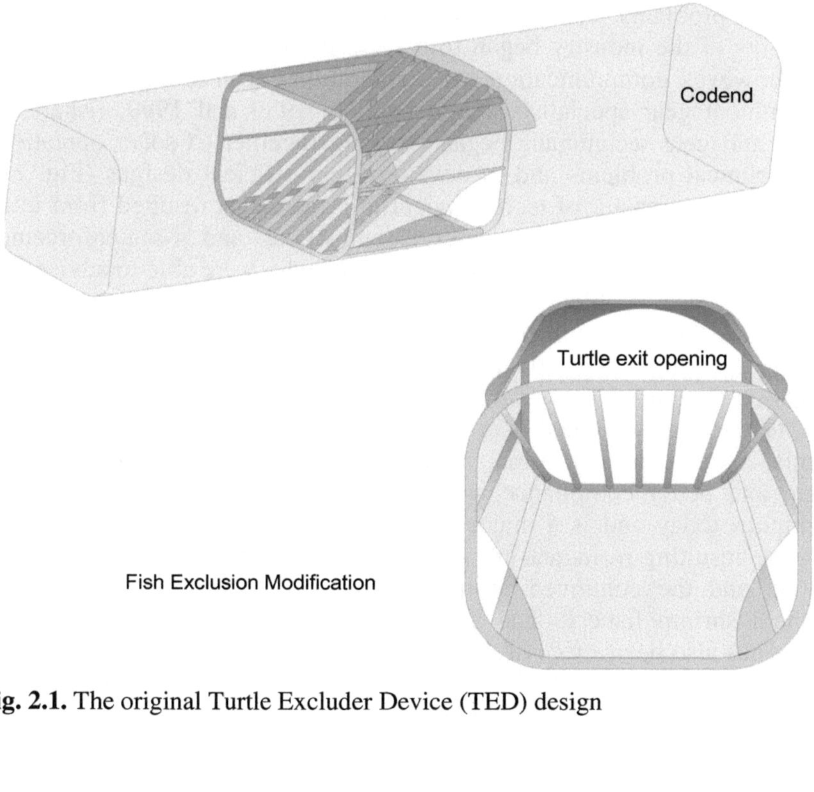

Fig. 2.1. The original Turtle Excluder Device (TED) design

In response to this, segments of the United States shrimp industry began a campaign to resist the requirement of mandatory use of the gear. Others began independent research to develop alternative designs when it became apparent that mandatory regulations would be enacted and enforced anyway. The new designs that were developed by the industry were cheaper, less complicated and easier to use than the design developed by the government researchers. These designs were certified by the government as effective in reducing sea turtle capture, but were less effective than the original design in retaining shrimp catch. Resistance from the industry to mandatory regulations was intense and included political pressure, litigation, adversarial confrontation, and civil and criminal disobedience. The intensity of the industry resistance resulted from fear of reduced revenue and economic hardship, denial and disbelief of the magnitude of the sea turtle problem, general opposition to regulations, distrust of federal regulators precipitated by an effective campaign by some segments of the industry, and ineffective communication between industry and government.

In 1989, after three years of litigation and opposition, mandatory regulations requiring TEDs were fully implemented. This implementation required intensive enforcement efforts and prosecution. The widespread use of certified industry designs resulted in the identification of operational and technical problems resulting from poor construction and installation. As segments of the industry began to accept the inevitability of the technology, however, communication and cooperation began to improve between government gear specialists and fishers. In 1989 and 1990, fishers, net shops and gear technicians began working together to solve operational and technical problems and to develop more efficient designs (Fig. 2.2). The effective transfer of technological improvements resulted from intensive technical training of law enforcement officers and at-sea enforcement assistance from government gear technicians who were able to advise fishers on technical problems. In addition, effective technical manuals, summary placards, and an intensive program of technical training workshops for fishers (including effective multimedia training presentations and hands-on demonstrations) were initiated. Cooperation between fishers and gear technologists resulted in efficient and effective technological improvements, better communication, a more effective technology-transfer program and better compliance with mandatory regulations. This program continues today and is a major component in sea turtle recovery efforts that are resulting in increasing populations of sea turtles in United States waters and the continued promulgation of the valuable United States penaeid shrimp fishery. The TED technology developed in the United States has also been successfully exported to 21 other countries around the world whose shrimp fisheries affect sea turtle populations through a foreign technology transfer program (Epperly 2003).

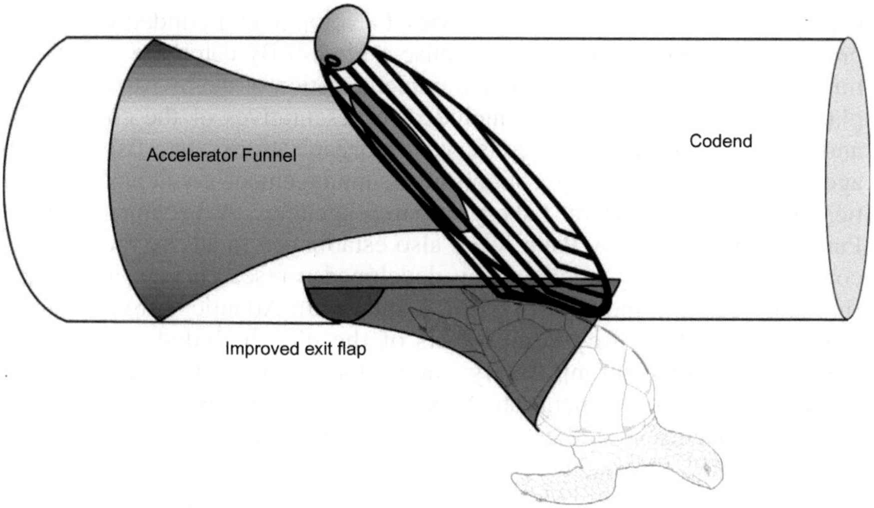

Fig. 2.2. An improved industry design (Super Shooter TED)

2.3 Development of By-catch Reduction Devices (BRDs)

In 1990, amendments to the Magnuson Fishery and Management Act (Public Law 101-27) required the United States Secretary of Commerce to conduct a 3 year research program to determine the impacts of shrimp-trawl by-catch on federally managed fishery resources. This legislation addressed the effect of by-catch mortality on fishery resources in addition to the protection of threatened and endangered species under the endangered species and marine mammal protection acts. The amendments required the secretary to establish a cooperative program to design and evaluate approaches for reducing the mortality of incidentally harvested fishery resources in shrimp fisheries.

A comprehensive approach was adopted to plan and implement a regional by-catch research program which had the advantage of the experience gained from the development and implementation of TEDs. In 1991, the National Marine Fisheries Service's Southeast Regional Office and Fisheries Science Centre developed and published a document entitled 'Shrimp Trawl By-catch Research Requirements' (NMFS 1991). This document established research protocols based on proven scientific methods which were subjected to peer review by an industry-organised panel of researchers and statisticians. To ensure effective communication and participation of all affected parties, the NMFS established cooperative agreements with the Gulf and South Atlantic Fisheries Development Foundation, a non-profit industry organisation, to organise a Finfish By-catch Steering Committee to guide the development and implementation of the by-catch research plan. The steering committee included representatives of the commercial and sport-fishing industries, conservation organisations, state fishery management agencies, fishery commissions, management councils, universities, and state and federal fishery research agencies. A Technical Review Panel and Gear Review Panel were also established to advise the Steering committee. Working together they developed a research plan addressing finfish by-catch in the Gulf of Mexico and South Atlantic shrimp fisheries (Hoar et al. 1992). Key components of the plan included: cooperative efforts between the shrimp fishery, states, universities, conservation groups and the federal government, strict adherence to stringent scientific protocols, and concurrent studies on social and economic impacts. One of the objectives of the by-catch research program was to identify, develop and evaluate gear options for reducing by-catch in the Gulf and South Atlantic shrimp fisheries. The gear review panel was responsible for selecting the best prototype of gear modifications for commercial evaluations, monitoring tests of the gears in different shrimping areas and prioritising options of gear modifications for consideration by management. The goal of the gear development project was to develop gear modifications to shrimp trawls and/or fishing practices that were capable of reducing the by-catch of finfish with minimal loss of shrimp catches. The research plan identified a 4-phase gear development plan:

1. **Initial Design and Prototype Development** – The full range of technical approaches to the modification of trawl designs was identified. Industry-based techniques, ideas solicited from fishers, designs from net shops and studies conducted by various research groups were evaluated. Studies of fish behaviour, gear instrumentation and gear performance were done on each design using SCUBA, remote video cameras and other techniques. This work evaluated fish behaviour and the feasibility of various prototypes. The results of this phase were subjectively evaluated based on the experience and expertise of the gear designer and research team. Operational data were taken on modified gears, and preliminary data on catching performance were obtained from comparative gear trials. One hundred and forty five gear modifications were evaluated by commercial fishers, universities and state and federal research agencies under this program between 1992 and 1996. The next phase of development was initiated once a design was determined to offer potential for by-catch reduction. This next phase involved integrating the design into the construction of nets.

2. **Proof-of-Concept** – Objectives during this phase were to evaluate effects of prototype modifications on key species, determine total rates of finfish reduction and establish catch rates of shrimp. Proof-of-concept testing was designed to evaluate the adequacy of the design for its safety features and any problems with the operational use of the design. Proof-of-concept testing was conducted under a specific scientific protocol developed under the 'Shrimp Trawl By-catch Research Requirements' (NMFS 1991). The most successful designs were prioritised based on the proportion of by-catch reduction and shrimp retention and were reviewed by a technical review panel for their progression to Phase 3 – operational evaluation by the commercial shrimp industry throughout the Southeast United States.
3. **Operational Evaluation** – The objective in this phase was to test each BRD-equipped net against a standard net under conditions encountered during commercial shrimping operations. Trained observers were placed onboard co-operating commercial vessels to collect data on both shrimp and finfish catch rates as well as species composition. Testing was conducted over a wide range of geographic areas, seasons and conditions.
4. **Industry Evaluation** – This phase involved widespread commercial evaluations of BRD designs. The research program was successful in developing and testing gear modifications to shrimp trawls that were capable of producing significant reductions in finfish by-catch with minimal reduction in shrimp catch rates. The most effective designs included the 'fisheye' BRD (Fig. 2.3) which is a simple metal, cone-shaped device inserted into the trawl codend to create an escape opening, the extended funnel BRD (Fig. 2.4) which consists of a large square-mesh section with a small mesh funnel inside of the square-mesh, and the Jone/Davis BRD (Fig. 2.5) which is a modification the expanded mesh BRD. The expanded mesh and Jones/Davis BRDs are installed between the TED and the codend. Approved BRD designs were made mandatory using management regulations in all state and federal waters in the southeastern United States in 1998 and 1999 (Watson et al. 1999).

The effectiveness of these technologies was monitored by fishery observers between 1998 and 2003. In 1998, under the Red Snapper Initiative Project, research was done to provide data on the effectiveness of the mandatory use of by-catch reduction devices by the commercial shrimp fishery in the north central and western Gulf of Mexico. The 1998 results for the Jones-Davis BRD were similar to results obtained during the development

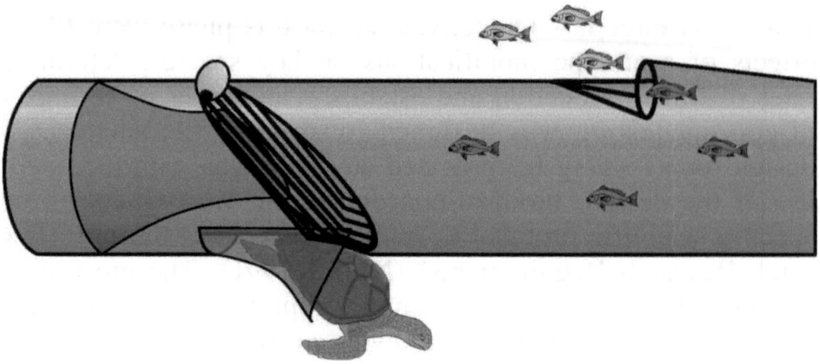

Fig. 2.3. The Fisheye BRD

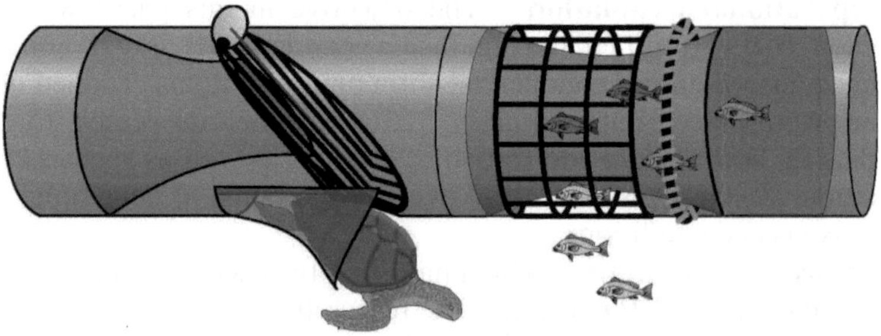

Fig. 2.4. The Extended Funnel BRD

Fig. 2.5. The Jones/Davis BRD

and certification of the device, but the results for the Fisheye BRD showed a lower finfish reduction rate. Regulations were modified to restrict the placement of the fisheye BRD in an attempt to improve its performance and continued monitoring of the fishery indicated that these changes to the regulations had little effect on the performance of the fisheye BRD. It is not clear what factors may have caused these results but video observations of fish behaviour associated with fisheyes indicated a large portion of fish escape through the device occurs at the surface when the net is hauled back. Observations also indicated that this escape at the surface is lowest when the fisheye is installed in the anterior section of the codend. When the fisheye is installed farther back in the codend, more fish escape through the device during the tow. Video observations of shrimp behaviour and fisheyes obtained by Texas A&M University and Georgia Sea Grant specialists indicate that the majority of shrimp loss through the device also occurs at the surface during haulback – when the fisheye is placed farther back in the codend. This information suggests that any operational efforts to reduce shrimp loss with fisheyes during haulback may have an adverse impact on reducing fish by-catch. In this case, the potential negative impact on fish reduction would be greatest for the fisheye installed in the forward position. Observations of the Jones-Davis BRD and other BRD designs using funnels indicated that the majority of fish escape occurs during the tow and is not subject to the above problems associated with haulback procedures.

Other variables identified by gear technologists during this work that could affect fisheye BRD performance included: the length of the codend, the location of tie-off rings in codends, the location and length of the lifting lines (triangular sections of webbing attached to the codend to which the codend haulback line is attached), the circumference of the codend and orientation of knots in the codend. Fishing practices that may affect the performance of fisheyes include: towing speed, winch retrieval speed, codend hauling procedures, hauling direction and frequent turning. Research is continuing in attempts to develop more effective technologies to reduce the by-catch associated with the shrimp trawl fishery while maintaining effective and efficient harvesting gear.

2.4 Development of Sea Turtle Mitigation Technologies in the Pelagic Longline Fishery

Pelagic longline gear is used throughout the world to catch widely dispersed species. The gear is very efficient at catching large pelagic fishes, such as bluefin, bigeye *Thunnus obesus*, yellowfin *T. albacares* and albacore *T. alalunga* tunas, broadbill swordfish *Xiphus gladius*, and the istiophorid billfishes. There are many possible variations in the configurations of this gear, but in general, when compared with such gears as trawls or pelagic gillnetting, pelagic longlines are considered highly selective for large target species (Yamaguchi 1989). However, the by-catch of protected species including sea birds, sea turtles and marine mammals by pelagic longline gear is considered a global problem. Loggerhead *Caretta caretta*, leatherback *Dermochelys coriacea* and Olive ridley *Lepidochelys olivacea* sea turtles are captured in longline fisheries, and although recorded mortalities are very low, the injuries sustained during interactions with hooks and lines are of concern as little data are available regarding post-release mortality (Watson and Kerstetter – in press).

Estimates of catches of turtles in the United States pelagic longline fishery have raised concerns that this fishery may be affecting the potential recovery of loggerhead and leatherback sea turtle populations in the Atlantic Ocean. In 2001, the NOAA Fisheries Service closed a large portion of the North Atlantic to longline fishing by United States fishers and initiated a cooperative research effort between fishers, universities and the federal government to investigate potential gear modifications and/or fishing practices to reduce sea turtle catch rates. Components of the project included the active participation of fishers in the design and implementation of the research, adherence to stringent scientific protocols and significant input from fishers on experimental designs. During initial meetings between industry and other partners, various problems were identified in relation to the need for stringent scientific protocols and the practical aspects of fishing practices.

A concerted effort was made to improve communications and trust among the various groups involved in this project, and to reach acceptable compromises on disagreements. The result was that fishers were able to educate scientists with their knowledge of the fishery, turtle interactions, and potential practical solutions while scientists were able to educate fishers on scientific research methodology and protocols. This cooperation, trust and effective communication resulted in an extremely effective and successful research project. The program tested potential sea turtle mitigation measures developed by an informal steering committee of fishers, research biologists, and fishery management personnel. Mitigation measures were evaluated on commercial longline fishing vessels in the Western Atlantic Ocean using appropriate experimental designs. Between 2001 and 2003, this program developed sea turtle mitigation techniques utilising circle hooks and fish bait that demonstrated reductions in interactions of between 58% and 94% for loggerhead turtles and 44% and 86% for leatherback turtles, without significantly affecting catch rates of the target species (Watson et al. 2005). Results from Western Atlantic studies determined that the use of circle hooks instead of traditional J hooks reduced the proportion of hard-shelled turtles swallowing hooks from 68.8 to 27.3% (Watson et al. 2005). Turtles that were hooked by circle hooks tended to be hooked in the mouth, where hooks could be safely removed, reducing the potential for post-hooking mortality. These studies determined that the use of large circle hooks 4.9 cm or larger in width significantly reduced turtle captures compared to 4.0 cm or smaller J and tuna hooks. Furthermore, large circle hooks were determined to be commercially viable for some target species. The catch rate for swordfish was increased by 30% when circle hooks were used with large (300 – 500 gram) mackerel bait and reduced by 33% when used with squid bait. Catch rates for bigeye tuna were increased by 26% when circle hooks were used with squid bait but reduced by 81% when used with large mackerel bait (Watson et al. 2005).

Laboratory studies with captive 45 – 65cm length loggerhead turtles indicated that the proportion of turtles that attempted to swallow circle hooks varied according to circle hook size, bait type and baiting technique (Watson et al. 2003). As circle hook size increased from 14/0 to 18/0, a smaller proportion of turtles tested attempted to swallow the hooks regardless of bait type or baiting technique. Fewer turtles attempted to swallow single hooked, sardine baits compared to threaded sardine baits, single hooked squid baits or threaded squid baits (Stokes et al. 2006). Threaded baits using both squid and sardine had a higher proportion of turtles attempt to swallow the hooks than single hooked baits.

In this project, gear technicians and fishers also developed effective tools and techniques to safely remove hooks from sea turtles and other pelagic by-catch species which has the potential to improve post-release survival (Watson et al. 2005). This pelagic longline sea turtle mitigation technology has been implemented in the pelagic longline swordfish and tuna fishery in the Atlantic and Gulf of Mexico and in the Pacific swordfish fishery, and is being evaluated in other fisheries around the world (Watson and Kerstetter – in press; Gilman 2006).

2.5 Discussion

The above examples of cooperative fishing gear research projects in the United States indicate some general characteristics of successful projects to reduce by-catch in commercial fisheries. Although the TED development project was ultimately successful, it was a very difficult, lengthy and expensive process and was not successful until a truly cooperative effort between government and the fishing industry was attained. These experiences identified several factors that are important to successful fishing gear research projects. First and foremost, the affected industry or constituents should be active participants in every aspect of research planning, technology development and evaluation. Secondly, planning for the development and implementation of new technologies needs to include a major long-term commitment for technology transfer and assistance to industry. Thirdly, voluntary acceptance of new technologies may also require financial or other incentives. New technologies which result in increased costs and/or loss of revenue will tend to be resisted. Fourthly, mandatory requirements for new technologies must include effective enforcement but regulations should be as flexible and as easily modified as possible to allow successful implementation and improvement of new technologies as they are developed. Fifthly, and most importantly, successful cooperative research programs require effective communication and trust between all partners.

Effective leadership is also a vital component of successful cooperative research programs. The most effective programs include leadership that is effective in communicating among partners and the development of advisory groups with the right mix of expertise and skill to be effective. Leadership must be receptive and proficient in developing compromise solutions when partners have divergent points of view, or be firm and resolute when compromises are not an option. A representative industry organisation that can effectively communicate industry concerns, needs and opinions in negotiations, and that can form a technical advisory group, is especially beneficial in a successful cooperative by-catch reduction program.

Challenges that are commonly encountered in such programs include a lack of trust and understanding between program partners, and misinformation. Other significant problems in developing cooperative research projects include securing adequate funding, requirements for permits, environmental impact statements, and other regulatory requirements that can significantly delay and, in some cases, prevent such programs.

Cooperative research programs have proven to be highly effective in developing solutions to critical fisheries problems like by-catch issues. To be successful, they must have common goals and sincere commitment from all partners. Partners must be effective communicators and negotiators who are willing to compromise to move processes forward. Programs need effective and expert advisory personnel to assist with regulatory and permit requirements, ensure that local knowledge is integrated into the scientific process, maintain acceptable scientific standards and effectively develop and execute successful research.

References

Alverson D, Freeburg M, Murawski S, Pope J (1994) A global assessment of fisheries by-catch and discards. FAO Fish Tech Pap. No 339. Rome, Italy, 233 pp

Branstetter S (1997) By-catch and its reduction in the Gulf of Mexico and South Atlantic Shrimp fisheries. Gulf and South Atlantic Fisheries Development Foundation Inc., Tampa, Florida, USA

Epperly SP (2003) Fishery related mortality and turtle excluder devices (TEDs). In: Lutz PL, Musick JA, Wyneken S (eds) The Biology of Sea Turtles, vol II. CRC Press, Boca Raton, Florida, USA, pp 339–352

Gilman E, Zollet E, Beverly S, Nakano H, Davis K, Shiode D, Dalzell P, Kinan I (2006) Reducing sea turtle by-catch in pelagic longline fisheries. Fish Fish 7:2–23

Hoar P, Hoey J, Nance J (eds) (1992) A research plan addressing finfish by-catch the Gulf of Mexico and South Atlantic shrimp fisheries. Tampa, Florida, USA, 114 pp

National Academy of Sciences, National Research Council (1990) Decline of the sea turtles: causes and prevention. National Academy Press, Washington DC, USA, 189 pp

National Marine Fisheries Service (1991) Shrimp trawl by-catch research requirements. NMFS Southeast Regional Office, St Petersburg, Florida, USA, 66 pp

National Marine Fisheries Service (2006) Gulf of Mexico shrimp trawl by-catch reduction. Report to Congress. NMFS Southeast Fisheries Science Centre, Miami, Florida, USA, 126 pp

Ogren LH, Watson JW, Wickham DA (1977) Loggerhead sea turtles (*Caretta caretta*) encountering shrimp trawls. Marine Fisheries Review 39(11):15–17

Seidel WR, McVea Jr C (1982) Development of a sea turtle excluder shrimp trawl for the southeast US penaeid shrimp fishery. In: Bjorndal KA (ed) Biology and Conservation of Sea Turtles. Smithsonian Institution Press, Washington DC, USA, pp 497–502

Stokes L, Hataway D, Epperly S, Belskis L, Bergmann C, Watson J, Higgins B (in preparation) Evaluation of injury potential in incidentally captured loggerhead sea turtles (*Caretta caretta*) relating to hook size and baiting technique. Proceedings of the 26th Annual Symposium on Sea Turtle Biology and Conservation. April 2006, Crete, Greece. NOAA Technical Memorandum, NMFS-SEFSC. Poster presentation

Watson JW, Kerstetter DW (in press) Pelagic longline fishing gear a brief history and discussion of research efforts to improve selectivity and sustainability. Mar Technol Soc J

Watson JW, Seidel WR (1980) Evaluation of techniques to decrease sea turtle mortalities in the southeastern United States shrimp fishery. ICES CM. 1980/B:31

Watson JW, Hataway BD, Bergmann CE (2003) Effect of hook size on ingestion of hooks by loggerhead sea turtles Report of NOAA Fisheries Service. Available online at http://www.sefsc.noaa.gov/seaturtleunpublishedreports.jsp

Watson JW, Mitchell JF, Shah AK (1986) Trawling efficiency device: a new concept for selective shrimp trawling gear. Marine Fisheries Review 48(1):1–9

Watson JW, Epperly SP, Shah AK, Foster DG (2005) Fishing methods to reduce sea turtle mortality associated with pelagic longlines. Can J Fish Aquat Sci 62:965–981

Watson J, Shah A, Nichols S, Foster D (1997) By-catch reduction estimates for selected species in the Gulf of Mexico for by-catch reduction devices evaluated under the regional by-catch program. NMFS Mississippi Laboratories, USA

Watson JW, Foster D, Nichols S, Shah A, Scott-Denton E, Nance J (1999) The development of by-catch reduction technology in the Southeastern United States shrimp fishery. Marine Technol Soc J 33(2):51

Watson JW, Workman IK, Foster D, Taylor CW, Shah A, Barbour J, Hataway D (1993) Status report on the potential of gear modifications to reduce finfish by-catch in shrimp trawls in the Southeastern United States. NOAA Technical Memorandum, NMFS-SEFC-327, 131 pp

Yamaguchi Y (1989) Tuna longline fishing (I-V): historical aspects, fishing gear and methods, selection of fishing ground, fish ecology, and conclusions. Mar Behav Physiol 15:1–81

3 By-catch Reduction Techniques in European Fisheries: Traditional Methods and Potential Innovations

PETRI SUURONEN AND FRANCESC SARDÀ

3.1 Introduction

Many commercial fish stocks in the Northeast Atlantic Ocean and the North, Baltic and Mediterranean Seas, are being exploited at levels considered by the International Council for the Exploration of the Sea (ICES 2005) and the General Fisheries Commission for the Mediterranean (GFCM 2005) to be beyond sustainable levels. In many of the fisheries that exploit these stocks, growth overfishing occurs. Discarding young fish is a common practice and discard mortality is a significant part of the fishing mortality for many commercially important species in European fisheries. The mortality of juveniles can lead to declines of future stocks and therefore to a substantial loss of potential income. In recent years, the incidental capture of endangered species or species that, for some other reasons should be avoided, has also become an important management issue in Europe.

In global terms, the Northeast Atlantic is an area with relatively high discard rates (average rate 13%), accounting for about 22% of the world's total discards (FAO 2005). In terms of fish biomass, this waste corresponds to a total of about 1.3 million tonnes yearly. It is noteworthy, however, that the North Sea bottom trawl fishery alone accounts for more than half of this amount. Although there have been several increases in minimum mesh sizes, it is obvious that for many key species the authorised mesh sizes remain far too small for the effective protection of immature fish. Moreover, it is obvious that a larger minimum mesh size alone would not provide a suitable tool for achieving maximum yield-per-recruit for each species in a mixed species fishery such as the North Sea demersal

S. J. Kennelly (ed.), By-catch Reduction in the World's Fisheries, 37–74.

trawl fishery which targets haddock, whiting, cod and many other species. Using one mesh size to catch several species of various sizes usually results in discarding or significant losses in catches. Overfishing, strict quota restrictions and minimum landing size (MLS) regulations on various species greatly contribute to high levels of discarding. For example, substantial quantities of fish caught in the North Sea are discarded yearly just because fish are undersized or the vessel has no quota for a particular species.

It is apparent that the multi-species nature of many European trawl fisheries greatly contributes to the capture of juveniles and individuals of many non-target species. The valuable fishery for *Nephrops* (Norway lobster) is a typical example, and demonstrates the problems faced in many fisheries. *Nephrops* is a widely distributed high-value commercial species in Europe and is a significant component of the catch from Iceland to the Mediterranean Sea. Due to the smaller mesh size used in *Nephrops* trawls compared to demersal fish trawls, the amount of by-catch and discards of other species and small *Nephrops* can be high (ICES 2004). FAO (2005) ranks the *Nephrops* trawl fishery as a high-discard-rate fishery. As a consequence, there has been considerable research into the fishing gear used in this fishery to improve species and size selection for both target and by-catch species.

In the Mediterranean Sea, a large variety of fish and shellfish species support a predominantly small-scale fishery, mostly operating close to the coast. Fisheries are more diverse in terms of fleet structure, species caught and the methods used than those in northern Europe. In many areas the catches mostly consist of juvenile fish. Regulations, particularly those applied to fishing gears and minimum landing sizes, are inconsistently enforced. It is noteworthy, though, that in the Mediterranean Sea the average discard rate is assessed as only about 5% (FAO 2005). This is largely because there are many pelagic fisheries with low by-catch rates and because undersized fish caught in the demersal trawl fishery are marketed effectively. There are, however, many multi-species bottom trawl fisheries conducted on the continental shelf of the Mediterranean Sea where discards can be up to 70% of the catch (e.g., Stergiou et al. 1997; Machias et al. 2001; Martin et al. 2001; Sànchez et al. 2004).

In recent decades, ICES and GFCM have consistently recommended a large reduction in the capture and discarding of juvenile fish throughout Europe, in particular for cod, haddock and hake. During the same period, the European Union financially contributed about €8 million per year for over 400 projects on gear selectivity, discard reduction and the quantification of impacts of fishing gears on habitats (Fischler 2004). At least the same amount of money was used by the national fisheries research laboratories and the European fishing industry to develop more selective gears and operations. There is no doubt that this effort has resulted in progress. The list of various gear solutions tested is exhaustive. These modifications include features such as square-mesh panels, exit windows, grids, fish excluders and separator panels and many of these modifications have been proven to reduce levels of by-catch (reviewed by van Marlen 2000; Valdemarsen and Suuronen 2003; ICES 2004). Many of them have been introduced into the fisheries through legislation but generally they are not yet being effectively used throughout all European fisheries. There is still substantial potential for adopting more responsible fishing gears and operations in European commercial fisheries.

This chapter attempts to pull together the lessons learned in by-catch management in European fisheries in order to draw relevant conclusions of the applicability, acceptance and efficiency of these techniques. The chapter also attempts to identify the most relevant issues that require further research to develop more responsible fishing methods and operations. The major focus is in the management of trawl fisheries because trawling is by far the most important fishing method in Europe, and is responsible for a major proportion of the by-catch and discards.

3.2 By-catch Reduction Technologies Tested in Active Fishing Gears Europe

The large variety of fishing gears used throughout the world can be classified as either active or passive gears. These divisions are not simply those gears that move and those that are stationary. Rather, for active fishing gears – discussed below (e.g., trawls, dredges, seine nets), fishers usually guide the gear into the path of the fish. Capture success depends largely on the fishers' skills and resources and often a substantial amount of power is needed. For passive gears – discussed in Section 3.3 (e.g., gillnets, traps, some types of hooks) the fish has to come to the gear. Detailed knowledge of fish behaviour is needed to successfully construct and locate passive fishing gears and often bait is used to attract the fish to the gear.

3.2.1 Improving Size-Selectivity

3.2.1.1 Codend Mesh Size

It is well demonstrated in many trawl fisheries that improvements in size-selectivity can be obtained with relatively simple constructional changes, such as modifying the size of codend meshes. During the last two decades, the minimum size of the conventional diamond meshes in the codends of North Sea demersal trawls has gradually been increased from 80 mm to

120 mm. Substantial increases in mesh size have also been made in many other European fisheries. Nevertheless, despite such changes relatively little progress has been noted in the reduction of by-catch and discards in these fisheries. One obvious reason it is the fact that the selectivity of a diamond-mesh codend depends on factors other than mesh size that can be relatively easily manipulated by fishers. The codend circumference, twine material and diameter, catch size, towing speed, season, weather, structure of the strengthening bag, vessel type and many other factors can affect the selectivity of a diamond-mesh codend (e.g., Ferro and Robertson 1988; Reeves et al. 1992; Tschernij et al. 1996; Ferro and Graham 1998; Dahm et al. 2002; Özbilgin and Wardle 2002; Hermann 2005). It is also generally considered that a diamond-mesh codend does not exhibit a sharp selectivity but has a wide selection range. This has stimulated research to find modifications that exhibit sharper selectivity and are less sensitive to the above-mentioned factors.

It is noteworthy, however, that a diamond-mesh codend can be highly size-selective and commercially applicable when certain factors are taken into account. At times, just small alterations in the codend construction can improve the selectivity of a traditional diamond-mesh codend. One of the simplest measures is to increase the diamond-mesh opening by either restricting the number of meshes in the circumference or by hanging the codend netting on ropes shortened relative to the stretched length of the codend (e.g., Robertson and Shanks 1989). Restricting the twine thickness and stiffness also reduces the mesh resistance to opening, and may thereby markedly improve selectivity.

3.2.1.2 Square-Mesh Codends

A substantial amount of work has been done in Europe on square-mesh codends (e.g., Robertson and Stewart 1988; Dahm 1991; Suuronen and Millar 1992; Campos et al. 2002; Bahamon et al. 2006). In square-mesh codends the meshes stay open during the tow irrespective of the tension placed on the meshes during hauling. In many trials, square-mesh codends have been shown to exhibit somewhat sharper selection than the corresponding diamond-mesh codends although this effect has not been found in all trials. A potential advantage of a square-mesh codend is that its selectivity is less affected by the type of twine used than a diamond-mesh codend; hence it may be less susceptible to measures by fishers to reduce improvements in selectivity. However, a square-mesh codend may not be efficient for all commercial species, such as many dorsally and vertically compressed fish. Moreover, altering the orientation of the mesh may create problems in terms of the strength of meshes and distortion. Square-mesh codends may also suffer from knot slippage that is difficult to repair. Moreover, a full square-mesh codend may be difficult to handle on deck when full of catch because the codend can become rigid and for some species, meshing problems may be substantial.

Square-mesh codends are unpopular among commercial fishers in Europe and have not been widely legislated in European fisheries. It is notable, however, that square-mesh codends have been voluntarily used in some specific European fisheries. For instance, in Norway the use of square-mesh codends has markedly improved the selectivity in the seine-net fishery for cod and haddock. Finnish inland water trawlers that are targeting adult vendace (*Coregonus albula*) have successfully used square-mesh codends since the mid 1990s.

It is notable that square-mesh codends have recently been tested in some multi-species fisheries for size sorting (e.g., Petrakis and Stergiou 1997; Campos et al. 2003a, 2003b; Guijarro and Massuti 2006; Bahamon et al. 2006). Generally, these studies showed that a square-mesh codend offers substantial improvement in size-selectivity for many important species, but it is obvious that a certain mesh opening is not optimal for all species; it will always be too large for some species and too small for others. Nevertheless, we believe that the full potential of the square-mesh codend principle has not yet been realised.

3.2.1.3 Turned Mesh Codend (T-90)

A square-mesh configuration is not the only alternative to the traditional diamond-mesh configuration. Some tests have been done with hexagonal mesh codends (e.g., Suuronen et al. 1991) and recently with turned mesh codends (Moderhavk 1997). A turned mesh codend is a codend where the netting has been turned 90 degrees (T-90). It is based on the observation that conventional knotted netting has more open meshes when turned 90 degrees. The construction is simple and low-cost, and easy to repair in case of damage. T-90 codends have recently been tested in the Baltic cod demersal trawl fishery (e.g., Dahm and Wienbeck 2000). According to these tests, this codend type has a good and stable selectivity and the whole area of the codend is selective. The debris gets out from the T-90 codend more easily than from the traditional diamond-mesh codend, resulting in less invertebrate discards. Codends tested so far have been made of polyamide (PA) and polyethylene (PE), and of single and double twine netting (of twisted, hollow braided and core-sheath braided twines). As in diamond-mesh codends, the number of meshes in the circumference of these codends has a marked effect on the selectivity of T-90 codends (Dahm and Wienbeck 2000).

There have been some concerns whether the selective properties of T-90 codends would last over time. It is notable that the recent

EU-regulation for technical measures in the Baltic cod fishery (enforced from January 2006) includes the 110 mm T-90 codend (50 meshes in circumference) as well as the 110 mm square-mesh window. The T-90 principle is currently being tested in other northern European fisheries. It is notable that positive results have not been found in all experiments. High variability in selectivity and relatively high selection ranges have been a cause of concern. Recently there have been tests of using T-90 netting in codend extension and trawl bellies.

3.2.1.4 Selection Panels and Escape Windows

Since late 1980s, substantial interest has been shown in Europe in the potential of using various types of selection panels (escape windows) attached in or near the codend to reduce the capture of juvenile fish (e.g., Arkley 1990; Briggs and Robertson 1993; Tschernij et al. 1996; Madsen et al. 1998, 1999, 2002; Graham and Kynoch 2001; Tschernij and Suuronen 2002; Graham et al. 2003, 2004). A selection panel installed in a standard diamond-mesh codend is often a flexible and practical means of excluding undersized roundfish. There are number of modifications (Fig. 3.1), and they are usually based on the square-mesh principle. A square-mesh panel maintains an open mesh structure irrespective of longitudinal tension in netting, providing improved chances for undersized fish to escape during the tow. An advantage with a selection panel is the ease by which selectivity can be changed, instead of manufacturing a whole new codend, only the panel needs to be replaced.

The optimal position and size of a selection panel are crucial for its proper performance. From underwater observations it has been shown that the majority of finfish escape attempts tend to be in the upper part of the codend, just in front of the catch (e.g., Main and Sangster 1981; Wardle 1989, 1992). This supports the use of panels in the upper codend where another advantage is that the panel is somewhat protected from stones and other larger debris that enter into the codend. The location of the panel in relation to the rear part of the codend is also important for effective selection performance and may vary according to the species (e.g., Madsen et al. 1999; Graham and Kynoch 2001; Graham et al. 2003; O'Neill et al. 2006). For the Baltic cod, to be effective, the panel has to extend into the rearmost part of the codend (Fig. 3.2; Madsen el al. 2002; Tschernij and Suuronen 2002). Moreover, to be effective with large codend catches, the panel should extend several metres along the codend.

Despite many positive features, selection panels have often been resisted by industry in Europe. This has mainly been due to disagreements concerning appropriate mesh size, type and positioning of the panel. Under

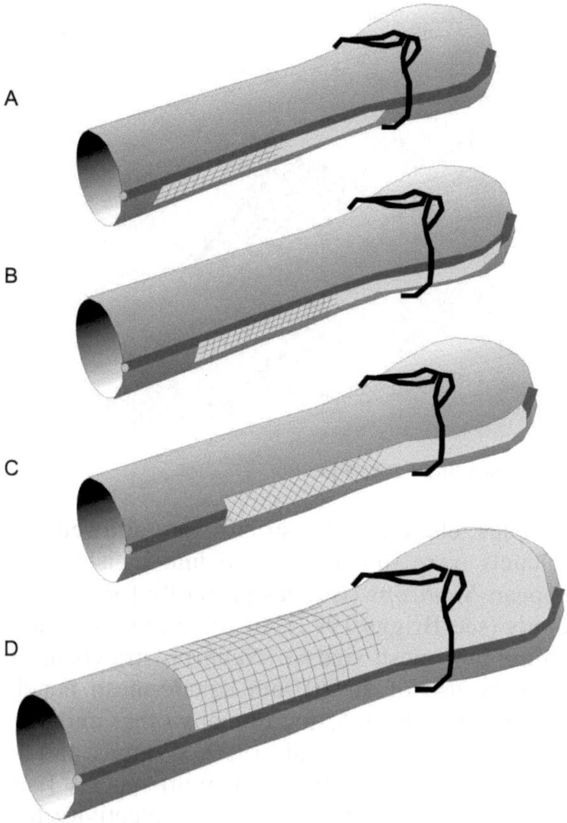

Fig. 3.1. Various escape panels enforced in the Baltic cod fishery in the 1990s and 2000s. A and B. two types of Danish exit window; C. the Swedish exit window and; D. the Bacoma panel

heavy force the distortion of panel meshes can be a serious problem. The weaker construction of a panel codend has also been a cause of concern. Many of these problems, however, can be solved by proper design. Although special knotless netting (e.g., Ultra-cross) is more suitable in a square-mesh panel, the practical constraints (price, net availability) are often forcing fisheries to use conventional knotted netting where knot slippage may be a problem.

Nevertheless, square-mesh windows have been the preferred choice of fisheries legislators in the European Union (EU); they have been legislated in several EU fisheries. These regulations include specific information regarding window size, construction, location, mesh size, twine type, and

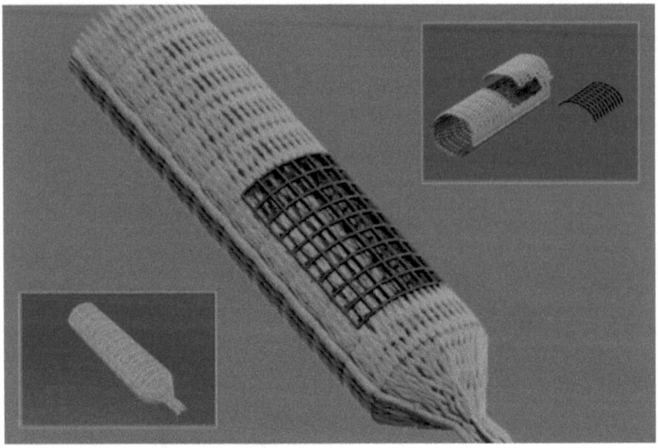

Fig. 3.2. The Bacoma square-mesh panel

whether lifting bags, chafers and restrictors are allowed. It is notable that square-mesh panels were first introduced into legislation in 1992 in the Northern European *Nephrops* fisheries primarily for improving the release of young gadoids (see Briggs 1992). In some regions, current EU regulations set the minimum mesh size for *Nephrops* trawls at 70 mm coupled with the mandatory use of an 80 mm square-mesh panel. Square-mesh panels have also recently been tested in many other multi-species fisheries for size sorting (e.g., Madsen et al. 1999; Campos and Fonseca 2004). Generally, these studies show that in a multi-species fishery, a square-mesh panel offers some improvement in size-selectivity for many species but a particular mesh opening will not be suitable for all species. Clearly, for different target species, different solutions are usually necessary.

3.2.1.5 Size Sorting Grids

Rigid sorting grids inserted into, or in front of, the trawl codend have been intensively studied and tested in many European fisheries for fish size sorting (i.e., for releasing juveniles – Larsen and Isaksen 1993; Suuronen et al. 1993; ICES 1996; Anon. 2002; Graham et al. 2004; Sardá et al. 2005). An advantage of a sorting grid is that the bar spacing of a grid is constant throughout the tow, regardless of towing speed and catch size. Another advantage is that practically all the fish can be forced to come into contact with a grid because it can be installed so that it completely blocks their way. These factors are considered to allow an effective and stable selection performance.

There are some indications that a codend equipped with a sorting grid has a sharper selectivity than a conventional codend with corresponding mean selection length (e.g., Larsen and Isaksen 1993; Isaksen and Valdemarsen 1994; Kvalsvik et al. 2002; Graham et al. 2004). However, few studies have directly compared the selectivity properties of a combined grid and codend to that of the codend only. Kvamme and Isaksen (2004) observed similar selection ranges for grid and codend selectivity for the north-east Arctic cod. When the grid was mounted in the codend, the mean selection length of the fish caught by the trawl increased by about 4 cm. Likewise, Jørgensen et al. (2006) presented experiments where the combined selectivity of a Sort-V sorting grid and codend were directly comparable to the selectivity of a conventional diamond-mesh codend in the demersal cod trawl fishery in the Barents Sea; there was no statistical difference in selection range. They concluded that the introduction of a 55 mm sorting grid in the Barents Sea demersal trawl fishery in 1997 only increased the mean selection length of cod; the same effect would have been attained with a traditional diamond-mesh codend made of 155 mm mesh. Their study also showed that the mean selection length of the grid codend was inversely related to catch rate when large catches were taken. This suggests that the sorting capacity of the grid is limited when large numbers of fish arrive at the grid simultaneously. The codend selection did not show such an effect. Grid selection, however, appeared less affected by seasonal variations in fish condition than the mesh selectivity. Whilst the results of these two studies are interesting, they may be only relevant for cod. Caution is still needed for any wider conclusions.

Rigid grids have been tested for improving *Nephrops* size selection in the North Sea and Skagerrak (Robertson and Shanks 1994; Valdemarsen et al. 1996; Anon. 2001) but so far very few commercial applications exist.

The survival of some fish that escape from a sorting grid may be slightly higher than that of escaping through a conventional codend mesh (e.g., Suuronen et al. 1996a), although this has not yet been demonstrated conclusively. Ingolfsson (2006) observed no significant difference in the mortality of haddock that escaped through a sorting grid and codend meshes. Soldal and Engås (1997) recorded no mortality in gadoid fish (cod and saithe) escaping through a sorting grid at low towing speed.

Relatively high prices, handling and safety problems, and the occasional blocking of grids by debris have resulted in industry resistance to grids, especially on vessels where the deck space behind the net-drum is small. Problems encountered in the installation and repair of sorting grids have also been a cause for concern. In the early 2000's an attempt was made to develop an user-friendly sorting grid system (EUROGRID) for the bottom trawl and seine net fisheries in the North Sea and adjacent waters with the aim of reducing the by-catch of juvenile gadoid fish (Anon. 2002). The 'Eurogrid' is made of polyamide; a material that has a high elasticity and strength. The grid is constructed of two hinged sections to enable it to be wounded around the net drum during haul-up. The grid is mounted into a separate net section that can easily be inserted between the codend extension and belly section of the trawl. This allows a damaged grid or netting section to be quickly replaced. The grid measures 1.5 x 0.75 m for average-sized North Sea trawlers and 1.2 x 0.6 m for the smaller vessels. The light weight of the grid (9 kg) allows its use on vessels without a ramp. The grid angle is 35 ± 5° and the bar spacing 40 mm (bars are teardrop-shaped in cross-section). The performance of the grid is highly sensitive to its rigging and the construction of the guiding funnel (Fig. 3.3).

In spite of the extensive amount of scientific work done and promising results obtained in many trials, few fisheries use grids for size sorting. Commercial applications remain few in Europe, the most remarkable being the sorting grids developed in Norway and used in the Barents Sea since the mid 1990s. Recent developments of flexible, lighter and more user-friendly sorting grids offer opportunities for more practical operation. In particular, the new polyurethane polymers offer substantial flexibility (e.g., Loaec et al. 2006) and the grids are light and can easily be wounded onto net drum (Fig. 3.4), but the long-term mechanical strength and durability for repeated bending may be a concern. Excessive flexibility and lower stiffness of the grid bars may also result in changes to bar spacings. To prevent this, there may be a need to add supporting structures or shorten the openings; this could reduce the selectivity performance of the grid. It appears, however, that polyurethane polymers facilitate a wide range of flexible, light weight and inexpensive sorting grids that allow easy handling.

Fig. 3.3. The principle of a size-sorting grid and the effect of a guiding panel. A. a side and an aft-view of a grid; B. this type of rigging creates a homogenous water flow pattern where the speed of the water in front of the grid is more or less constant. If the water cannot easily pass through the grid, the main stream of the water and fish will be directed downwards, and not towards the grid. C. one way to prevent this is to use a panel to guide the fish upward so that they meet the grid at its front part, thus enabling an efficient size-selection along the whole area of the grid. Guiding panels, however, often cause problematic water turbulence in front of the grid. D. a grid installed so that the net section around the grid takes a conical shape which, D. allows a more favourable water flow pattern where the fish are guided towards the whole grid

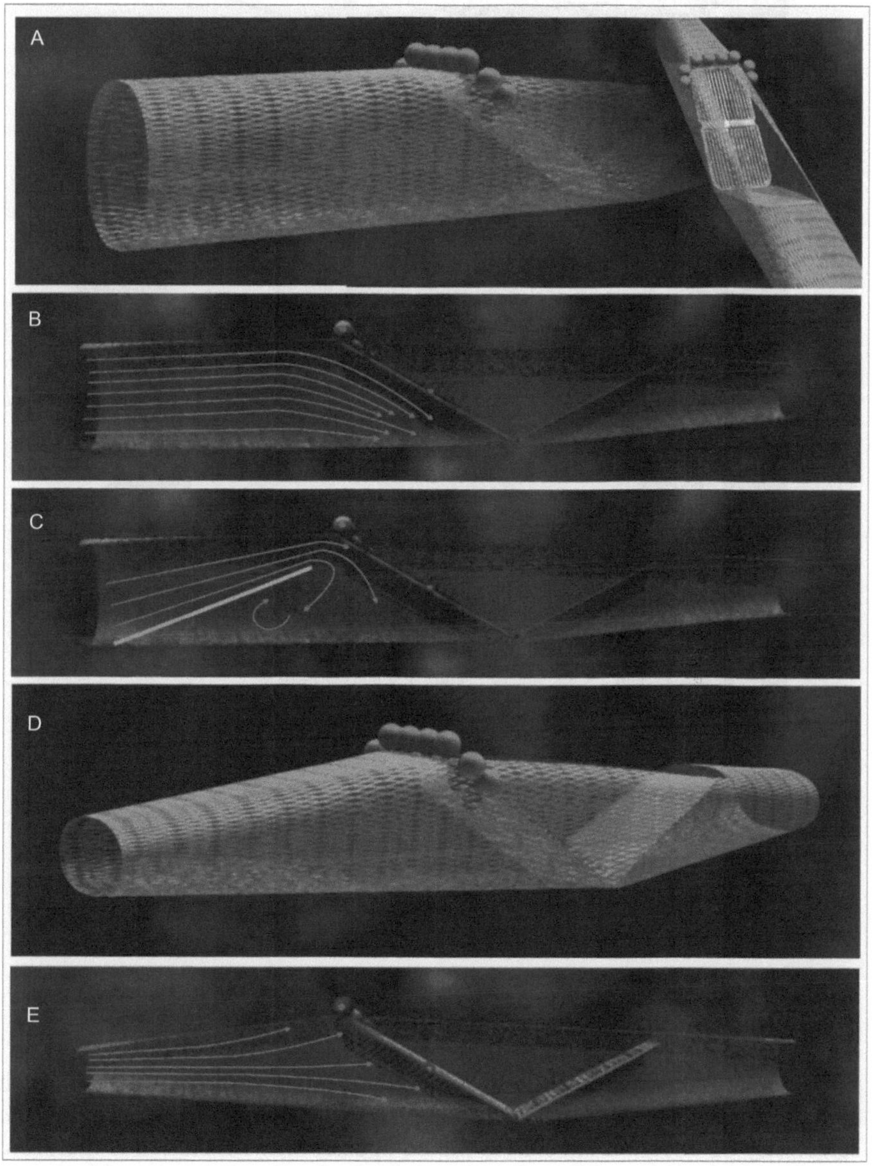
A
B
C
D
E

Fig. 3.4. A flexible sorting grid

3.2.2 Improving Species Selection

In demersal fisheries, catches are typically a mixture of roundfish, flatfish and shellfish, varying widely in shape, size and behaviour. Improving the selectivity of different species in mixed species trawl fisheries is therefore not easily achieved by simple gear modifications. Alternative solutions to this problem involves separating the species during fishing and ensuring suitable selection by the appropriate mesh size for each species, or simply excluding unwanted species by guiding them out of the gear before they enter the codend. Various types of species-selective gear modifications and devices have been developed and some designs are successfully being deployed. The well-known examples are turtle excluder devices and the Nordmøre grid.

3.2.2.1 Nordmøre Grid

The Nordmøre Grid (Fig. 3.5) is based on a rigid filtering system; the device comprises a series of parallel bars spaced to allow shrimp to pass through the grid into the codend whereas fish and other unwanted organisms are guided by the bars out of the trawl (e.g., Isaksen et al. 1992). The device is inserted in front of the codend. It was developed in Norway in the late 1980s to reduce the capture of non-wanted by-catch of juvenile finfish in northern water shrimp fisheries. It was originally invented by a Norwegian fisher who initially wanted to sort out unwanted jellyfish in his shrimp trawl fishery. This device proved to be an effective fish excluder, whilst simultaneously retaining the targeted shrimp. The grid was made mandatory in early 1990s in the Norwegian shrimp trawl fishery, and its use spread quickly in many shrimp fisheries around the world. The major disadvantage of early grids were their heavy and rigid construction and the tendency to become blocked by various objects, resulting in a loss of shrimp catch. However, the designs have been constantly improved and many practical problems have been overcome.

Fig. 3.5. The Nordmøre grid in action

The rigging angle of the grid has a marked effect on the Nordmøre Grid's performance. Grimaldo (2006) studied the effect of Nordmøre grid angle on its efficiency in releasing unwanted fish (cod, haddock, halibut, red fish) and on any loss of shrimp. He used a grid design where the bars were made of a combination of polyester and fibreglass and the frame made of Polyamide 6; this type of grid is light and has low deformation under stress. The bars were tear-drop shaped to improve water flow through the grid. The grid angle affected the percentage of shrimp loss and the escape of unwanted fish. When the angle was 33°, shrimp loss increased to 12.3% (from the original 5.6%) and the escape of fish reached 86% (originally 72%). When the grid angle was increased to 39 degrees, shrimp loss decreased on average to 4.2% and the escape of fish was reduced to 73%. Grimaldo notes that in addition to grid angle, the towing speed is an important parameter in the functioning of a grid; when towing speed is reduced the escape of unwanted fish increases considerably. The effective sorting area in relation to the area covered by the bars is also an important design factor because it directly affects water flow patterns.

3.2.2.2 Modified Nordmøre Grids and Other Excluding Grids

Various modifications of the Nordmøre grid principle have been tested for species selection in many fisheries other than shrimp fisheries. However, relatively little species selection work has been done on the use of grids in the *Nephrops* fishery (but see Catchpole et al. 2006; Graham and Fryer 2006). One reason for this is that finfish by-catch in most *Nephrops* fisheries has a relatively high commercial value; these fish are usually retained and marketed (at least those fish larger than the minimum landing size). This makes their total exclusion often unacceptable for fishers. However, in Swedish national waters it is now mandatory for the inshore *Nephrops* fleet to fish with species-selective trawls by having a Nordmøre type of grid with a bar spacing of 35 mm to exclude fish from the catch.

Fonseca et al. (2005) tested a modified Nordmøre grid for by-catch reduction in the Portuguese multi-species trawl fishery targeting three crustaceans (rose shrimp, *Nephrops* and red shrimp). They obtained an encouraging exclusion (50–70%) of non-target fish species (blue whiting, boardfish) but the average loss of 15% of large-size *Nephrops* raised some concern. In the North Sea shrimp fishery, Madsen and Hansen (2001) tested a flexible grid system made of polyamide with a fish escape hole at the top and a *Nephrops* escape hole at the bottom. Their results showed that the grid achieves a substantially greater reduction of by-catch (cod, whiting, *Nephrops*) than the square-mesh panel. However, significant work remains to be done before commercial applications are ready including work on the grid angle, clogging, handling, weight, material, price and maintenance.

Polet (2002) and Graham (2003) tested a modified Nordmøre Grid for by-catch reduction in the North Sea brown shrimp (*Crangon crangon*) beam trawl fisheries. The catch composition had a strong effect on the reduction of by-catch, but clogging of the grid was a major problem. When there were no clogging problems, the reduction of fish (> 70%) and benthos (65%) by-catch was quite high (Polet 2002) whilst the commercial brown shrimp catch was reduced by 15%.

Kvalsvik et al. (2006) tested grids in the North Sea industrial Norway pout (*Trisopterus esmarki*) trawl fishery to separate by-catch species like haddock and whiting from the smaller sized target species. They observed a significant reduction of by-catch; all haddock longer than 24 cm were sorted by the grid (at 22 mm bar spacing). However, there was also a substantial loss of target species. The rigging of the grid and the guiding funnel had a marked effect on the sorting efficiency. Eigaard and Holst (2004) tested the selectivity of a composite gear that consisted of a sorting grid and square-mesh window that retained the larger marketable by-catch fish in the fishery for Norway pout. The reduction of undersized haddock and whiting was 37–57%.

Zachariassen and Thomsen (2006) tested a range of rigid and flexible grids for reducing the by-catch of cod and saithe in the semi-pelagic blue whiting (*Micromesistius poutassou*) fishery in the North East Atlantic. The best results were obtained using flexible grids made of plastic tubes. By-catch has been reduced by 95% without losing more than 1% of the targeted catch.

It is notable that grids can have a dual function in selectivity. They may be used simply to sort individuals by size. However, they may also be used to guide unwanted organisms to an area where escape can occur by simultaneously allowing size sorting of the target species, i.e., they can function as a guiding and sorting device. Clearly, species separation by a grid is easier when the species are very different in size or shape, or when there are large differences in behaviour between the species.

3.2.2.3 Selective Ring Device

Some entirely new design principles for species-selection in shrimp trawl fisheries have recently been tested in Norway (Valdemarsen 2005). One of these designs is the selective ring device. It is based on an assumption that nearly all shrimp that enter the trawl mouth will hit the netting and thus be guided along it towards the codend, whereas fish will react to approaching netting by trying to avoid it, and thus stay at some distance away from the netting when they are on their way towards the narrow aft belly of the trawl. The ring device (diameter 1.6 m) is inserted in the aft belly of the shrimp trawl. The 10 cm slot around the ring perimeter is divided into a lower and upper half, each covered with a collecting bag. The central hole was covered with a third collecting bag. Most of the shrimp pass through the lower half of the ring slot whereas roundfish pass through the central hole of the ring. For shrimp that are mainly distributed in the vicinity of the bottom this means that the bottom panel of a trawl can be the main guiding panel. Clearly, this principle is based on behavioural differences between fish and shrimp; unwanted fish will swim out of the trawl through the outlets whilst shrimps passively swim into the codend.

3.2.2.4 Horizontal Separator Panel

Behavioural observations of groundfish have demonstrated that species like haddock, saithe and whiting may swim upwards when entering the trawl mouth whereas cod, many flatfish and *Nephrops* have less tendency to do so (e.g., Main and Sangster 1982a, 1982b, 1985; Wardle 1992). One practical application of this difference is to insert a horizontal dividing panel inside the trawl and have upper and lower codends with different mesh sizes, providing the opportunity to manipulate the overall selectivity to suit the species entering each codend. In general, cod, flatfish and *Nephrops* fall back under the separating panel, while haddock and whiting tend to pass over the panel and into the upper codend (e.g., Main and Sangster 1985; Moth-Poulsen 1994; Arkley et al. 1995; Engås et al. 1998). Experiments have shown that separation efficiency is strongly affected by the vertical height of the panel´s front edge and visibility. Engås et al. (1998) showed that the separation of cod, haddock and saithe by a horizontal panel was 60–90% in most cases. It is interesting that in their underwater video observations, haddock appeared to often enter the trawl at all levels, but many of those in the lower half swam upwards and through the separating panel as they passed towards the trouser codends; the final separation was best with haddock.

It is noteworthy that full-length horizontal separator panels have proved difficult to rig, repair and maintain, and therefore have often been resisted by fishing industries. Simpler and more robust constructions are now being designed for use in north east Atlantic fisheries. Ferro and Kynoch (2006) described a new horizontal panel design that is inserted only in the aft part of the tapered body of the trawl. This horizontal panel has been successfully tested in the northern North Sea mixed-species whitefish fishery. In general, more than 70% of haddock, whiting and saithe entered the upper compartment whereas more than 70% of cod, monkfish and flatfish entered the lower compartment. For haddock, some length-related effects in separation were observed. Towing speed was found to have no significant effect on separation (range 2.6–4.4 knots). Ferro and Kynoch (2006) stress that a horizontal panel helps to maintain suitable selection processes for different commercial species caught in a mixed fishery.

3.2.2.5 Inclined Separator Panel

Inclined separator panels made of netting inserted in front of the trawl codend can be used as an alternative to grids and can assist in species selection in *Nephrops* fisheries. They have already been introduced in the Irish Sea to reduce the by-catch of cod and other valuable whitefish species. Inclined separator panels are easier to handle than grids and are relatively inexpensive. The panel is fitted into the modified extension piece of a standard *Nephrops* trawl at approximately 30° angle to divert whitefish species towards an escape hole in the top of the trawl (see ICES 2004). The panel starts 50 meshes before the codend with the leading edge approximately 30 cm above the bottom sheet, allowing the passage of *Nephrops* and species such as monk and flatfish into the codend, while guiding cod, haddock and whiting out of the escape hole. In fishing trials, about 90% of whiting and 77% of cod have been released whereas the majority of *Nephrops* have been retained.

3.2.2.6 Sieve Net

In the commercial brown shrimp fishery, substantial progress in discard reduction has achieved by the use of sieve-nets (Revill and Holst 2004). The problem in the brown shrimp fishery is the substantial by-catch of young fish that are discarded. EC Council Regulation 850/98 requires that all vessels engaged in brown shrimp fisheries in European waters must have sieve nets or separator grids fitted into their trawls (for species selection). However, a sieve net (veil net) reduced the capture of small discard-sized brown shrimps but its effectiveness varied considerably between fishing fleets. In the Wadden Sea fishing grounds discarding of small commercial fish species such as juvenile plaice remained substantial despite the widespread use of sieve nets. In other grounds that have larger fish as discards, the introduction of this measure was predicted to result in larger benefits. Clearly, some technical solutions are simply not applicable in all regions.

3.2.2.7 Set-back Headline (Cut-Away-Trawl)

Modified trawls with set-back (cut-away) headlines allow certain fish species to escape before they enter the trawl. As *Nephrops* tend to keep low as they enter the trawl mouth, the only reason for having an upper panel above the groundgear is to prevent the escape of fish that tend to rise as they fall back into the trawl. This behaviour pattern can be used for species-selection by extending the headline downwards. These types of trawls are called cut-away-trawls. Recent trials by the Sea Fish Industry Authority in the United Kingdom have shown that by extending the headline in conjunction with a large mesh panel behind, it is possible to exclude some finfish species (haddock and whiting) from a *Nephrops* trawl but the results were inconclusive for cod (Arkley and Dunlin 2003a, 2003b). These designs tend to mimic the operation of traditional low headline *Nephrops* trawls that have low finfish by-catches (ICES 2004). Few commercial applications exist yet.

3.2.2.8 Large Mesh Top Panel Net

Thomsen (1993) demonstrated that gadoid by-catch in otter trawls can be reduced by placing large meshes on the top panel of the trawl body. Madsen et al. (2006) further developed the concept. A species-selective trawl that would catch flatfish such as plaice and flounder but would release all cod and other gadoid fish would be desirable for exploiting flatfish stocks without affecting the depleted cod stocks in the North Sea. To allow gadoid fish to swim out of the trawl, a large part of trawl upper panels, extending from the headline to the first belly section, was made of large meshes (400 mm full mesh). The vertical opening of the trawl was kept low by using little flotation on the headline. In addition, a 130 mm square-mesh panel was attached in front of the codend to further improve the exclusion of gadoids. The trawl caught more plaice and flounder, and markedly reduced the catch of cod, particularly those smaller than the minimum landing size, when compared to a conventional flatfish trawl. Van Marlen (2003) obtained a reduction of 30–40% for cod and whiting with no loss in flatfish when using a large-meshed top panel in a beam trawl.

Present European Union regulations require a large-mesh (140 mm) escape panel to be inserted in the upper belly of *Nephrops* trawls, directly behind the headline. The operating principle is to provide an escape opportunity for roundfish as they rise and fall back into the trawl. An advantage with these types of panels is that they are cheap and easy to install, and they maintain the original trawl geometry and stability. However, no adequate assessment exists of the efficiency of such panels. It is known, however, that loss of some target species can be significant with incorrect rigging of the panel. It is likely that these designs could be further improved.

3.2.2.9 Selective Ground-gear

It has often been discussed that ground gear equipped with special escape gaps may allow some species to escape under the trawl but this solution is largely untested. Ingolfsson and Jørgensen (2006) demonstrated and quantified the escape of Atlantic cod, haddock and saithe beneath a commercial bottom-rigged trawl with rockhopper groundgear in the Barents Sea. To collect the escapees, three sampling bags were attached behind the goundgear. Approximately one third of the cod and a quarter of the haddock available to the trawl escaped beneath the trawl. The escape of saithe was substantially less: such an effect could be used in species selection. Furthermore, the escape of cod was length-dependent, the smaller cod escaping more frequently below the trawl. The estimated length at 50% escape was 38.5 cm but the selection range was high (34.1 cm). This length-dependence was less pronounced in haddock, and the escape of saithe exhibited no length-dependence. Fish abundance had no observable effect on escape.

This study clearly demonstrates that a substantial part of the species- and size-selection may take place in front of the trawl and this information could be used when developing more selective trawl gears. It is interesting that about 50% of fish caught in the collecting bags showed scale abrasion, suggesting that many of them had been overrun or contacted by the gear. Little is known whether these damages are fatal for these fish and this should be explored in future studies.

3.2.2.10 Benthic Release Panels and Electric Stimuli in a Beam Trawl

Beam trawls are commonly used in the North Sea on flat bottoms, mainly to catch flatfish such as plaice and sole, but also for shrimp. The beam is supported at each end by a trawl head that has a steel plate (shoe) welded to the bottom of the beam. The steel plates are in direct contact with the seabed when fishing. Beam trawls are usually equipped with tickler chains to dig the flatfish off the seabed, and, on rougher grounds, to prevent boulders from being caught. By-catch in these gears can be high and possible modifications in beam trawls and their operations to reduce by-catch are currently being explored. Reducing the amount of chain or modifying the chain design (e.g., using parallel tickler chains) can reduce benthic by-catch but may also markedly reduce the catching efficiency of target species (van Marlen et al. 2005). Benthic release panels and drop-out panels in the belly of beam trawls have been tested with up to 80% release of benthic by-catch but often with some loss of target species (e.g., van Marlen et al. 2005; Revill and Jennings 2005). Fonteyne and Polet (2002) obtained promising results in the reduction of benthic organisms with square-mesh windows inserted in the belly just in front of the codend. The use of electric stimuli as an alternative to chains for digging out flatfish (pulse-beam trawl) is promising but requires further testing (van Marlen 2000). Polet et al. (2005) tested electric pulses in the North Sea brown shrimp beam trawl fishing to reduce finfish by-catch. The basic idea was to selectively invoke a startle response with shrimp without stimulating by-catch species. A selective ground-gear could then be used in combination with the electric pulses to obtain cleaner catches. The preliminary results showed that the use of such electric stimuli has substantial potential.

3.2.3 Do Fish that Escape from a Trawl Codend Survive?

It has been demonstrated that most roundfish that are discarded from a vessel deck do not survive and are often taken by seabirds following fishing vessels (reviewed by Chopin and Arimoto 1995; ICES 2000; Suuronen 2005). Further, those fish that escape a trawl codend during a haul may not always survive. Selective fishing can be justified only if significant numbers of escaping animals survive. If most of them die, selective devices are of little conservation value. In towed fishing gears, escape often occurs after the fish have been subjected to a wide variety of stressors and possible damage through contact with other fish, debris or the gear itself. Nevertheless, experiments conducted in Scotland, Norway and Finland have shown high (80–100%) survival likelihood for many gadoid fish (e.g., cod, saithe) that escape from trawl codends (Main and Sangster 1990, 1991; Soldal et al. 1993; Sangster et al. 1996; Suuronen et al. 1996b, 2005; Soldal and Engås 1997; Wileman et al. 1999). For haddock and whiting, observed survival rates have been somewhat lower and more variable; around 60–90% (e.g., Sangster et al. 1996; Ingolfsson 2006). Substantially lower survival rates (10–50%) have been recorded for some pelagic species such as herring and vendace (e.g., Suuronen et al. 1995, 1996a, 1996c). Clearly, the robustness and ability of various species to withstand physical injury and fatigue associated with capture and escape vary markedly. Moreover, the smallest escapees often appear the most vulnerable. There are some indications that escape at night can result in higher mortality than escape in daylight conditions (Suuronen et al. 1995) but this has not been demonstrated conclusively.

It is notable that very little is known about the survival of fish that escape during the haul-up of a trawl; survival may not be as high among fish that escape near the surface than among those that escape at the fishing depth during towing because the former are also vulnerable to predation by sea birds. Nevertheless, the survival likelihood of fish escaping from a fishing gear, whether it takes place during the capture or haul-up process is, in practice, always higher than survival of fish that are discarded from a vessel deck.

When developing selective fishing gears and practices, it is important to address the whole range of stressors caused by the capture and selection process and there are various options available to improve survival

(Suuronen 2005). Firstly, fish that escape from a fishing gear should do so quickly and should not enter into the aft part of the codend, where the risk of serious injury is greatest. Installing escape panels or other sorting devices at strategic positions in a fishing gear can enhance escape and the survival of juveniles and non-target species. Furthermore, facilitating voluntary escape through various constructional and operational solutions would increase the likelihood of survival. The use of non-abrasive netting materials, the exclusion of debris and large objects from codends, and better design, operations and rigging of nets could further improve survival. In some cases, however, the use of alternative fishing methods (such as pots and seine nets) may be the only appropriate approach to reduce unaccounted mortality.

3.2.4 Potential Benefits and Costs of Improved Selectivity

3.2.4.1 Accounting Unaccounted Mortality

Relatively few studies have been done to investigate the costs and benefits of selective fishing gears. Most quantitative assessments of such impacts focus on the medium- and long-term effects and are usually predicted to be positive. For instance, Kvamme and Frøysa (2004) assessed the effects of the sort-X grid system that became mandatory in 1997 in the demersal trawl fisheries of northeast Arctic cod for size sorting. Their simulations showed that there would be substantial long-term gains, in terms of both stock size and catches, from increasing the mean retention length by 5 to 8 cm (from the present 47 cm). Catches of three- to four-year-old fish would decrease, while catches of fish of six years and older would increase within a few years. It is notable that north-east Arctic cod reach maturity when they are 6 to 12 years of age and 65 to 105 cm long. Hence, immature fish would be the most affected by the change in selectivity. Kvamme and Frøysa (2004) pointed out that the change in selectivity would lead to a more efficient exploitation of the stock's growth potential, and more fish would have a chance of growing to mature size and spawn. This would increase the spawning biomass and result in greater and more stable catches within a few years. They noted, however, that the total catch would decrease during the first three years following the implementation of a mesh size increase. It is worth noting, however, that in this work they assumed that all escapees would survive.

Ingólfsson (2006) assessed the effect of escape mortality on the Barents Sea haddock stock. His analysis indicates that, with the present fishing mortality, the annual escape mortality at the stock level is about 3% for 20 cm haddock and declines with length to 1% and 0.3% for 30 cm and 40 cm haddock respectively. Escape mortality corresponds to a removal of

6.5 million 0.1 kg individuals, weighing about 650 metric tons. If the fishing mortality is as high as in the North Sea, the effects of escape mortality would be larger.

Breen and Cook (2002) analysed the potential impacts of selective fishing on North Sea haddock assessments by taking into account the unaccounted mortality. Their simulation was run with discard mortality set at zero (no discards) and one (all discards die) and with varying escape mortalities. The simulation showed that including discard mortality significantly increased fishing mortality estimates, particularly for ages one and two, and including escape mortality (assuming that 25% of escaping fish die) produced less significant but still substantial increases in fishing mortality (38% at age one; 7% at age two). That is, their analyses showed that compared to escape mortality, discarding has a far more profound effect on the fishing mortality of haddock. Furthermore, the relative importance of escape mortality decreases as age increases. Their analyses provide a useful insight into the relative importance of the different components of fishing mortality (landing, discard and escape mortality) for the stock-assessment process. They also assessed the long-term benefits of increasing the minimum legal mesh size. Their analyses showed that this benefit is greatly reduced if, for instance, only 25% of escaping fish die. Significant benefits would be obtained only if most escapees survive.

Rahikainen et al. (2004) applied length-specific selection and escape-mortality functions to estimate the total quantity of escapees that die and the actual removals from Baltic herring stock in the northern Baltic Sea. They assumed that the smallest (< 12 cm) escapees have 100% escape mortality and that herring of 12 to 17 cm would have an escape mortality of 90%. Their analyses suggested that more age 0 to one year old herring die as a result of escape from trawl codends than are landed. Their analyses showed that the effect of fishing-induced escape mortality decreases as a function of age and size, so that the impact on estimated recruitment and fishing mortality at age one is considerable, while it is almost irrelevant at age two and older. The actual fishing mortality at age one was estimated to be more than twice as high as estimates of fishing mortality based on unadjusted data. Rahikainen et al. (2004) emphasized that correct catch and mortality data are necessary for age-structured assessment models and such data may be biased due to unaccounted mortality associated with escape from trawl gears.

Kuikka et al. (1996) assessed the effect of mesh size increases on the economic value of the annual herring catches in the northern Baltic Sea. Their results showed that under the conditions prevailing in 1974 to 1992, the increase in codend mesh size would have led to reduced catches and lower yield-per-recruit values. The magnitude of the estimated reduction of catches varied greatly, according to the growth and natural mortality of the population. The calculation suggested that, in order to make an increase in mesh size profitable for this fishery over the long term, the price of large herring processed for human consumption would have to be approximately six times greater than that of smaller herring, or the survival of codend escapees would have to be increased to 80% from its current estimated level of about 15%.

Some of these studies clearly demonstrated that for species incurring a high post-escape mortality, there may be no biological or economic justification for a mesh size increase. Clearly, unless the level of escape mortality is known, the benefits of a change in selectivity could be largely overestimated. In the worst case, this type of unaccounted mortality can have a negative effect on fish stocks because overall fishing mortality may be underestimated. The problem of poor survival after escape may be a common characteristic of many pelagic fisheries and therefore mesh size management may not be the most appropriate tool to manage them (Suuronen et al. 1997).

3.2.4.2 Potential Short-term Losses

The short-term effects of increases in selectivity may require special attention. The following examples from the Baltic cod and Mediterranean mixed-species trawl fisheries demonstrate the importance of understanding and addressing short term effects.

In 2002, a highly size-selective 120-mm square-mesh panel (the Bacoma panel; see Madsen et al. 2002; Tschernij and Suuronen 2002) was enforced in the Baltic cod demersal trawl fishery. The decision to do this was based on long-term projections that suggested that there would be a substantial increase in spawning stock size and a marked reduction in discards if a larger panel mesh size was enforced (Kuikka et al. 1999; Suuronen et al. 2000). The short-term effects of a new selectivity pattern were modelled with a stochastic size-selective simulation model (Tschernij et al. 2004). Selectivity estimates based on vessel-type and catch-per-unit-of-effort (CPUE) data from the Baltic cod demersal trawl fishery were used to estimate catch losses. The simulations suggested that when the codend selectivity is increased due to an increase in the Bacoma panel from 105 to 120 mm, the overall loss in catch of fish of marketable size during the first month would be around 40 to 50% (with the same fishing effort). The discarding of undersized cod would decrease by about 70%. If fishers decided to compensate their loss in marketable catch by increasing their fishing effort, they would have to increase it by 55 to 90% and Tschernij et al. (2004) suggested that fishers were unlikely to increase their efforts to such a large extent. Instead, they might try to circumvent the regulations by intentionally decreasing the selectivity of their gear. In fact, widespread gear manipulation – legal and illegal – was observed in 2002 and 2003 in the main fishing grounds (Suuronen and Tschernij 2003). Fishers were not able to adapt to heavy losses in catches, which apparently were often larger than predicted by the simulations. The overall fleet selectivity did not improve; instead, it may have got worse than it was before the decision. Consequently, in September 2003, the minimum mesh size of the Bacoma window was reduced from 120 to 110 mm, leading to a greater compliance.

This example demonstrates that even in a case where fishing targets almost exclusively one species, increasing mesh size may be very complex even though the biological preconditions appear favourable. This case also demonstrates that too large an increase in selectivity may not be commercially acceptable. Gears will be manipulated and rules will be circumvented if the losses are too large (see also Ferro and Graham 2000; Halliday and Pinhorn 2002). Clearly, short-term effects should be addressed in management plans; it is not enough to assess only the long-term effects of a mesh size increase.

Relatively high short-term economic losses (12–33%) were also estimated by Bahamon et al. (2006) for the 40 mm square-mesh codend in shallow shelf fishing grounds (depths < 100 m) in the Mediterranean; this was due to the escape of a high number of accompanying species with a relatively high commercial value. These losses may cause substantial resistance by the fishing industry to accept such a codend although, for many species, it would reduce discards and the sorting work on deck. In deeper slope fishing grounds (approximately 400 m depth), short-term economic losses would be substantially smaller than in the shallower shelf grounds due to a smaller number of commercial species. In any case, an increase in fleet selectivity would increase the average age-at-first-capture for most species and therefore would improve the overall situation and should increase long-term total yield from the fishery even if a precise optimum is not achieved for all species. Clearly, there is a general need to identify the potential long term benefits and short term losses of any changes in selectivity.

3.3 Solving By-catch Problems in Fisheries with Passive Gears

3.3.1 Gillnetting

Bottom-set gillnets are widely-used throughout European fisheries and improved materials and techniques have allowed their expansion into rougher grounds and deeper waters. In general, gillnets are considered highly size-selective (Bjordal 2002), however, species selectivity can be poor, and in particular the entanglement of sea birds, turtles and marine mammals in pelagic gillnets has aroused significant concerns. Several potential solutions have been explored. Acoustic scaring devices (pingers) for deterring cetaceans have been introduced in many fisheries, but pingers are considered labour-intensive and expensive by fishers. Other solutions being explored include improved setting of nets and the use of netting materials in which cetaceans are not easily entangled. In deeper water gill-netting, reef-forming organisms and other sessile epibenthic organisms frequently become entangled in gillnets and are damaged when they are hauled. These problems apparently can be reduced by raising the ground-line of gillnet above the bottom, but this may reduce the catching efficiency of certain target species.

Ghost fishing by lost gillnets can be a significant problem – especially in deep, low-current and cold-water grounds; gillnets may continue to catch target and non-target species for up to two years (e.g., Humborstad et al. 2003; Tschernij and Larsson 2003; Ayaz et al. 2006). So, lost nets can cause a substantial unaccounted fishing mortality. Techniques to collect lost nets have therefore been developed. Netting material has a substantial effect on ghost fishing efficiency and the longevity of a gillnet (Ayaz et al. 2006); hence, by proper choice of material, ghost-fishing mortality can be reduced. Another issue with these ghost nets is that, by accumulating detritus and biofouling on the lost nets, they gradually decrease their capture efficiency (e.g., Revill and Dunlin 2003).

In southern European countries, trammels nets are among the most important gears and are widely used in the coastal fishery to catch a variety of demersal species such as sole, sea bream, red mullet, shrimp, lobster and cuttlefish (e.g., Erzini et al. 2006). Trammel nets generally catch a wide size range of many species and, compared to gill-nets, they are less selective. Therefore, by-catch and discards can be substantial and modifications are needed for reducing unwanted by-catch.

3.3.2 Longlining

Longlines are considered a relatively selective gear although in some cases by-catch can be high and the fish released may suffer high mortality. Species- and size-selectivity of a longline gear can be affected by bait size and type (Bjordal 2002). Artificial baits that target particular species and sizes offer a promising area of research as does the new area of bait-odour release technology.

The design and size of hooks can affect selectivity in longlines but substantially less than bait. Baited lines can be hazardous to seabirds when they try to eat the bait on the hooks while these are near the surface behind the vessel. A solution to this problem is to make the baited hooks less accessible for seabirds. This can be achieved by using bird-scaring lines above the longline when setting, or by setting the longline through a tube that leads the lines directly underwater, thus making the baited hooks invisible or inaccessible to birds (e.g., Løkkeborg 1998, 2003). A range of other options have been developed, including setting longlines during darkness, and adding extra weight to lines so that they sink faster. Many of the solutions that have been developed to address this problem also reduce the loss of baits, and thereby increase the fishing efficiency of the gear. The incidental capture of sea turtles on longline hooks is a problem in certain European fisheries – in particular in the Mediterranean – but satisfactory technical solutions have not yet been found. Apparently, the wider the hook, the less likely it is that a turtle will swallow it. Hence, a circle hook may cause fewer 'deep-hookings' than the conventional J-hook. More research is needed to produce hooks and baits that reduce the capture of turtles and facilitate their release in these fisheries.

3.3.3 Trap and Pot Fishing

Fishing with traps and pots normally results in catches that are alive and uninjured, so in most cases unwanted by-catch organisms can be released with a good chance of survival (e.g., Siira et al. 2006). However, factors such as on-deck injury, barotrauma and thermal shock may jeopardize the survival of released organisms. Despite this, traps and pots offer the potential for low by-catch mortality in comparison with many other fishing methods. By-catches from traps and pots can be minimized by design elements of the gear, including appropriate mesh sizes, materials and twines, and choosing the correct size, shape, location and design of entrances and escape openings as well as excluder devices. Often the major problem with traps and pots is their low catching efficiency compared to many other fishing gears. The use of various types of baits in traps and pots has the potential to attract the target species and/or repel unwanted species. A substantial amount of research has been undertaken in the Faroe Islands to improve the catching efficiency of fish traps with the aim to make them an alternative fishing gear for traditional species such as cod and haddock (Thomsen 2006). One task has been to develop a long lasting and effective bait for the trap and promising results have been obtained with frozen bait soap.

Traps and pots are often lost at sea and they may continue to catch fish or other organisms via ghost fishing. Bio-degradable materials, galvanic timed releases and various escape vents can be used to reduce this ghost-fishing capacity. It is worth noting, however, that in environments where there is little natural structure or complexity, lost traps may add to habitat complexity and offer refuges for various species, thus functioning in the same manner as artificial reefs. Traps and pots offer marked potential to decrease habitat impacts in fragile grounds where active fishing methods may cause severe damages to benthic ecosystems. The potential for catching new target species that are not currently pursued with traps should be investigated in order to facilitate the movement away from gear that lead to greater levels of impact.

3.3.4 A Seal-safe Trap-net

The current yearly growth rate of the grey seal population in the northern Baltic Sea is about 10%. In a traditional salmon trap-net, seals can readily enter all parts of the gear and eat and/or damage the catch (Kauppinen et al. 2005). A seal can also tear a hole in the fish bag, allowing fish to escape through the hole and fishers have to waste substantial time at sea to repair gear damages due to seals. There is therefore a major conflict between coastal fisheries and seal protection policies. Practices that effectively minimize seal-induced damage, as well as reduce the incidental mortality of seals caught in gears (i.e., by-catch mortality of seals), are needed.

The potential of scaring away those seals that had learnt to feed on fish caught in fishing gear has been explored intensively in the Baltic Sea. It is clear, however, that scaring seals away from fishing gear – for instance with an acoustic device – is not an easy task, especially in remote and exposed off-shore areas where it is difficult to get electric power for the devices. Development of this technology will therefore take time. Meanwhile, to reduce seal-induced damage to gears and catches, modifications to gears that prevent seals from entering the fish bag of a trap-net have been developed (e.g., Lehtonen and Suuronen, 2004; Suuronen et al. 2006). Modifications that have been tested include a wire grid installed in the funnel to prevent seals (but not fish) from entering the fish bag and various types of fish bags made of extra-strong polyethylene netting to prevent seals from ripping through the netting. Lunneryd et al. (2003) tested a large-mesh middle-chamber to allow fish to escape the gear when chased by a seal. Substantial progress in protecting catches from seals has been achieved with the so-called pontoon trap (Fig. 3.6) that is now in wide use in Swedish and Finnish salmon fisheries. It is equipped with a fish bag made of double-layer netting held under tension.

Net materials also have a major impact on seal protection. The use of thick and stiff polyethylene netting in the wings and middle chambers of nets effectively prevents entangling of fish and thereby reduces their vulnerability to seal predation (Suuronen et al. 2006). More work, however, is needed, however, to find effective and acceptable methods to resolve this issue.

Fig. 3.6. A seal-safe pontoon trap that was invented in the late 1990s in Sweden

3.4 Conclusions

This chapter has shown that modifications to fishing gears and operations may significantly help to reduce by-catch and discards in European fisheries and that there has been substantial progress although a significant amount of work remains to be done. A large number of juvenile fish continue to be caught and discarded in European waters but there are possibilities to improve size and species selection in the fishing gears used. To enable the best possible selectivity, a combination of various modifications may be required. In general, the capture of a fish is not a passive process but depends largely on how fish react to the gear so to be efficient, selection systems have to be adapted to the specific behaviour characteristics of particular species. In most fisheries there is still insufficient quantitative information on the capture behaviour of key species and more work is obviously needed in this field.

It is unlikely that gear modifications alone will eliminate all adverse effects of fishing but there has been, and will continue to be, progress in this area (see also Hall et al. 2000; Cook 2003). Many current problems in the implementation of more responsible fishing gears and operations could be avoided with better dialogue and relations between policy makers and those fishers who have to live and work by those policies. Obviously, as an alternative to large-scale closures to fisheries as a means of reducing by-catch, the adoption of fishing gears that are more selective and minimise effects on benthic habitats are better options for the fishing industry.

One of the obstacles in the adoption of more selective gears is the general observation that they usually result in a loss of landings and revenue, at least in the short-term. Techniques that are not practical or increase costs will be resisted by users and will most likely fail. Success in implementing more sustainable fishing gears and practices therefore depends somewhat on the economic forces to which fishers are subjected. In fact, for commercial fishers, there are often strong economic incentives to use relatively non-selective gears in order to maximize short-term profits. Incentives that take into consideration short-term influences are often necessary to get fishers to adopt gears that meet discard-reduction objectives. For example, in Europe, such incentives as additional fishing days or improved access to fishing grounds for those who use selective gears have been developed and tested (i.e., extra days at sea permitted, extra quotas of fish, subsidies for initial costs, etc.).

It is also necessary that fishers can trust that there will be positive effects in the near future by switching to selective gears. Because of this, gear changes have to have predictable effects and to ensure confidence in projected medium and long-term gains. Moreover, objectives for by-catch reduction should be realistic and the regulations enforcing new technologies need to be consistent among jurisdictions. The same solutions are not effective in all fisheries so regional flexibility is necessary.

In many fisheries, enforced management measures have unintended effects on by-catches (these are regulatory-induced discards). For example, fish are discarded when a vessel has no quota for that species, or when fish are below minimum landing (or market) size. Failures of the quota system enforced in many European fisheries and in particular in mixed-species fisheries have contributed greatly to a large volume of fish that must be discarded. Inconsistent minimum landing size regulations among jurisdictions further influence wide-spread discarding practices.

It is important to recognize that economic pressures often magnify by-catch and discards problems. As exploitation increases, the average size of fish is reduced, and fishers tend to reduce further the selectivity of their gears, causing an increase in by-catch and discards. There are many fisheries where it is unlikely that success in reducing by-catch can be achieved without additional measures such as reducing fishing effort. Reducing exploitation improves the status of stocks and increases the abundance of large fish. This in itself provides an incentive for fishers to adopt more selective gears.

The ultimate success of technical measures largely depends on the willingness of the fishing industry to accept them. The fishing effectiveness and practicality of new designs are important because an inefficient gear will not be used, may be 'sabotaged', or may require so much additional fishing effort that overall impacts on ecosystems could actually be increased. Individual fishers often feel that they have little stake in contributing to by-catch reduction, and may require significant incentives to be motivated to reduce his by-catch. However, in order for a by-catch reduction system to be successful, all individuals in a fishery should participate. In short, individual fishers must not only understand the basic nature and magnitude of the problem but also believe that the resulting measures are effective and fair. Close cooperation between the fishing industry, scientists, managers and other stakeholders is necessary to develop and introduce environmentally friendly fishing technologies. Innovative management plans that offer positive incentives for the effective use of such fishing techniques should have a high priority. In conclusion, new responsible fishing technologies in Europe (and indeed throughout the world) should be enforceable, practical, acceptable to fishers and acceptable to management.

References

Anon (2001) *Nephrops* trawl discard reduction using activating selection grids (NETRASEL). Final Report, Project FAIR-CT-98-4164

Anon (2002) EUROGRID: Selective whitefish grid system for demersal towed gear fisheries in the North Sea and adjacent waters. Final Report, EC Project FAIR CT-98-3536, 45 pp

Arkley K (1990) Fishing trials to evaluate the use of square mesh selector panels fitted to *Nephrops* trawls. Seafish Report no. 383

Arkley K (1996) Separator trawls as a tool for improving selectivity for cod. Seafish Report no. 464

Arkley K, Dunlin G (2003a) Commercial proving trials of a new prawn trawl design. Seafish Report no. SR551

Arkley K, Dunlin G (2003b) Improving the selectivity of towed fishing gears. New Prawn Trawl Designs to Avoid Capture of Unwanted By-catch. Seafish Report no. SR542

Arkley K, Lart W, Swarbrick J (1995) Improved selectivity in towed gears – sea trials to evaluate selectivity of separator trawls used to catch mixed whitefish species. Seafish report no. 441

Ayaz A, Acarli D, Altinagac U, Ozekinci U, Kara A, Ozen O (2006) Ghost fishing by monofilament and multifilament gillnets in Izmir Bay, Turkey. Fish Res 79:267–271

Bahamón N, Sardà F, Suuronen P (2006) Improvement of trawl selective in the NW Mediterranean demersal fishery by using a 40 mm square mesh codend. Fish Res 81:15-25

Bjordal Å (2002) The use of technical measures in responsible fisheries: Regulation of fishing gear. In: Cochrane KL (ed) A fishery manager's guidebook:

Management measures and their application. FAO Fish Tech Pap no. 424, pp 21–47

Breen M, Cook R (2002) Inclusion of discard and escape mortality estimates in stock assessment models and its likely impact on fisheries management. ICES CM 2002/V:27

Briggs RO (1992) An assessment of nets with a square mesh panel as a whiting conservation tool in the Irish Sea *Nephrops* fishery. Fish Res 13:133–152

Briggs RP, Robertson JHB (1993) Square mesh panel studies in the Irish Sea *Nephrops* fishery. ICES Fish Capture Committee, CM 1993/B:20

Campos A, Fonseca P (2004) The use of separator panels and square mesh windows for by-catch reduction in the crustacean trawl fishery off the Algarve (South Portugal). Fish Res 69:147–156

Campos A, Fonseca P, Erzini K (2002) Size selectivity of diamond and square mesh cod ends for rose shrimp (*Parapenaeus longirostris*) and Norway lobster (*Nephrops norvegicus*) off the Portuguese south coast. Fish Res 58: 281-301

Campos A, Fonseca P, Erzini K (2003a) Size selectivity of diamond and square mesh codend for four by-catch species captured in the crustacean fishery off the Portuguese south coast. Fish Res 60:79–97

Campos A, Fonseca P, Henriques V (2003b) Size selectivity for four fish species of the deep groundfish assemblage off the Portuguese southwest coast: Evidence of mesh size, mesh configuration and cod end catch effects. Fish Res 63:213–233

Catchpole TL, Revill AS, Dunlin G (2006) An assessment of the Swedish grid and square-mesh codend in the English (Farn Deeps) *Nephrops* fishery. Fish Res 81:118-125

Chopin FSM, Arimoto T (1995) The condition of fish escaping from fishing gears – a review. Fish Res 21:315–327

Cook R (2003) The magnitude and impact of by-catch mortality by fishing gear. In: Sinclair M, Valdimarsson G (eds) Responsible Fisheries in the Marine Ecosystem. FAO and CABI International Publishing, pp 219–234

Dahm E (1991) Doubtful improvement of the selectivity of herring midwater trawls by means of square mesh codends and constructional modifications of diamond mesh codends. ICES CM 1991/B:2, 8 pp

Dahm E, Wienbeck H (2000) New ways for an improvement of the selectivity of trawl codends in the Baltic cod fishery. Meddelande fran Havsfiskelaboratoriet, Lysekil pp 80–93

Dahm E, Wienbeck H, West CW, Valdemarsen JW, O'Neill FG (2002) On the influence of towing speed and gear size on the selective properties of bottom trawls. Fish Res 55:103–119

Eigaard OR, Holst R (2004) The effective selectivity of a composite gear for industrial fishing: a sorting grid in combination with a square mesh window. Fish Res 68:99–112

Engås A, Jørgensen T, West CW (1998) A species-selective trawl for demersal gadoid fisheries. ICES J Mar Sci 55:835–845

Erzini K, Goncalves JMS, Bentes L, Moutopoulos DK, Herdando Casal JA, Soriguer MC, Puente E, Errazkin LA, Stergiou KI (2006) Size selectivity of trammel nets in southern European small-scale fisheries. Fish Res 79:183–201

European Commission (2001) Green Paper. The future of the common fisheries policy. Volume I. Luxembourg: Office for Official Publication of the European Communities, 47 pp

FAO (2005) Discards in the world's marine fisheries: an update (by Kelleher K). FAO Fisheries Technical Paper No. 470, Rome. 134 pp

Ferro RST, Graham N (1998) Recent Scottish data on demersal fish selectivity. ICES CM 1998/Open:03

Ferro RST, Graham GN (2000) A recent UK joint initiative to revise technical conservation measures regulating the design of mobile gears. ICES CM 2000/W:05, 6 pp

Ferro RST, Kynoch RJ (2006) A horizontal panel to separate species in a mixed whitefish fishery. EU 'Recovery' Project Status Report 1. Fisheries Research Services Report (in preparation)

Ferro RST, Robertson JHB (1988) Mesh selection within the cod-end of trawls. The effect of narrowing the cod-end and shortening the extension. Scott Fish Res Rep no. 39

Fonseca P, Campos A, Larsen RB, Borges TC, Erzini K (2005) Using a modified Nordmøre grid for by-catch reduction in the Portuguese crustacean-trawl fishery. Fish Res 71:223–239

Fischler F (2004) Fast-tracking the development of environmentally-friendly fishing methods. Conference on environmentally-friendly fishing methods. Dundalk, 11–12 March 2004. Available online at http://europa.eu.int/comm/fisheries/news corner/discours/speech46_en.htm

Fonteyne R, Polet H (2002) Reducing the benthos by-catch in flatfish beam trawling by means of technical modifications. Fish Res 55:219–230

GFCM (2005) Recommendations of Mediterranean Management. General Fisheries Commission for the Mediterranean. Rome, 21 – 25 February 2005, 12 pp

Graham N (2003) By-catch reduction in the brown shrimp, *Crangon crangon*, fisheries using a rigid separation Nordmøre grid (grate). Fish Res 59:393–407

Graham N, Fryer RJ (2006) Separation of fish from *Nephrops norvegicus* into a two-tier cod-end using a selection grid. Fish Res 82:111–118

Graham N, Kynoch RJ (2001) Square mesh panels in demersal trawls: some data on haddock selectivity in relation to mesh size and position. Fish Res 49: 207–218

Graham N, Kynoch RJ, Fryer RJ (2002) Square mesh panels in demersal trawls: further data relating haddock and whiting selectivity to panel position. Fish Res 62:361–375

Graham N, O'Neill FG, Fryer RJ, Galbraith RD, Myklebust A (2004) Selectivity of a 120 mm diamond cod-end and the effect of inserting a rigid grid or a square mesh panel. Fish Res 67:151–161

Grimaldo E (2006) The effects of grid angle on a modified Nordmøre-grid in the Nordic Shrimp Fishery. Fish Res 77:53–59

Guijarro B, Massuti E (2006) Selectivity of diamond- and square-mesh codends in the deepwater crustacean trawl fishery off the Balearic Islands (western Mediterranean). ICES J Mar Sci 63:52–67

Hall MA, Alverson DL, Metuzals KI (2000) By-catch: Problems and Solutions. Mar Pollut Bull 41:204–219

Halliday RG, Pinhorn AT (2002) A review of the scientific and technical bases for policies on the capture of small fish in North Atlantic groundfish fisheries. Fish Res 57:211–222

Herrmann B (2005) Effect of catch size and shape on the selectivity of diamond mesh cod-ends. II. Theoretical study of haddock selection. Fish Res 71:15–26

Humborstad O-B, Lokkeborg S, Hareide N-R, Furevik DM (2003) Catches of Greenland halibut (*Reinhardtius hippoglossoides*) in deep-water ghost-fishing gillnets on the Norwegian continental slope. Fish Res 64:163–170

ICES (1996) Report of the Study Group on grid (grate) sorting systems in trawls, beam trawls, and seine nets. ICES Fish Capture Committee CM 1996/B:1, 87 pp

ICES (2000) Report of the FTFB Topic Group on Unaccounted Mortality in Fisheries. ICES FTFB Working Group, Haarlem, The Netherlands, 10 – 11 April 2000

ICES (2004) The *Nephrops* fisheries of the Northeast Atlantic and Mediterranean – A review and assessment of fishing gear design. Graham N and Ferro RST (ed). ICES Cooperative Research Report no. 270, 40 pp

ICES (2005) Report of the Advisory Committee for Fisheries Management. Available online at http://www.ices.dk/ICES advice

Ingólfsson O (2006) Size selectivity and escape mortality of gadoid fish in the Barents Sea trawl fishery. PhD thesis, University of Bergen, 56 pp

Ingolfsson OA, Jørgensen T (2006) Escapement of gadoid fish beneath a commercial bottom trawl: relevance to the overall trawl selectivity. Fish Res 79: 303–312

Jørgensen T, Ingólfsson OA, Graham N, Isaksen B (2006) Size selection of cod by rigid grids – Is anything gained compared to diamond mesh codends only? Fish Res 79:337–348

Isaksen B. and Valdemarsen JW (1994) By-catch reduction in trawls by utilizing behaviour differences. In: Ferno A, Olsen S (eds) Marine Fish Behaviour in Capture and Abundance Estimation. Fishing News Books, Blackwell Science, Oxford, pp 69–83

Isaksen B, Valdemarsen JW, Larsen RB, Karlsen L (1992) Reduction of fish by-catch in shrimp trawl using a rigid separator grid in the aft belly. Fish Res 13:335–352

Kauppinen T, Siira A, Suuronen P (2005) Temporal and regional patterns in seal-induced catch and gear damage in the coastal trap-net fishery in the northern Baltic Sea: Effect of netting material on damages. Fish Res 73:99109

Kuikka S, Suuronen P, Parmanne R (1996) The impacts of increased codend mesh size on the northern Baltic Sea herring fishery: ecosystem and market uncertainties. ICES J Mar Sci 53:723–730

Kuikka S, Hildén M, Gislason H, Hansson S, Sparholt H, Varis O (1999) Modelling environmentally driven uncertainties in Baltic cod (*Gadus morhua*) management by Bayesian influence diagrams. Can J Fish Aquat Sci 56:629–641

Kvalsvik K, Misund OA, Engås A, Gamst K, Holst R, Galbraith D, Vederhus H (2002) Size selection of large catches: using sorting grid in pelagic mackerel trawl. Fish Res 59:129–148

Kvalsvik K, Huse I, Misund OA, Engås A, Gamst K (2006) Grid selection in the North Sea industrial trawl fishery for Norway pout: Efficient size selection reduces by-catch. Fish Res 77:248–263

Kvamme C, Frøysa KG (2004) Assessing the effects on stocks of selectivity changes in a fishery. Fish Res 69:283–293

Kvamme C, Isaksen B (2004) Total selectivity of a commercial cod trawl with and without a grid mounted: grid and codend selectivity of north-east Arctic cod. Fish Res 68:305–318

Larsen R, Isaksen B (1993) Size selectivity of rigid sorting grids in bottom trawls for Atlantic cod (*Gadus morhua*) and haddock (*Melanogrammus aeglefinus*). ICES Mar Sci Symp 196:178–182

Lehtonen E, Suuronen P (2004) Mitigation of seal-damages in salmon and white-fish trap-net fishery by modification of the fish bag. ICES J Mar Sci 61: 1195–1200

Loaec H, Morandeau F, Mellita M, Davies P (2006) Engineering development of flexible selectivity grids for *Nephrops*. Fish Res 79:210–218

Løkkeborg S (1998) Seabird by-catch and bait loss in longlining using different setting methods. ICES J Mar Sci 55:145–149

Løkkeborg S (2003) Review and evaluation of three mitigation measures – bird-scaring line, underwater setting and line shooter – to reduce seabird by-catch in the north Atlantic longline fishery. Fish Res 60:11–16

Lunneryd S-G, Fjälling A, Westerberg H (2003) A large-mesh salmon trap: a way to mitigate seal impact on a coastal fishery. ICES J Mar Sci 60:1194–1199

Machias A, Vassilopoulou D, Vatsos D, Bekas P, Kallioaniotis A, Papaconstantinou C, Tsimenides N (2001) Bottom trawl discards in the northeastern Mediterranean Sea. Fish Res 53:181–195

Madsen N, Hansen KE (2001) Danish experiments with a grid system tested in the North Sea shrimp fishery. Fish Res 52:203–216

Madsen N, Holst R, Foldager L (2002) Escape windows to improve the size selectivity in the Baltic cod trawl fishery. Fish Res 57:223–235

Madsen N, Moth-Poulsen T, Lowry N (1998) Selectivity experiments with window codend fished in the Baltic Sea cod (*Gadus morhua*) fishery. Fish Res 36:1–14

Madsen N, Moth-Poulsen T, Holst R, Wileman DA (1999) Selectivity experiments with escape windows in the North Sea *Nephrops* (*Nephrops norvegicus*) trawl fishery. Fish Res 42:167–181

Madsen N, Tschernij V, Hansen K, Larsson P-O (2006) Development and testing of a species-selective flatfish ottertrawl to reduce cod by-catches. Fish Res 78:298–308

Main J, Sangster GI (1981) A study of the fish capture process in a bottom trawl by direct observations from a towed underwater vehicle. Scott Fish Res Rep 23:1–23

Main J, Sangster GI (1982a) A study of separating fish from *Nephrops norvegicus* L. in a bottom trawl. Scott Fish Res Rep no. 24, 8 pp

Main J, Sangster GI (1982b) A study of a multi-level bottom trawl for species separation using direct observation techniques. Scott Fish Res Rep no. 26, 16 pp

Main J, Sangster GI (1985) Trawling experiments with a two-level net to minimise the undersized gadoid by-catch in a *Nephrops* fishery. Fish Res 3:31–145

Main J, Sangster GI (1990) An assessment of the scale damage to and survival rates of young fish escaping from the cod-end of a demersal trawl. Scott Fish Res Rep 46/90, 28 pp

Main J, Sangster GI (1991) Do fish escaping from cod-ends survive? Scott Fish Res Rep 18/91

Martín P, Carbonell A, Belcari P (2001) Estimation of trawl discards in the western Mediterranean. European hake (*Merluccius merluccius*) as case study. Final Report, EC DG Fisheries Study 00/009, 135 pp

Moderhak W (1997) Determination of selectivity of cod codends made of netting turned through 90°. Bull Sea Fish Inst 1(140):1–24

Moth-Poulsen T (1994) Development of a species selective whiting trawl. ICES CM 1994/B:22

O'Neill FG, Kynoch RJ, Fryer RJ (2006) Square mesh panels in North Sea demersal trawls: Separate estimates of panel and cod-end selectivity. Fish Res 78:333–341

Özbilgin H, Wardle CS (2002) Effect of seasonal temperature changes on the escape behaviour of haddock, *Melanogrammus aeglefinus*, from the cod-end Fish Res 58:323–331

Petrakis G, Stergiou K (1997) Size selectivity of diamond and square mesh codends for four commercial Mediterranean fish species. ICES J Mar Sci 54:13–23

Polet H (2002) Selectivity experiments with sorting grids in the North Sea brown shrimp (*Crangon crangon*) fishery. Fish Res 54:217–233

Polet H, Delanghe F, Verschoore R (2005) On electrical fishing for brown shrimp (*Crangon crangon*). I. Laboratory experiments. Fish Res 72:1–12

Rahikainen M, Peltonen H, Pönni J (2004) Unaccounted mortality in northern Baltic Sea herring fishery – magnitude and effects on estimates of stock dynamics. Fish Res 67:111–127

Reeves SA, Armstrong DW, Fryer RJ, Coull KA (1992) The effects of mesh size, codend extension length and codend diameter on the selectivity of Scottish trawls and seines. ICES J Mar Sci 49:279–288

Revill AS, Dunlin G (2003) The fishing capacity of gillnets lost on wrecks and on open ground in UK coastal waters. Fish Res 64:101–113

Revill A, Holst R (2004) The selective properties of some sieve nets. Fish Res 66:171–183

Revill AS, Jennings S (2005) The capacity of benthos release panels to reduce the impacts of beam trawls on benthic communities. Fish Res 75:73–85

Robertson JHB, Shanks AM (1989) Further studies of the size selection of *Nephrops* by different cod-ends. Scottish Fisheries Working Paper no. 1/89

Robertson JHB, Shanks AM (1994) Experiments with rigid grids in the *Nephrops* and whitefish fisheries. Marine Laboratory Report no. 4/79

Robertson JHB, Stewart PAM (1988) A comparison of size selection of haddock and whiting by square and diamond mesh codends. J Cons Cons Int Explor Mer 44:148–161

Sánchez P, Demestre M, Martin P (2004) Characterisation of the discards generated by bottom trawling in the northwestern Mediterranean. Fish Res 67: 71–80

Sangster GI, Lehmann KM, Breen M (1996) Commercial fishing experiments to assess the survival of haddock and whiting after escape from four sizes of diamond mesh cod-ends. Fish Res 25:323–346

Sardà F, Bahamón N, Sardà-Palomera F, Molí B (2005) Commercial testing of a sorting grid to reduce catches of juvenile hake (*Merluccius merluccius*) in the western Mediterranean demersal trawl fishery. Aquat Living Resour 18:87–91

Siira A, Suuronen P, Ikonen E, Erkinaro J (2006) Survival of Atlantic salmon captured in and released from a commercial trap-net: Potential for selective harvesting of stocked salmon. Fish Res 80:280–294

Soldal AV, Engås A (1997) Survival of young gadoids excluded from a shrimp trawl by a rigid deflecting grid. ICES J Mar Sci 54:117–124

Soldal AV, Engås A, Isaksen B (1993) Survival of gadoids that escape from demersal trawl. ICES Mar Sci Symp 196:122–127

Stergiou KI, Politou C-Y, Christou ED, Petrakis G (1997) Selectivity experiments in NE Mediterranean: the effect of trawl codend mesh size on species diversity and discards. ICES J Mar Sci 54:774–786

Suuronen P (2005) Mortality of fish escaping trawl gears. FAO Fish Tech Pap no. 478, 72 pp

Suuronen P, Millar RB (1992) Size-selectivity of diamond and square mesh codends in pelagic herring trawls: only small herring will notice the difference. Can J Fish Aquat Sci 49:2109–2117

Suuronen P, Tschernij V (2003) The problems encountered in the adoption of improved selectivity in the Baltic cod demersal trawl fishery. Theme Session Z: The Historical and Current use of Technical Conservation Measures and the Evaluation of their Effectiveness. ICES CM 2003/Z:07, 14 pp

Suuronen P, Erickson D, Orrensalo A (1996c) Mortality of herring escaping from pelagic trawl cod-ends. Fish Res 25:305–321

Suuronen P, Lehtonen E, Jounela P (2005) Escape mortality of trawl-caught Baltic cod (*Gadus morhua*) – the effect of water temperature, fish size and cod-end catch. Fish Res 71:151–163

Suuronen P, Lehtonen E, Tschernij V (1993) Possibilities to increase the size-selectivity of a herring trawl by using a rigid sorting grid. NAFO SCR Doc. 93/119, serial no. N2313, 12 pp

Suuronen P, Millar RB, Järvik A (1991) Selectivity of diamond and hexagonal mesh codends in pelagic herring trawls: evidence of a catch size effect. Finnish Fish Res 12:143–156

Suuronen P, Lehtonen E, Tschernij V, Larsson P-O (1996b) Skin injury and mortality of Baltic cod escaping from two codends equipped with exit windows. Arch Fish Mar Res 44:165–178

Suuronen P, Peres-Comaz JA, Lehtonen E, Tschernij V (1996a) Size related survival of Baltic herring (*Clupea harengus*) escaping through a rigid sorting grid and trawl codend meshes. ICES J Mar Sci 53:691–700
Suuronen P, Turunen T, Kiviniemi M, Karjalainen J (1995) Survival of vendace (*Coregonus albula* L.) escaping from a trawl codend. Can J Fish Aquat Sci 52(12):2527–2533
Suuronen P, Siira A, Kauppinen T, Riikonen R, Lehtonen E, Harjunpää H (2006) Reduction of seal-induced catch and gear damage by modification of trap-net design: Design principles for a seal-safe trap-net. Fish Res 79:129–138
Suuronen P, Kuikka S, Lehtonen E, Larsson P-O, Tschernij V, Madsen N, Holst R (2000) Improving technical management in Baltic cod fishery (BACOMA). Final Report EC 4th Framework Programme, Contract FAIR CT 96-1994. Brussels, EC, 106 pp
Thomsen B (1993) Selective flatfish trawling. ICES Mar Sci Symp 196:161–164
Thomsen B (2006) Long lasting bait for fish pots. ICES Symposium on Fishing technology in the 21st Century – Integrating Fishing and Ecosystem Conservation, Abstract ICES BOS06 2:136P
Tschernij V, Larsson P-O (2003) Ghost fishing by lost gill nets in the Baltic Sea. Fish Res 64:151–162
Tschernij V, Suuronen P (2002) Improving trawl selectivity in the Baltic – Utökning av trålselektion i Östersjön. TemaNord 2002, 512, 56 pp
Tschernij V, Suuronen P, Jounela P (2004) A modelling approach for assessing short-term catch losses as a consequence of a mesh size increase. Fish Res 69:399–406
Tschernij V, Larsson P-O, Suuronen P, Holst R (1996) Swedish trials in the Baltic Sea to improve selectivity in demersal trawls. ICES 1996 CM/B:25
Valdemarsen JW (2005) Selective shrimp trawl development using behaviour differences between fish and shrimp. ICES-FAO FTFB WG meeting, Rome 18 – 22 April 2005, 8 pp
Valdemarsen JW, Suuronen P (2003) Modifying fishing gear to achieve ecosystem objectives. In: Sinclair M, Valdimarsson G (eds) Responsible Fisheries in the Marine Ecosystem. FAO and CABI International Publishing, pp 321–341
Valdemarsen JW, Ulmestrand M, West CW (1996) Experiments on size-selectivity for Norway lobster using sorting grids in the aft trawl belly. In: Report of the Study Group on Grid (Grate) sorting systems in trawls, beam trawls and seine nets, Appendix VI. ICES CM 1996/B:1
van Marlen B (2000) Technical modifications to reduce the by-catches and impacts of bottom-fishing gears. In: Kaiser MJ, de Groot SJ (eds). The effects of fishing on non-target species and habitats: Biological conservation and socio-economic issues. Blackwell Science Ltd, Oxford, pp 253–268
van Marlen B (2003) Improving the selectivity of beam trawls in the Netherlands. The effect of large mesh top panels on the catch rates of sole, plaice, cod and whiting. Fish Res 63:155–168
van Marlen B, Bergman MJN, Groenewold S, Fonds M (2005) New approaches to the reduction of non-target mortality in beam trawling. Fish Res 72:333–345
Wardle CS (1989) Understanding fish behaviour can lead to more selective fishing gears. In: Campbell CM (eds) Proceedings of the World Symposium on

fishing gear and fishing vessel design. Marine Institute, St John's, Newfoundland November, 1988

Wardle CS (1992) Fish behaviour and fishing gear. In: Pitcher TJ (ed) The behaviour of teleost fishes, 2nd edition, London, Chapman and Hall, pp 606–641

Wileman DA, Sangster GI, Breen M, Ulmestrand M, Soldal AV, Harris RR (1999) Roundfish and *Nephrops* survival after escape from commercial fishing gear. Final report, EC Contract FAIR-CT95-0753. Brussels, EC

Zachariassen K, Thomsen B (2006) Sorting grids in large blue-whiting trawls. ICES Symposium on Fishing Technology in the 21st Century – Integrating Fishing and Ecosystem Conservation, Abstract ICES BOS06 2:137V

4 By-catch Reduction in Wire-mesh Fish Traps

JOHN STEWART

4.1 Wire-mesh Fish Traps

The use of traps to catch fish is one of the most primitive and oldest forms of fishing. Traps are passive fishing gears that have evolved from simple barriers made of rocks in flowing rivers, lakes and along coasts, through portable baskets made from woven sticks or palm leaves (which are still used today in some artisanal fisheries), to modern-day traps that are mostly covered with wire-mesh. Fish traps have been developed independently in many regions throughout the world and their designs are varied and numerous. Different trap designs are generally unique to particular locations or groups of fishers, and today, the use of wire-mesh fish traps is an established method of fishing in many parts of the world (see review by Mahon and Hunte 2001).

In general, wire-mesh fish traps are timber or steel framed, can be as large as 3 m x 2 m x 2 m, and are typically covered with galvanised hexagonal-shaped wire-mesh (Slack-Smith 2001). The mesh size and shape of such traps varies significantly throughout the world. Hexagonal mesh (sold as chicken mesh in many countries) is probably the most commonly used product (Fig. 4.1), of mesh sizes generally ranging between 3 cm (in the Caribbean – Ward 1988) and 5 cm (in Australia – Stewart and Ferrell 2003). These hexagonal meshes are measured as the shortest distance between the centres of wire strands. Some fisheries use traps that are covered with welded mesh – e.g., 50 x 50 mm welded mesh is used in South Australia (Grove-Jones and Burnell 1990) and 50 x 75 mm welded mesh is used in northern Australia (Whitelaw et al. 1991). Wire-mesh fish traps, regardless of the design, work by allowing fish to enter through funnels that taper from a wide entrance to a narrow opening within the trap. Fish are not

S. J. Kennelly (ed.), By-catch Reduction in the World's Fisheries, 75–93.

prevented from leaving the trap after entering, but their escape is inhibited by the small funnel size inside the trap. Traps may be baited or not – depending on the fishery.

The most common trap designs are the Antillean 'Z' trap, 'S' trap and arrowhead or chevron trap (all generally used in the Caribbean), rectangular traps (used in the Caribbean and Australia), round or 'O' traps (northern Australia) and 'D' traps (called 'Gargoor' in middle-eastern fisheries) that are used throughout the world (Fig. 4.2). These traps are all demersal fishing gears, however pelagic wire-mesh fish traps have previously been used in Australia to target species such as *Seriola lalandi*.

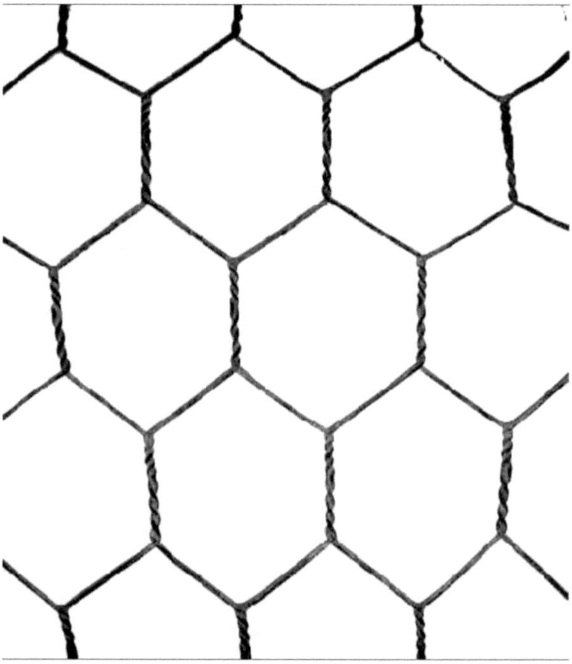

Fig. 4.1. An example of hexagonal wire-mesh typically used to cover fish traps

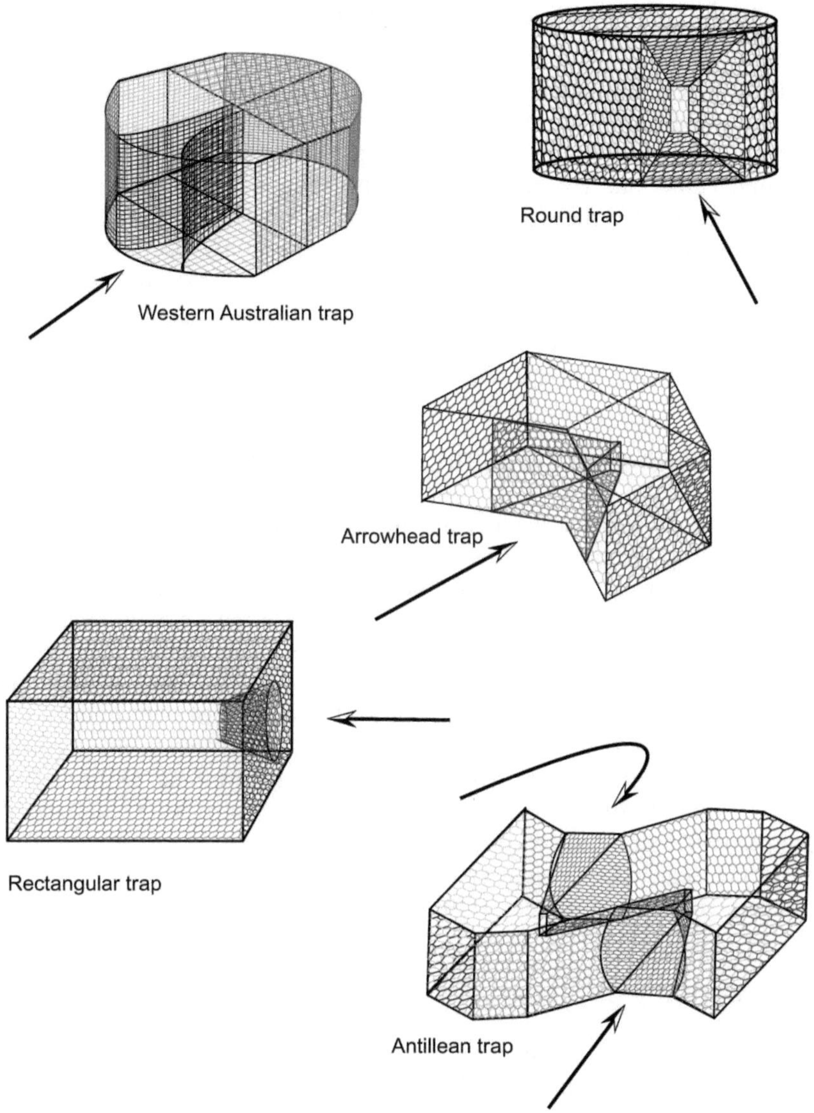

Fig. 4.2. Common wire-mesh fish trap designs. Arrows indicate position of trap entrances NB. Some sections of the mesh have been omitted to show internal structures and the mesh shown is not to scale

4.2 By-catch in Wire-mesh Fish Traps

By-catch in wire-mesh trap fisheries is defined as that portion of the catch that is not retained, and can be divided into two categories: (a) small individuals of target species that cannot be landed due to regulations or very low economic value; and (b) unmarketable species. Quantification of the catch composition of wire-mesh trap fisheries, using observers onboard commercial vessels and fishery-independent surveys, has only occurred relatively recently compared to fisheries that have had greater perceived by-catch issues such as trawl fisheries. There exist three main areas of concern regarding by-catch in wire-mesh fish traps: (i) discarding of large proportions of by-catch; (ii) the mortality of discards; and (iii) ghost-fishing.

4.2.1 Discarding of Large Proportions of the Catch

Studies have shown that usually a significant proportion of the catch from wire-mesh trap fisheries is discarded. Up to 50% of the catch from fish traps used in Florida, U.S.A. was reported to be non-commercial species in addition to significant quantities of sub-legal target species (Sutherland and Harper 1983; Taylor and McMichael 1983). Harper et al. (1994) in Florida, using Antillean 'Z' traps and 2.5 x 5.1 cm PVC coated rectangular mesh and 3.8 cm hexagonal mesh, showed that 14% of captured lutjanids and serranids were undersized, and recorded 79 species as discarded. Similar catch compositions are reported from the Caribbean (Munro 1983). In New South Wales, Australia, Stewart and Ferrell (2003) showed that more than 30% of the catch of the target species, *Pagrus auratus*, was sub-legal and discarded. A recent (2001) increase in legal minimum length for *P. auratus* has further increased this discard rate to be greater than 50%.

4.2.2 Mortality of Discards

Clearly, the risks to the sustainability of populations and fisheries due to discarding from fish traps depend on the survival rates of the discards. As in all fisheries, it is extremely difficult to estimate the survival of fish that are discarded after being caught in fish traps. We know that undersized commercial species and non-commercial species discarded from fish traps may suffer injury and mortality from: (i) attempting to escape from the traps; (ii) barotrauma caused by being hauled to the surface in the traps from depth; (iii) handling and stress onboard before release; and (iv) predators feeding on them after release. Fishers often argue that the survival of discarded fish is high because they are released quickly after capture and most are seen to swim away. This is certainly true for many species in the fishery in New South Wales, Australia, (pers. obs.) and is also reported in the Caribbean (Taylor and McMichael 1983; Sutherland and Harper 1983). However, there have been very few studies that have attempted to quantify the survival rates of discards in these fisheries. Survival rates of discards in the commercial trap fishery in Florida have been estimated by recording injuries and fish behaviour on release (Taylor and McMichael 1983; Sutherland and Harper 1983; Harper et al. 1994). These studies reported that the two most common injuries to fish were from barotrauma (evidenced by bulging eyes, protruding intestines or everted stomachs) and snout damage due to abrasion against the trap mesh. The majority of injured fish were the deep bodied chaetodontids and pomacanthids. Discarded fish were categorised into: (i) those that immediately swam down on release; (ii) those that died either from barotrauma, physiological stress or being eaten by birds or sharks; and (iii) unknown – fish that neither swam down nor died. The vast majority of fish (nearly 80%) were observed to swim down on release, however some usually died regardless of apparent injuries or handling time. At times, up to 30% of discarded fish were observed to die and 20% neither swam down nor died. These studies concluded that rates of discard mortality were unacceptably high and that an increase in mesh size in these fisheries would reduce discarding while providing some protection against overfishing.

The mortality of discards from wire-mesh fish traps is likely to be significantly lower than that from other fisheries (like trawling) because of the relatively short time that they are in physical contact with the fishing gear and the quick discarding process. Survival is also likely to be variable and species- and depth-specific. The over-fished status of many target species in these fisheries means that even relatively low levels of discard mortality may be very detrimental to the stock and increase the risk of its collapse. It is therefore extremely important for these wire-mesh trap fisheries to minimise their by-catch in order to: (i) reduce the risk of stock collapse of target species; (ii) decrease any potential impacts on broader ecosystems; and (iii) prevent societal disapproval over perceived waste caused by killing unmarketable fish.

4.2.3 Ghost-fishing

It has been generally assumed and publicly stated by opponents to fish trapping that ghost fishing by traps is a problem, and that lost traps continue to kill fish indefinitely (www.reefguardian.org). The potential for ghost-fishing to be problematic depends on three factors: (i) the number of traps that are lost; (ii) the ability of fish to escape from traps once they have entered them; and (iii) the durability of lost traps. Many fishery managers believe that ghost-fishing is a problem and mandatory sacrificial panels have been either implemented or recommended in several wire-mesh trap fisheries, e.g., in the Middle-East (Al-Masroori et al. 2004), the South Atlantic (Sheperd et al. 2002), the Caribbean (Kumpf 1994) and Canada (Scarsbrook et al. 1988).

The number of traps lost will vary enormously between different fisheries, is extremely difficult to quantify, and is often estimated by fishers' self-reporting. Some studies indicate that the number of trap losses and mortality of fish within these traps results in unacceptably high levels of ghost-fishing. Taylor and McMichael (1983) reported annual trap losses off Florida averaged 63%. They observed 27% of fish from traps soaked for 20 days to be dead or injured. They concluded that ghost-fishing in this fishery can cause high mortality rates and was therefore a significant problem. Munro et al. (1971) observed that fish that did not escape from traps lived for varying periods but showed injuries after two weeks, and concluded that these fish died and were eaten by eels. Al-Masroori et al. (2004) investigated ghost-fishing of traps near Oman and concluded that it decreased exponentially through time but was still problematic. An average of 70 kg of fish per trap was estimated to be killed during the first 3 months of such ghost-fishing. They concluded that escape gaps to release undersized fish and sacrificial panels should be implemented.

Conversely, there is considerable evidence in some trap fisheries that ghost-fishing is unlikely to be problematic. Video footage from Australia (of round and rectangular traps – Whitelaw et al. 1991) and from New Zealand (blue cod – Cole et al. 2004) has shown high rates of escape of fish from traps. Whitelaw et al. (1991) used video and baited traps off the north-west shelf of Australia to observe very high rates of ingress and egress (i.e., 1 to 2 hours soak times only retained 60% of the fish that actually entered the trap). Luckhurst and Ward (1987) documented high egress rates for tropical snappers (Lutjanidae).

There remains considerable work to be done before the impacts of ghost-fishing of wire-mesh traps is fully understood. It appears as if ghost-fishing may be problematic for some species in some fisheries, but is unimportant for others. Until definitive research is done, it may be wise to take the precautionary approach and implement escape gaps to allow undersized fish to escape and sacrificial panels to prevent any potential for ghost-fishing.

4.3 Overfishing of Target Species

Most wire-mesh trap fisheries are used in tropical and temperate waters and catch a large array of species. For example, more than 100 species have been recorded in the Caribbean trap fishery alone (Munro et al. 1971; Taylor and McMichael 1983; Sutherland and Harper 1983). Despite the species diversity in catches, trap fisheries around the world have traditionally targeted the larger, long-lived, slow-growing, reef-associated, top-order species. These groups of fish are generally the large groupers and cods (epinephilids and serranids), tropical snappers (lutjanids), emperors (lethrinids) and temperate snappers (sparids).

Unfortunately, these groups of fish are all vulnerable to overfishing. They are not only highly-prized food and trophy fish, but also have life-histories that are characterised by slow growth rates and the formation of spawning aggregations that are easily targeted. There is now widespread recognition that in places where wire-mesh fish traps have been used extensively, the target species are currently over-exploited (Mahon and Hunte 2001; Stewart and Ferrell 2003). Further, in areas where long-term research has been done on fish trapping, the trends in the composition of landings of target species have shown consistent patterns of decline. The most noticeable trends have been declines in both the sizes of target individuals and their proportions in landings. For example, the Nassau grouper spawning aggregation in the Virgin Islands has been fished to virtual extinction, with a decline in the average size and proportion of target species landed – most of the catch are now juveniles. Gobert (1994), in the Antilles, showed that the trap fishery now depends on small-sized species and also on small to medium sized individuals of these species. The larger species have been eliminated through overfishing. The trap fishery in the Middle-East has reported a drastic decline in the major target species (*Epinephalus coiodes*) since 1996 and the catch is now dominated by secondary species (Al-Husaini et al. 2002). Many *E. coiodes* retained are also juveniles.

The large tropical snappers, cods, groupers, emperors and temperate sparids targeted by trap fisheries are also highly prized by recreational fishers, and there is little doubt that increased pressure by recreational fishers is a contributing present-day factor in observed declines of such species. However, it is also widely recognised amongst scientists who have studied these fisheries that the wire-mesh used to cover these fish traps is often too small, and that the capture of small individuals of target species is wasteful, almost certainly leading to growth overfishing and potentially recruitment overfishing.

4.4 Factors Affecting the Catch Composition of Wire-mesh Fish Traps

Before solutions to by-catch problems in wire-mesh trap fisheries can be developed, it is important for scientists to understand the factors that affect catch rates and the species and size compositions of fish retained in these gears. Such an understanding is not only vitally important to the operation of commercial trap fisheries but also when traps are used for scientific sampling (Sheaves 1995; Cappo and Brown 1996). Discussion of factors known to influence catch rates and species composition has been done previously (Mahon and Hunte 2001) and is outside of the scope of this chapter, but in practice, any modifications to trap design and the method of fishing them can influence the catch composition.

Generally, catch rates of traps have been found to be proportional to trap volume, with bigger traps having higher catch rates (Munro 1983). Further, traps baited with good quality fish bait (e.g., pilchards Sardinops Sagax) have better catch rates (Whitelaw et al. 1991), and may attract different species than traps that are unbaited (Newman 1990). For example, unbaited traps may catch more herbivorous fish than baited traps (Newman 1990). Many artisanal trap fisheries either use no bait, broken shells or pottery as attractants, or bait of very poor quality (offal or trash fish), and they have relatively poor catch rates. It is likely that if these fisheries used good bait such as pilchards, their catch rates would improve.

Traps have their best catch rates when set with the entrance funnel facing down current. Video footage has shown that the bait burley plume is the major attractant to fish and they move up-current to the bait, congregating down-current of the trap and entering through the funnel to feed (Whitelaw et al. 1991). Some predatory fish such as serranids are thought to enter traps to feed on captive fish (Whitelaw et al. 1991), while the presence of conspecifics in traps may increase catch rates of some species (Sheaves 1995; Santurtun 1995). Soak-time is a major determinant of catch rate and several studies have demonstrated that catch rates decline after several days soak-time (Munro et al. 1971; Munro 1974; Stevenson and Stuart-Sharkey 1980; Wolf and Chislet 1974; Luckhurst and Ward 1987). Video observations and experiments with varying soak-times have shown that fish will enter a trap at a certain rate (governed by factors such as bait, fish abundance and behaviour) and also escape through the entrance funnel at a certain rate. Catch is determined by the balance of ingress and egress at the time the trap is hauled (Whitelaw et al. 1991).

The most important factors controlling the sizes of fish caught in fish traps, and therefore the quantities of fish that are generally unmarketable and are discarded, are the sizes of fish available to be caught and the size of the mesh that covers the trap. Assuming there is no size selectivity operating on fish that escape through the entrance funnel, then fish escaping through the trap meshes will determine the size-composition of the catch. Studies that have examined mesh selectivity in wire-mesh fish traps have been reviewed in Mahon and Hunte (2001). Overall, results have been intuitively obvious, in that traps covered with small mesh retain more small fish than those covered with larger mesh. These small fish may be either small-sized species or smaller individuals of larger growing species. Fish body shape is also important in determining retention sizes, with slender fishes like eels being more likely to escape through meshes than compressed fishes like triggerfishes or depressed fishes like flatfishes (Sutherland et al. 1991).

The factors affecting wire-mesh trap selectivity are reasonably well understood and are species-specific. It has generally been assumed that the sizes of fish retained in a trap is a direct function of their body size and the trap mesh size (Munro 1983; Ward 1988). Stewart and Ferrell (2003) showed that the sizes of most target species retained in traps off New South Wales, Australia could be accurately predicted from the maximum mesh aperture and the body-depth of the fish (Fig. 4.3). However, Stewart and Ferrell (2002) found that species-specific behaviour affected selectivity, with silver trevally (*Pseudocaranx dentex*) being selected at considerably smaller sizes than would have been predicted based on fish body-depth and mesh aperture alone. In contrast, several studies have shown that models based only on mesh size and fish size may under-estimate the sizes of fish retained (Ward 1988; Gobert 1998; Robichaud 1999). Hypotheses explaining this under-estimation have concentrated on the 'squeezability' of some fish, i.e., the ability of fish with body-depths slightly larger than the maximum mesh aperture to squeeze through the meshes (Ward 1988; Gobert 1998; Robichaud et al. 1999). Gobert (1998) showed that the ability to squeeze through trap meshes is affected by fish behaviour and may be density-dependent.

Fish may escape through trap meshes in two ways: (i) while the trap is passively fishing undisturbed on the sea floor; and (ii) while the trap is being hauled to the surface. Fish may escape through any part of the trap when it is on the sea floor; however observations that fish in traps tend to swim into the current (Whitelaw et al. 1991) suggest that when set properly (i.e., with the trap entrance facing down-current), fish would be more likely to encounter and escape through the back panel of the trap (see Fig. 4.4). Traps have been traditionally viewed as being passive fishing gears, however, while being hauled to the surface, they are actively forcing fish to make

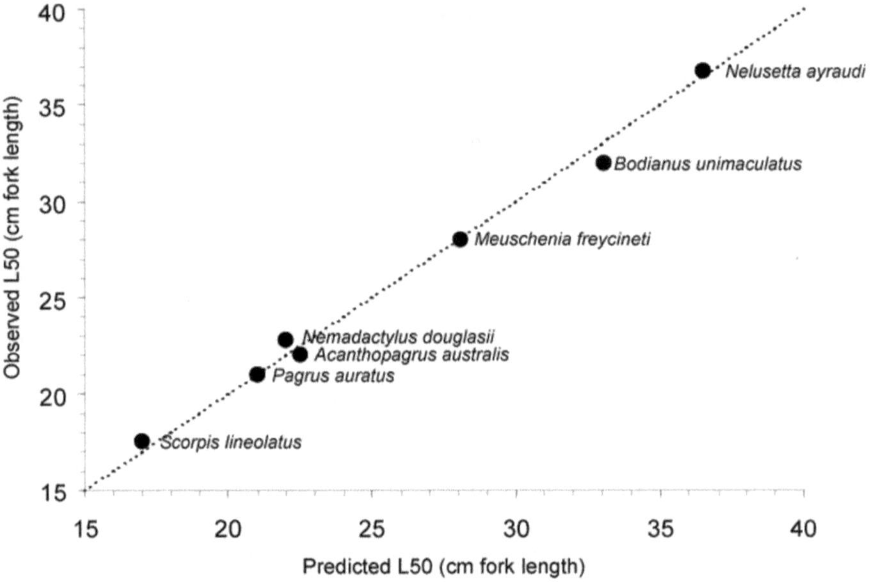

Fig. 4.3. Comparison of observed sizes at 50% retention (L50) in 50 x 75 mm welded mesh determined using selectivity models, with those predicted using fish body height and the maximum mesh aperture dimensions. Data from Stewart and Ferrell (2003)

contact with the trap meshes and mesh selectivity is an active process in this instance (akin to that occurring in trawls). Many species escaping as the trap is being hauled to the surface are also most likely to escape through the meshes in the back panel of the trap because of their behaviour in swimming downwards as the trap is being lifted (Stewart pers. obs).

There is widespread recognition amongst scientists who have studied catches from wire-mesh fish traps that the mesh sizes being used throughout the world in such traps are generally too small and are the major factor responsible for by-catch problems in these fisheries. The simplest solution to these by-catch problems is therefore to increase the size selectivity of traps by increasing the mesh size used in the back panel of traps. Such modifications create minimal changes for fishers who can still construct most of their traps from traditionally-used materials except for the back panel. The rigid nature of wire mesh means that fish that are physically capable of escaping through it often do so, unlike net fisheries where size-selectivity is largely influenced not only by mesh size but also by the hanging ratio of the net. The result is that the size selectivity of wire-mesh fish traps, while species-specific, is often nearly knife-edged. Knife-edge selectivity in any fishing gear is highly desirable because it can lead to the

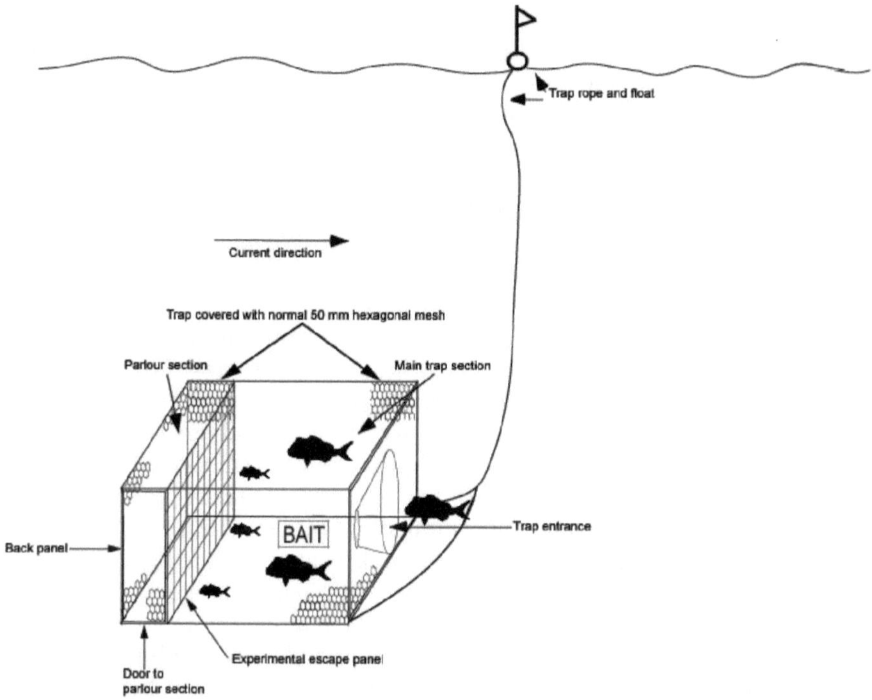

Fig. 4.4. Diagram of the parlour trap used to estimate selectivity in the trap fishery of New South Wales, Australia

development of gears that select fish at pre-determined sizes, therefore minimising both the retention of fish that are discarded and the loss of target fish. The following case study serves to illustrate one of the most recent examples of research into reducing by-catch in wire-mesh fish traps by changing gear selectivity.

4.5 Case Study: The Wire-mesh Trap Fishery in New South Wales, Australia

The wire-mesh trap fishery in New South Wales (NSW), Australia provides a recent example of a fishery that has been identified as having a significant by-catch problem and of research into effectively reducing this by-catch while maintaining sustainability.

The use of demersal, wire-mesh fish traps is an established method of fishing in NSW and forms the basis of a valuable fishery, landing more than 600 tonnes of finfish per year. This fishery has traditionally targeted mainly pink snapper (*Pagrus auratus*), using rectangular, timber-framed, traps approximately 2 x 1 x 1 m, with single entrance funnels on one side, baited and set on or near reefs at depths of between 10 and 150 m. These traps have a regulated minimum mesh size of 50 mm and fishers almost exclusively cover their traps with a galvanised hexagonal mesh (see Fig. 4.1). Like some other wire-mesh trap fisheries in the world, the NSW fishery is a multi-species one, with fishers landing, in addition to pink snapper, considerable quantities of rubberlip morwong (*Nemadactylus douglasii*), silver trevally (*Pseudocaranx dentex*), bream (*Acanthopagrus australis*) and ocean leatherjackets (*Nelusetta ayraudi*).

This fishery has recently been assessed in terms of reported landings and of discarded and retained catches (Stewart and Ferrell 2003). Patterns in landings from this fishery were found to be similar to those described for other trap fisheries that have been overfished. Landings of trap-caught pink snapper declined steadily from approximately 600 tonnes in 1992/93 to approximately 128 tonnes in 2004/05. The proportion of pink snapper in landings has declined from approximately 50% in 1992/93 to approximately 20% in 2004/05. The sizes of pink snapper landed are generally close to the minimum legal size limit, with approximately 75% being within the first 5 cm of the legal limit. Pink snapper are an iconic species in this part of the world and are also highly prized by recreational fishers – the most recent estimate of the recreational catch in NSW being around 120 tonnes per year (Henry and Lyle 2003). Pink snapper have been designated as being growth overfished in NSW and a recovery program is being developed for them.

Information from observers onboard commercial vessels and from voluntary fisher logbooks during 1999/00 showed that approximately 30% of the pink snapper retained in fish traps covered in 50 mm hexagonal mesh were below the minimum legal size limit of 28 cm total length (TL) and were discarded with unknown mortality. Since this time, the minimum legal size limit for pink snapper has been increased to 30 cm TL and the discard rate is now approximately 50% – again with unknown mortality. The selectivity of 50 mm mesh was found to be inappropriate for all of the species caught in this fishery that had minimum legal size limits, so significant numbers of sub-legal fish were caught and subsequently discarded (Stewart and Ferrell 2003). Large numbers of unmarketable species such as *Atypichthys strigatus* and *Enoplosus armatus*, were also observed to be retained in traps covered with 50 mm mesh.

Observations of fish behaviour in these traps being hauled to the surface, and video footage of fish behaviour in traps on the sea floor, showed that those fish which escaped through the trap meshes did so almost exclusively through the back panel (that side of the rectangular trap that is opposite to the trap entrance and is up-current of the bait when set correctly with the trap entrance facing down-current – see Fig. 4.4). This back panel was colloquially called the 'escape panel' by fishers. Comparative experiments using traps covered with small (37 mm) hexagonal mesh and back panels of either 37 or 50 mm hexagonal mesh or a 50 x 75 mm rectangular welded mesh, were done to determine the selectivity of these meshes for the major species. The results showed that size selectivity occurred over a small range with selectivity ogives for commonly caught species having very small (< 2 cm) selection ranges (SR) (Fig. 4.5). The size-selection for most species was found to be a direct function of fish body-depth and the maximum aperture of the trap mesh in the back panel (see Fig. 4.3).

Having recognised that the traps used in the NSW fishery had poor selectivity, a series of trials were done using various mesh sizes as escape panels (i.e., in the back panel of the trap only – Stewart and Ferrell 2002). It was recognised that any increase in trap mesh selectivity would have some initial impacts on commercial fishers in terms of losses of secondary species, and industry were engaged in this research as collaborative investigators. An important component of this work was the development of a 'parlour trap' designed to allow fishers to test (for themselves) the impacts

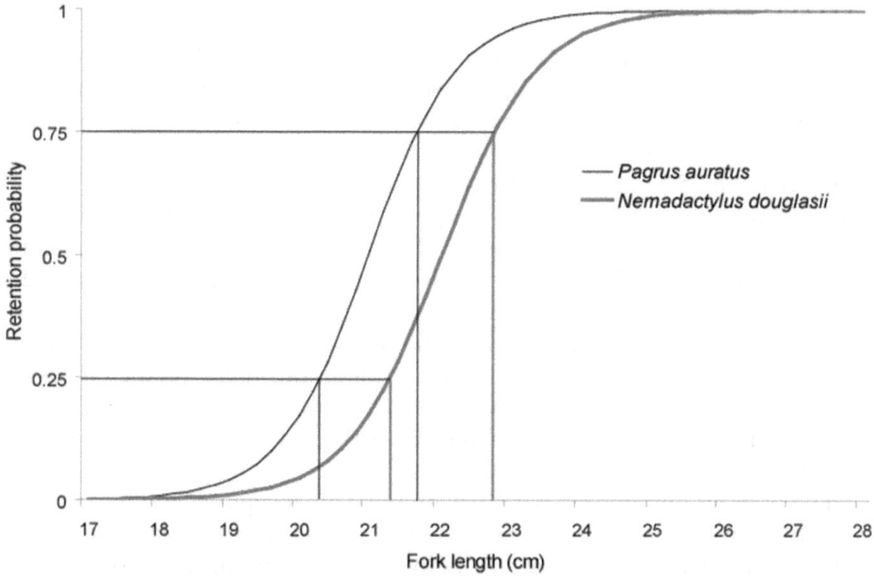

Fig. 4.5. Selectivity ogives showing narrow selection ranges for *Pagrus auratus* and *Nemadactylus douglasii* in traps with escape panels of 50 x 75 mm mesh in New South Wales, Australia

of changes in mesh size without losing marketable fish (Fig. 4.4). This was called a 'parlour type' trap because it had two sections, the main body of the trap and a back, 'parlour' section separated only by an experimental mesh panel approximately 50 cm from the back. The trap was wholly wrapped with standard 50 mm hexagonal mesh and had a second door into the parlour section through which fish could be removed. Analogous to the covered codend experiments used to test the selectivities of trawls, the parlour traps allowed better estimates of selectivity from fewer trap lifts. The concept behind this system was that once a fish had entered the trap normally through the entrance, it would pass through, or be retained by, the experimental mesh panel when the gear was being lifted to the surface.

The successful prediction of selection size from fish body-depth and the maximum aperture of the trap mesh allowed the development of custom-made mesh designed to select pink snapper at their minimum legal size. This custom-made 50 x 87 mm mesh was tested along with two commercially-available products, 60 x 80 mm and 80 x 100 mm hexagonal, woven wire used to make gabions for rock retaining walls. Some parlour traps were built using the commercially available hexagonal meshes with the longest axis vertical, and others with the longest axis horizontal. These experimental parlour traps were used by commercial fishers during 2000 and scientific observers documented the sizes of fish retained in the main and parlour sections.

The selectivity of the custom-made 50 x 87 mm welded mesh was found to be appropriate for pink snapper, selecting fish at their minimum legal size. Predicted reductions in catches of undersized pink snapper by 77% and undersized rubberlip morwong by 97%, with losses of marketable fish being 0.9 and 25.5% by weight respectively, indicated that this mesh was a suitable product for many fish trappers. However, this mesh resulted in relatively moderate predicted losses of some species that are known to be important to some fishers (e.g., yellowfin bream – 39%, ocean leatherjackets – 55% and pigfish *Bodianus unimaculatus* – 69%). These predicted effects were initial losses and did not account for increases in yields from harvesting them at larger sizes. In addition, these predicted losses were for the smaller fish that often receive low prices at market, so any financial losses would be less than those predicted from the loss in weight of fish by an increase in mesh selectivity.

The 60 x 80 mm gabion wire proved to be an acceptable, commercially-available product suitable for use as escape panels in fish traps in NSW. Pink snapper were selected at just below their minimum legal size and the catch of undersized pink snapper reduced by around 61% with almost no losses of legal sized fish. However, using 60 x 80 mm gabion reduced catches yellowfin bream, ocean leatherjackets and pigfish by similar amounts to the 50 x 87 mm mesh.

The 80 x 100 mm gabion wire selected pink snapper at 2 cm above their minimum legal size and it could be a viable mesh in future for some fishers if the size limit is increased further. The selectivity of the 80 x 100 mm gabion wire was found to be inappropriate for all other important species in the fishery, however, with up to 100% of some species escaping through this mesh.

Fish behaviour was found to be important in determining selectivity in this study. Most species had similar predicted selection sizes in meshes oriented either vertically or horizontally, but silver trevally had significantly smaller selection sizes in the mesh oriented horizontally. It is thought that species such as pink snapper were able to turn on their sides to escape through the longest mesh aperture, whereas silver trevally did not. In traps being lifted to the surface, silver trevally were observed to form a tight school and to swim in the same direction that the trap was moving, only encountering the back of the trap as it was being lifted onto the vessel.

Silver trevally also provided an example of a problem when using larger trap meshes and, through their behaviour, a simple solution. In meshes oriented vertically, 48 silver trevally (approximately 5% of those caught) were observed to be meshed (being stuck between their anterior dorsal and protruding anal spines in the longest axis of the mesh). These fish were damaged to such an extent that they were unmarketable. No silver trevally were meshed in wire that was oriented horizontally. As above, this was believed to be a result of silver trevally not being able to turn on their sides to escape through meshes. The solution to meshing large numbers of silver trevally, whilst maintaining the desired selectivity for pink snapper, was therefore to place the wire meshes horizontally.

The study was one of the most comprehensive to be done on the selectivity of wire-mesh fish traps and, by engaging industry to assist in the project, provided the following conclusions and recommendations:

(i) The fishery currently has an unacceptably large by-catch problem. Having approximately 50% of pink snapper caught in traps being discarded is likely to be ecologically unsustainable and socially unacceptable.

(ii) This by-catch problem could be simply solved by introducing 'escape panels' of larger mesh in the traps.

(iii) A range of custom-made and commercially available alternatives to 50 mm hexagonal wire-mesh were tested as escape panels. Fishers and fishery managers were provided with tables of the predicted reductions in by-catch achieved for each mesh type and also initial impacts from the losses of small marketable fish. Potential longer-term increases in yields for these species were expected from harvesting them at larger sizes.

(iv) To reduce by-catch and to minimise short-term losses to fishers, recommendations were made to fishery managers to implement 50 x 75 mm weldmesh as escape panels in fish traps in this fishery. They should then periodically increase this mesh size until the by-catch of sub-legal pink snapper is minimised.

These recommendations were made in 2001 and many fishers who were involved in the study now voluntarily use escape panels of larger mesh to reduce their by-catch. However, after 5 years following these recommendations, there have been no mandated changes to the mesh size regulations in this fishery although a management strategy being developed has made a commitment to implement 50 x 75 mm escape panels. The lesson learnt here is that even involving industry in developing solutions to by-catch problems doesn't always guarantee timely or effective management changes. In this case study, submissions from a small number of fishers against the implementation of any escape panels, due to perceived short-term losses of income, led to a slowing of management action. It is hoped that managers eventually will implement the recommendations of this research to minimize by-catch. The worst-case scenario, however, is that by the time this happens, stocks of pink snapper in this region may have declined to commercial extinction.

4.6 Conclusions

Wire-mesh trap fisheries around the world have been subjected to increasing scrutiny in recent years, and most have been shown to have unacceptably high rates of by-catch. Generally, the wire mesh being used to cover these fish traps is too small, resulting in large levels of by-catch of unmarketable fish and potentially reducing long-term yields of these species. The survival of this by-catch when discarded is largely unknown and should be the focus of further research in these fisheries.

One of the simplest solutions to reduce by-catch in these gears is to increase the trap mesh selectivity. In one documented case where increases in mesh size were implemented, significant reductions in by-catch and increases in fish stocks were detected after just 3 years (Sary et al. 1997). Elsewhere, panels of large square-mesh (110 mm diagonally) have recently been regulated in the trap fishery in the Emirate of Abu Dhabi to enable juvenile fish to escape. In Australia, there remains a recommendation to implement escape panels of larger mesh in the pink snapper trap fishery in NSW.

Such efforts to reduce by-catch appear to be vital to the long-term continuation of wire-mesh fish trap fisheries throughout the world. These are uncertain times for the world's wild harvest fisheries, and it is important that these fisheries are seen to be as environmentally sound as possible. The general public and environmental groups are well-informed of unacceptable by-catch issues, and have considerable political influence. Commercial fishers and the managers who oversee their fisheries increasingly need to answer to non-consumptive users of the resource (e.g., environmentalists and conservationists) to prove that their activities are environmentally sound. The worst-case scenario for fisheries that cannot demonstrate acceptable practices and levels of by-catch is that they are closed, and there are several examples of wire-mesh fish trap fisheries where this has already occurred. For example, the South Atlantic Fishery Management Council and the Florida Marine Fisheries Commission both banned the use of fish traps in the early 1990's and the Gulf of Mexico Fishery Management Council has announced that all fish trapping within their jurisdiction will be banned after 2007.

It is hoped that co-operation among fishers, fishery managers and scientists to develop and implement solutions to by-catch problems will ensure the long-term viability of wire-mesh fish trapping. This fishing method is a simple, yet very effective one, that can be used to harvest a wide variety of fish. Once changes to the gears have been tailor-made to harvest only fish at suitable sizes, it will be important for these fisheries to be pro-active and to educate society as to their sustainable and environmentally-friendly nature.

References

Al-Husaini M, Al-Baz A, Safar S, Al-Wazan Z, Al-Jazzaf S (2002) Age, growth, mortality, and yield-per-recruit for nagroor, *Pomadasys kakaan*, in Kuwait's waters. Fish Res 59:101–115

Al-Masroori H, Al-Oufi H, McLean E (2004) Catches of lost fish traps (ghost fishing) from fishing grounds near Muscat, Sultanate of Oman. Fish Res 69:407–414

Cappo M, Brown IW (1996) Evaluation of sampling methods for reef fish populations of commercial and recreational interest. CRC Reef Research Centre Ltd. Technical Report no. 6. CRC Reef Research Centre, Townsville, Australia, 72 pp

Cole RG, Alcock NK, Tovey A, Handley SJ (2004) Measuring efficiency and predicting optimal set durations of pots for blue cod *Parapercis colias*. Fish Res 67:163–170

Gobert B (1994) Size structures of demersal catches in a multispecies multigear tropical fishery. Fish Res 19:87–104

Gobert B (1998) Density-dependent size selectivity in Antillean fish traps. Fish Res 38:159–167

Grove-Jones R, Burnell A (1990) Ocean jacket traps assessed. SAFISH 15(1): 10–11

Harper DE, Bohnsack JA, McClellan DB (1994) Investigation of bycatch from the wire fish-trap fishery in federal waters off southern Florida. Proceedings of the 43rd Annual Gulf and Caribbean Fisheries Institute, pp 3–25

Henry GW, Lyle JM (2003) The National Recreational and Indigenous Fishing Survey. Final Report to the Fisheries Research and Development Corporation and the Fisheries Action Program. Project no. 1999/158. NSW Fisheries Final Report Series no. 48. ISSN 1440-3544. 188 pp

Kumpf HE (1994) Practical considerations and testing of escape panel material in fish traps. Proc Gulf Caribb Fish Inst 43:211–214

Luckhurst B and Ward J (1987) Behavioral dynamics of coral reef fishes in Antillean fish traps at Bermuda. Proc Gulf Caribb Fish Inst 38:528–546

Mahon R, Hunte W (2001) Trap mesh selectivity and the management of reef fishes. Fish Fish 2:356–375

Munro JL (1974) The mode of operation of Antillean fish traps and the relationship between ingress, escapement, catch and soak. J Cons Cons Int Explor Mer 35:337–350

Munro JL (1983) Caribbean coral reef fishery resources. ICLARM Studies and Reviews 7, 276 pp

Munro JL, Reeson PH, Gaut, VC (1971) Dynamic factors affecting the performance of the Antillean fish trap. Proc Gulf Caribb Fish Inst 23:184–194

Newman SJ (1990) Effects of depth of set, mesh size, bait and lunar phase on the performance of fish traps on the Great Barrier Reef. Honours Thesis, James Cook University of North Queensland, Townsville, Australia, 112 pp

Robichaud D, Hunte W, Oxenford HA (1999) Effects of increased mesh size on catch and fishing power of coral reef fish traps. Fish Res 39:275–294

Santurtun M (1995) Catch structure and dynamics of Antillean 'Z' traps on a coral reef: Interaction between trap characteristics and fish behaviour. MSc Thesis, Department of Marine Biology, James Cook University, North Queensland, Australia, 72pp

Sary Z, Oxenford HA, Woodley JD (1997) Effects of an increase in trap mesh size on an overexploited coral reef fishery at Discovery Bay, Jamaica. Mar Ecol Prog Ser 154:107–120

Scarsbrook JR, McFarlane GA, Shaw W (1988) Effectiveness of experimental mechanisms in sablefish traps. N Amer J Fish Manag 8:158–161

Sheaves MJ (1995) Effect of design modifications and soak time variations on Antillean-Z fish trap performance in a tropical estuary. Bull Mar Sci 56(2): 475–489

Sheperd GR, Moore CW, Seagraves RJ (2002) The effect of escape vents on the capture of black sea bass, *Centropristis striata*, in fish traps. Fish Res 54: 195–207

Slack-Smith RJ (2001) Fishing with traps and Pots. FAO Training Series no. 26. Rome, Italy, 61pp

Stevenson DK, Stuart-Sharkey P (1980) Performance of wire fish traps on the western coast of Puerto Rico. Proc Gulf Caribb Fish Inst 32:172–193

Stewart J, Ferrell DJ (2002) Escape panels to reduce by-catch in the New South Wales demersal trap fishery. Mar Freshw Res 53:1179–1188

Stewart J, Ferrell DJ (2003). Mesh selectivity in the New South Wales demersal trap fishery. Fish Res 59:379–392

Sutherland DL, Harper DE (1983) The wire fish-trap fishery of Dade and Broward counties, Florida December 1979 – September 1980. Florida Marine Research Publications 40. Florida, USA, 21pp

Taylor RG, McMichael Jr RH (1983) The wire fish-trap fisheries in Monroe and Collier counties, Florida. Florida Marine Research Publications 39. Florida, USA, 19pp

Ward J (1988) Mesh size selection in Antillean arrowhead fish traps. In: Venema S, Moller-Christensen J, Pauly D (eds) Contributions to Tropical Fisheries Biology. Fisheries Report no. 389. Food and Agriculture Organisation. pp 455–467

Whitelaw AW, Sainsbury KJ, Dews GJ, Campbell RA (1991) Catching characteristics of four fish-trap types on the north west shelf of Australia. Aust J Mar Freshwater Res 42:369–382

Wolf RS, Chislett GR (1974) Trap fishing explorations for snapper and related species in the Caribbean and adjacent waters. Marine Fisheries Review 36:49–61

5 Modifying Dredges to Reduce By-catch and Impacts on the Benthos

MIGUEL B. GASPAR AND LUÍS M. CHÍCHARO

5.1 Introduction

The sustainable existence of particular fishing activities in a certain area depends on the maintenance of the stock of the target species in that area. The cumulative effect of fishing can lead to overfishing with a consequent decrease in the abundance of targeted species, the fisheries that depend on them and significant impacts on ecosystems. Ecosystem changes caused by fishing are mostly associated with mobile bottom gears, especially dredges, which impact the benthic habitat and associated assemblages of species. The magnitude of impacts from such fishing depends on factors such as the fishing frequency, towing speed, gear type, gear penetration into the sediment, time of year, local environmental conditions (such as water depth, tidal strength and currents), nature of the substratum and the structure of the benthic communities affected (de Groot 1984; Churchill 1989; Mayer et al. 1991).

The environmental effects of shellfish dredging have received special attention throughout the world in recent decades (e.g., Caddy 1973; Conner and Simpson 1979; McLoughlin et al. 1991; Eleftheriou and Robertson 1992; Dare et al. 1993; Hall et al. 1993; Jennings and Kaiser 1998; Hall-Spencer and Moore 2000). These gears re-suspend and rework bottom sediments, move and bury boulders, reduce microtopography and may leave long-lasting grooves (e.g., Caddy 1973; Churchill 1989; Mayer et al. 1991). Sediment re-suspension by towed gears can alter the composition of sediments (usually to coarser grain sizes), reduce chemical exchanges in the water-sediment interface and increase water turbidity with deleterious effects on planktonic productivity (Hayes et al. 1984; LaSalle 1990; Coen 1995). Along with the target species, dredges also catch algae and other epifauna and infauna, many of which have no commercial value and are therefore discarded either alive or dead.

S. J. Kennelly (ed.), By-catch Reduction in the World's Fisheries, 95–140.

In recent times, one of the most significant issues affecting the management of marine fisheries is the high mortality associated with discarded fish after capture. In the long-term, high mortality rates of discards can have very significant impacts on the ecosystem by modifying benthic and demersal food-webs. Therefore, one of the main aims of fisheries management is the reduction of by-catch and, consequently, of discards. This goal can be achieved through establishing closures to fishing in areas of high rates of discard of key species – including juveniles of targeted species; and/or improving the selectivity of fishing gear to allow most of the non-targeted species to escape.

It is, however, important to emphasise that closures are only effective for the area that is closed (Kennelly 1999). Indeed, since whole fisheries usually are not closed, fishing effort is often simply redirected to other areas when small-scale closures are used. As a consequence, fishing effort increases outside the closure, which can cause negative environmental effects. Thus, fishery closures may result in fishers moving to areas that were previously only slightly impacted or not at all. Gear modifications are often a better strategy to use than closures to reduce discards. Moreover, a gear-based solution is also advantageous because adoption by the entire fleet that operates a specific fishing gear would lead to reductions of discards in all areas where the fishery occurs (Kennelly 1999).

Most studies of gear modifications in mobile bottom fishing gears are related to trawls (see Broadhurst 2000 for a review of this work in shrimp-trawl fisheries) and few have focused on dredges. However, these former studies tended to concentrate on reducing non-target and juvenile fish by-catch with few attempting to reduce the benthic by-catch (other than fish) nor the potential damage of mobile fishing gears to invertebrate benthic species. To minimise the adverse ecological effects of fishing gears, fishery managers and the fishing industry should promote modifications that enhance selectivity and reduce habitat damage and impacts on benthic communities (Morgan and Chuenpagdee 2003). In this chapter we describe the main types of fishing dredges used to harvest bivalves and the methods used by fishers to handle the catch. Afterwards, we review the mortality that may occur during the entire fishing process and the impacts of dredging on benthic habitats and communities. We also describe recent research by the Portuguese Fisheries Research Institute (IPIMAR) which aims to modify clam dredges in order to reduce by-catch, discards and environmental impacts. In this case study, we describe the stages involved in developing modifications that have been successful in reducing by-catch and ecosystem effects in the dredge fishery that occurs along the Portuguese coast. Finally, at the end of the chapter we identify how the solutions achieved in these studies can be applied to other dredge fisheries in the world.

5.2 Dredge Designs

The design of fishing dredges can vary greatly according to harvesting objectives. It is known that the impact of dredges on the seafloor varies greatly between gear types and therefore any assessment of the effects of those gears on habitats must consider their different specifications.

Dredging involves fishing techniques ranging from small and light gears towed by hand to large and heavy gears operated by large vessels. Dredges can be divided into three classes: manual, mechanical and hydraulic dredges. Hand dredges are adapted to scrape the smooth bottom in the intertidal and/or in very shallow waters. Mechanical dredges are those that scrape the surface of the seabed (including scallop dredges) and those that penetrate the substratum using a toothed bar to dig clams out of the sediment up to 60 cm in depth. Hydraulic dredges use water jets to fluidise the sediment and wash clams out of the seafloor.

5.2.1 Manual Dredges

Manual dredges (Fig. 5.1) are small and light, consisting of a mouth frame or a rigid metal cage (opened in its posterior part) with a digging blade or a

Fig. 5.1. Photograph of the manual dredge used in the *Donax trunculus* fishery along the Portuguese south coast

toothed lower bar, attached to a collecting bag constructed of metal rings or netting material. The mouth of the gear is fixed to a wood handle. The tooth length does not exceed 10 cm. These dredges are operated by hand or from small boats.

5.2.2 Mechanical Dredges

5.2.2.1 Scallop Dredges

Scallop dredges (Fig. 5.2) used in soft bottom habitats comprise a fixed digging blade with or without teeth and a collecting bag made from steel rings. In some scallop fisheries, the ringed bag is replaced by strong wire meshed boxes with depressor plates, which help keep the dredge on the seabed while it is being towed. On harder substrata, in order to avoid damage to the tooth bar, two absorbing springs are attached to the frame. The tooth length usually does not exceed 9 cm. In some scallop dredges, a tickler chain is used to induce scallops to propel from the seafloor so they are more easily caught. In rocky areas, chains are used to prevent large boulders from entering the bag. Small vessels work with up to 6 dredges, whereas large vessels can operate with up to 24 dredges. A wheeled towing beam is usually used when more than two dredges are towed simultaneously. Scallop dredges are towed along the seabed by vessels travelling at 2 to 6 knots.

Fig. 5.2. Dredge used to harvest scallops (Photo: Antonio Hervas)

Fig. 5.3. Photographs of a razor clam dredge (left) and a clam dredge (right) used by the Portuguese dredge fleet

5.2.2.2 Clam and Razor Clam Dredges

Clam and razor clam dredges are comprised of a metallic frame, a toothed lower bar and a mesh bag or a rectangular metallic grid box to retain the catch (Fig. 5.3). When a cage is used to collect the catch, a diving vane on the back can be used to maintain bottom contact when the digging blade encounters resistance. The length of the teeth used in dredges varies according to the target species and takes into account the maximum burrowing depth of the species being harvested. Usually, the length of the teeth used to catch clams does not exceed 20 cm, whilst in the case of the razor clam fishery, the tooth length may reach 60 cm. For clam dredges, small boats can work with up to 6 dredges, whereas large vessels can operate with up to 24 dredges. When razor clam dredges are used in a fishery, small boats can operate a single dredge only, while larger vessels work with two dredges that are deployed and hauled together or individually. Dredges are towed with a cable normally at 3:1 warp depth ratio. The duration of each tow varies between 1 and 20 minutes depending on the target species. In the case of razor clams, the number of damaged individuals increases with tow duration (Gaspar et al. 1998) and the tow is performed at a speed of 1 to 3 knots.

5.2.3 Hydraulic Dredges

A hydraulic dredge (Fig. 5.4) consists of a rectangular cage made of steel bars to retain the catch with a cutting blade and a system for delivering pressurised water through jets. Usually, in the anterior part of the cage, there are two adjustable running sledges to prevent the dredge both from

Fig. 5.4. Photograph of a hydraulic dredge used in the *Chamelea gallina* fishery in the southwestern coast of Spain

sinking in the substratum and from digging the sediment too deeply. In most European bivalve fisheries, the hydraulic dredge is secured to the boat by two towing ropes. As it is towed over the seabed, the sediment is dug by a cutting blade located in front of the dredge mouth. High-pressure water is pumped from the ship through a hose and delivered as a series of pressure jets at the mouth of the dredge and inside the dredge cage. The water expelled from the jets placed in front of the cage fluidises the sediment allowing the blade to cut it easily and therefore dig out the clams. The depth penetration of water jets used depends on the target species, type of sediment and water pressure. Hydraulic suction dredges are used in some cockle fisheries where the catch is continuously brought up onto the deck of the boat through a suction pipe instead of being retained in a rigid cage.

In Spain and Italy, hydraulic dredges are hauled from the bow of the vessel and towed astern either by warping on a big anchor using a winch or by moving backwards by means of the propeller. However, hydraulic dredges may also be hauled from the side (such as side rig dredges) or from the stern (stern rig dredges) of the boat. Each tow lasts 10–20 minutes depending on the density of the target species, the type of sediment and the amount of debris in the area. The towing speed may attain 3 knots but usually is lower as the dredges accumulate clams. Although large vessels may operate two hydraulic dredges simultaneously, most vessels operate only one dredge.

5.3 Catch Handling

Catch handling must be considered in any study of the fishing impacts of dredging, since the survival of discards can be affected by the onboard processing procedure of the catch and the time that organisms are exposed on the deck of the vessel.

In both scallop and clam fisheries, some large vessels are equipped with a conveyor belt to handle the catch. In this case, the dredges' catch is put into a tumbler (that sifts out empty shells and rocks) where a continuous waterflow leads the catch onto a conveyer belt that ends in a tray. Here, the bivalves are hand-sorted by fishers into heaps of small, medium and large individuals, which are stored in net bags or boxes. Non-commercial species are discarded immediately. In some vessels, the dredges are emptied directly onto the deck. The catch is then shovelled into rotary sieves to separate large individuals from empty shells and juveniles which pass through the grids of the sieve back into the sea. The remainder of the catch is collected in baskets or boxes that are emptied on a sorting table and hand-sorted by the crew. After sorting, the discards are thrown overboard. In small vessels, the dredge is brought aboard by hand or by a powered winch, and lifted from the rear so the catch is dumped out through the mouth. The catches are collected in boxes on the deck. During the next tow, fishers sort the catch manually or using manual sieves. In the razor clam fishery, catches are put into boxes placed on the deck. These boxes are then emptied on a sorting table and sorted by the crew. The discards are collected in baskets and then returned to the sea.

For hydraulic dredges, the rigid cages are periodically retrieved and their contents are spilled on a collecting table where a continuous water flow leads the catch to a mechanical vibrating sieve comprising two or three superimposed grids that sort clams by commercial size-classes. By-catch individuals that have not passed through the screens of this 'riddle' are manually sorted and thrown overboard while small individuals from target species and non-commercial species that pass through all the screens of the sieve fall on a tray that, using water jets, returns them into the sea. The individuals of commercial species are stored in boxes or net bags, on the deck, until landed.

5.4 Mortality Through the Entire Fishing Process

Any attempt to minimise fishing impacts on the benthos should consider the mortality that may occur during the entire fishing process (Fig. 5.5). Apart from landings, dredging also causes other kinds of mortality, either

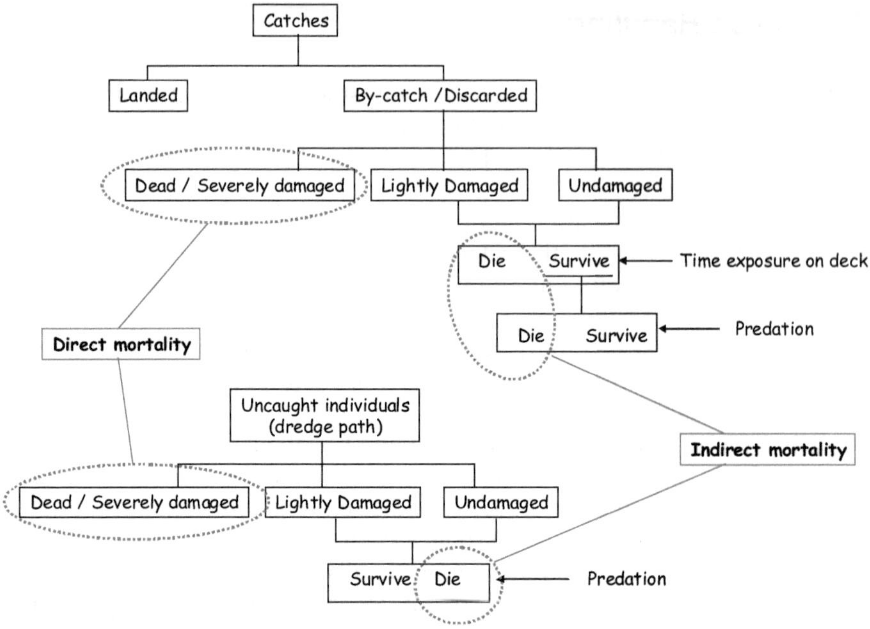

Fig. 5.5. Schematic representation of the mortality that may occur during the entire fishing process

directly or indirectly. To evaluate the total impact of such a fishery on clam beds, it is necessary to account for the specimens which have been discarded and killed afterwards, as well as those which have died because of contact with the gear, but released before the catch has been brought on deck. Indirect mortality of undamaged or slightly damaged individuals (either discarded or dislodged by the gear) through predation should also be considered. Therefore, the extent of this additional mortality must be quantified if we are to calculate the total mortality associated with fishing (Kaiser and Spencer 1995; Broadhurst et al. 2006).

The low selectivity of many dredging gears inevitability results in some level of unintended catch and not all individuals captured will be landed. Part of the catch will be returned to the sea-bed as discards, due to: (i) regulatory prohibitions (such as minimum landing sizes); (ii) captured species that have no current market; (iii) quota surpluses; and/or (iv) commercial specimens that are severely damaged.

The discarding of by-catch by commercial dredge fishing vessels is a common practice in many bivalve fisheries but should not be a major problem if the discarded individuals survive. However, the survival of discarded specimens will depend on their air exposure on deck, the degree of the damage suffered during the tow, the size of the specimens discarded, on-board sorting operations and their susceptibility to predation after discarding (Broadhurst et al. 2006). According to Medcof and Bourne (1964), only a small proportion of the damaged clams that are thrown back into the sea will survive if sorting times are long or conditions on deck are unfavourable. High rates of mortality due to desiccation may be observed during the summer, especially when air temperatures are high. Gaspar and Monteiro (1999) showed that the length of deck exposure is directly related with the mortality of *Spisula solida* juveniles. Indeed, all individuals that were transplanted to tanks immediately after being caught survived, while the survival rate of individuals that were exposed to air decreased dramatically with time (84%, 69%, 56% and 46% for deck exposure times of 1, 2, 3 and 4 hours, respectively).

In contrast, Gaspar and Monteiro (1998) found that undamaged juveniles of *Donax trunculus* could sustain long periods out of the water and these high survival rates can be explained by its ecology. *Donax trunculus* inhabits the high-energy environment of exposed sandy beaches, where it is the dominant macrobenthic organism. In such an environment, daily salinity changes and large temperature fluctuations occur year-round, chiefly during the summer. Thus, *Donax trunculus* has developed a natural resistance to high temperatures and can therefore support long exposure times (more than 8 hours) on the decks of vessels. Survival of long deck exposures therefore varies greatly between species, depending on their ecologies and life histories (Broadhurst et al. 2006).

Even when the catch is immediately sorted, the survival of discarded individuals is also species-specific. Fonds (1994) noted that the survival of flatfish and roundfish discarded from beam-trawls are very low. For non-commercial invertebrates, however, a major part of the discarded by-catch may survive. Both starfish and brittlestars autotomise their arms as an escape response to predators or as a result of gear damage, subsequently regenerating their arms. Hence the survival of these echinoderms may be high. Kaiser and Spencer (1995) reported that nearly 100% of the common starfish, *Asterias rubens*, and nearly 80% of the brittlestar, *Ophiura ophiura*, survive capture in the nets of beam trawls. Similarly, Bergman et al. (1990) stated that common starfish have a high chance of survival after returning to the sea. In contrast, mortality rates of ascidians returned to the seabed are likely to be high (Currie and Parry 1999).

The survival chance of an individual (discarded or dislodged) depends on the degree of damage suffered. The vulnerability of infauna and epifauna to dredging is highly variable. For example, in the Portuguese clam dredge fishery, robust bodied or thick shell species such as whelks and hermit crabs are unlikely to be affected, while fragile species such as heart urchins (*Echinocardium cordatum*) will suffer badly from impact with a passing clam dredge. Of the bivalves, there are species that are very resilient to fishing, such as *Donax* spp. and *Arcopagia crassa*, whilst thin-shelled species such as *Ensis siliqua*, *Pharus legumen*, *Tellina* spp., usually show high percentages (> 60%) of damaged individuals in the catches (Gaspar et al. 2001, 2002, 2003a).

The nature of the bottom can also affect the mortality induced by mobile fishing gears on benthic species. Several authors (e.g., Hall 1994; Currie and Parry 1996; Jennings and Kaiser 1998; Kaiser et al. 1998; Franceschini et al. 1999) have noted that the impact of towed gears is lower on mobile sandy sediments than on rocky or muddy bottoms, or those with a large amount of debris. On the latter kinds of grounds, the net is often filled with mud and/or stones, which damage the catch during fishing and sorting operations. Houghton et al. (1971) observed that in hauls on sandy grounds, the extent of damage inflicted on invertebrate species varied with the quantity of empty shells caught.

Independent of the survival rates of the different species to catch and sorting procedures, individuals returned to the seabed provide potential food for scavengers and predators. Therefore, their survival depends on the time needed to reach the bottom and to rebury (in the case of infauna) or to restart their normal activity (in the case of epifauna) (Gaspar et al. 2003a). Reburial time has been proposed as a valuable indicator of stress level in clams (Chícharo et al. 2002c, 2003a, 2003b). According to Phelps et al. (1983), burrowing behaviour is adaptive and allows clams to escape predation and, if stress conditions affect burrowing behaviour, this may increase mortality. According to Hauton et al. (2003), the heart urchin *Echinocardium cordatum* is a good example of a species that fails to rebury once exposed to the air. Furthermore, several authors (Minchin et al. 2000; Jenkins and Brand 2001; Maguire et al. 2002) observed that undersized scallop discards are more susceptible to predation due to a reduction in their response or inability to recess, indicating that indirect, fishing-induced mortality can be significant for this species. Similar results were found by Chícharo et al. (2003a, 2003b). These authors showed that the cumulative effect of mechanical stress and temporary exposure on deck affects the behavioural and physiological condition of juvenile *Spisula solida*, probably decreasing its ability to escape predation. However, there are interspecific differences in reburial behaviour as a response to stress conditions. In laboratory-based simulations of mechanical dredge stress and air exposure, a reduction of bivalve reburial capacity for juveniles of *Spisula solida* was recorded, but also an increase of reburial ability of undersized *Donax trunculus* (Chícharo, unpublished results). Dias (2004) also showed a significant decrease in the reburial time of *Solen marginatus* juveniles due to fishing stress and Robinson and Richardson (1998) found that undersized razor clams *Ensis arcuatus* returned to the seabed were slow to rebury, becoming highly vulnerable to predatory crabs' attacks. On the other hand, Hauton et al. (2003) observed that razor clams which had been brought to the surface in the hydraulic dredge collecting box maintain the ability to rebury rapidly. A significant rate of predation on undamaged discarded whelks by starfish in an area disturbed by scallop dredging was reported by Ramsay and Kaiser (1998). Their laboratory studies demonstrated that whelks that had been in contact with bottom gears and rolled over took longer to straighten themselves and were less likely to have an escape response than whelks that had not been under the influence of the dredge.

As mentioned above, predation and scavenging on uncaught but dislodged individuals may be an important source of indirect fishing mortality. In fact, several studies (e.g., Caddy 1973; Meyer et al. 1981; Wassenberg and Hill 1987; Kaiser and Spencer 1994; Lambert and Goudreau 1996; Ramsay et al. 1996, 1998; Fonds et al. 1998; Kaiser et al. 1998; Ramsay and Kaiser 1998; Bergman and van Santbrink 2000; Hall-Spencer and Moore 2000; Chícharo et al. 2002c; Gaspar et al. 2003b) reported that the numbers of scavengers and predators can be elevated in fished areas. As for discarded individuals, indirect mortality of uncaught bivalves attributable to predation will depend on the relation between the reburial time and the time needed for predators to reach the impacted area. Meyer et al. (1981) found that 80% of the *Spisula solidissima* dislodged by hydraulic dredges reburied within 2 hours. Furthermore, Michael et al. (1990) have shown that for the clam species *Paphies donacina*, *Spisula aequilateralis*, *Dosinia anus*, *Mactra discors* and *Mactra murchiosini*, most clams had reburied after 20 minutes. Similar reburrowing times were observed by Chícharo et al. (2002c) for *Spisula solida*. These authors observed that clams exposed in dredge tracks reburied within 19 minutes. Species' life cycles also seem to be important as the impact of predators on bivalves is probably greater during their post-larval phase, especially when the scavengers and predators in the impacted area are specialised meiofauna feeders.

Evidence of predation or scavenging of exposed invertebrates after dredging or trawling activities has been reported by several authors (e.g., Caddy 1968; Arntz and Weber 1970; Meyer et al. 1981; Michael et al. 1990; Kaiser and Spencer 1994; Hall-Spencer and Moore 2000; Jenkins et al. 2004). Hall-Spencer and Moore (2000) reported that after dredging, damaged *Limaria hians* left on the dredge track had attracted a dense aggregation of scavengers. These authors found that the flesh from the file shells was consumed within 24 hours by juvenile cod (*Gadus morhua*), dragonets (*Callionymus* sp.), dogfish (*Scyliorhinus canicula*), edible whelks (*Buccinum undatum*), brittlestars (*Ophiocomina nigra*), swimming crabs (*Liocarcinus depurator*) and hermit crabs (*Pagurus bernhardus*). Even though there is an aggregative behaviour leading to the increase of scavengers and predators, such actions are short-lived. They last a few minutes (Gaspar et al. 2003b) to a few days (Ramsay et al. 1996 1997; Demestre et al. 2000; Jenkins et al. 2004), depending on the density of predators and scavengers in area and on the number of individuals that were damaged and dislodged by the fishing gear.

It is important, however, to determine if predation on discarded or dislodged individuals really occurs, or if only scavenging takes place. In the latter case, only the damaged individuals with their survival definitely compromised would be consumed and, therefore, relatively little additional indirect mortality due to predation would occur. Predator and scavenger aggregation to simulated discards from a scallop dredge fishery were investigated by Jenkins et al. (2004) in the north Irish Sea to determine differences in response to varying levels of damage to the discards (undamaged, lightly damaged and highly damaged individuals of the king scallop). These authors found that there was no apparent aggregation near stressed but undamaged scallops, whilst highly damaged scallops provided an attractive food source that was readily available to all scavengers. There was a lack of aggregation by predators to undamaged individuals left on dredge tracks in the Portuguese clam and razor clam fisheries (Gaspar, personal observation).

From these results, therefore, in bivalve fisheries where no additional mortality due to predation occurs, assessing indirect mortality may be restricted to survival experiments of undamaged or slightly damaged organisms that are discarded or exposed on the dredge tracks.

A final point that may add to the indirect mortality of dislodged or discarded organisms concerns the location at which such organisms find themselves after returning to the bottom. The relocation of individuals into unsuitable habitat after being discarded may contribute to increased mortality (Gaspar 1996). That is, habitats that are not optimal for the particular species affected due to unsuitable bottom type, food resources, etc. may compromise a dislodged or discarded species' chances of survival.

5.5 Review of Dredging Impacts

The environmental impacts of mobile fishing gears has increasing attention worldwide in recent years as the extent and intensity of this type of fishing has increased. Potential effects of dredging on habitats include changes in: physical structures (seabed topography and sedimentary biogenic structures such as reef corals, tubes, shells, burrows, etc.); the chemistry of the environment; sediment suspension and its redistribution; the benthic community; and ultimately the structure and dynamics of the whole ecosystem. Whist such effects do not strictly reside under the definition of "by-catch" effects, the recent increase in importance of such issues, and the technical measures developed to ameliorate them, makes it appropriate to include some discussion of them in this chapter about dredging.

5.5.1 Physical Impact of Dredging on the Sea Bed

Dredges alter the topography of the seabed while being towed across the sediment. Immediate physical effects are evident, with the dredge leaving visible furrows in the seabed. The dimensions of the furrows will depend on the sediment type and the gear's specifications, especially, the width of the dredge, the length of the tooth, the cutting depth of the blade and/or the pressure of the water jets. The width of individual tracks ranges from 1–5 m based on dredge design, while the depth of the tracks can exceed 50 cm (Meyer et al. 1981; Hall et al. 1990). The filling rate of the dredge track by sediment will determine the longevity of the furrows. Several authors (e.g., Gaspar et al. 2003b; Rosenberg et al. 2003) observed that the sides of the trench start to erode as soon as it is cut, whereas Meyer et al. (1981) reported that the trench walls began slumping two hours after dredging. Gaspar et al. (2003b) found that the dredge path was deeper on sandy-mud sediments than in sandy sediments. They also observed that the dredge track persisted for a longer period in finer sediments, varying from a few hours to several days for tows performed in sandy and sandy-mud sediments, respectively. Similarly, DeAlteris et al. (1999), in a study in Narragansett Bay, observed that the tracks in shallow waters and sandy sediments completely faded within 1–4 days, while tracks produced in deeper water and in muddy sediments remained unchanged for a period greater than 60 days. The difference in track longevity in these studies reflects the difference in the sediment recovery time.

Long-lasting hydraulic dredge tracks were observed by Pickett (1973) in the Thames estuary, where they took 2 months to disappear. However, hydraulic dredge trenches may remain visible for longer periods as demonstrated by Gilkinson et al. (2003). These authors conducted a hydraulic clam dredging experiment on a deep (70–80 m) offshore sandy bank on the Scotian Shelf to examine the immediate impacts of hydraulic dredging on physical habitats. They analysed video images and sidescan sonograms showing that the relatively flat seabed was transformed into a series of irregularly spaced, deep (20 cm) and wide (4 m) furrows which remained for at least 3 years.

From the above studies, it can be concluded that the longevity of the dredge tracks on the seabed, depends on the grain size, depth and the hydrodynamics of the area where the fishery is conducted (i.e., the strength of tidal and bottom currents and the frequency of storms). That is, wherever currents are weaker, the dredge tracks may be recognisable for a much longer time and even a minor fishery may have a significant cumulative effect on the microtopography of the bottom (Caddy 1973).

Sediment fluidisation in fished tracks has also been reported (Lambert and Goudreau 1996; Tuck et al. 2000). In a study conducted in a shallow sandy area in the Outer Hebrides on the west coast of Scotland, the effects of water jet dredging for *Ensis* spp. on the seabed and benthos were assessed by Tuck et al. (2000). These authors found that after the passage of the gear, the sediment within the tracks was fluidised to a depth of approximately 0.3 m. Although the dredge tracks had disappeared after 11 weeks, the sediment (0.2 m) in fished tracks remained fluidised. This result suggests that dredging may break natural cohesive and biological bonds in the sediment (Black and Parry 1994). Dredging also eliminates natural bottom features such as sand ripples and dislodges large shell fragments, rocks and cobbles (Caddy 1973; Butcher et al. 1981; Eleftheriou and Robertson 1992; Auster et al. 1996; Curry and Parry 1999). Dredging also removes large sessile epifauna and alters seabed micro-habitats by reducing the density of polychaete tubes, bivalve burrows and empty shells due to a combination of direct destruction by the dredges and burial (Gilkinson, et al. 2003). In a manipulative experimental trawl study in Sweden, Rosenberg et al. (2003) found significant impacts to sediment profiles in trawled benthic habitats (73 – 93 m deep) when compared with reference areas. Generally, attached epifauna and polychaete tubes were either rare or not observed at all on trawled sediment surfaces. We believe that it is likely that these findings for trawl fisheries can be readily extrapolated to dredge fisheries.

Reducing habitat complexity by dredging can have large ecological implications. Benthic infauna can be considered as ecological engineers (Coleman and Williams 2002), since they bioturbate, burrow and irrigate the sediment (Francois et al. 2001). These activities produce significant effects on the sediment biogeochemistry by enhancing nutrient exchange and the recycling of organic matter (Aller 1988; Kristensen 1988). Moreover, seafloor structures provide refuges for both predators and prey (Tuck et al. 2000), so changing habitat complexity may influence predation rates (e.g., Persson and Eklov 1995; Rooker et al. 1998). Another important consequence of changing sediment topography is the alteration of the near-bed hydrodynamics (Thrush et al. 1992). Increased sediment relief and mud clasts that are created by mobile fishing gears like dredges can have similar ecological consequences as increased habitat roughness available for biota (Rosenberg et al. 2003). Thus, near-bed currents can accelerate and decelerate in response to small protrusions and depressions (Rosenberg et al. 2003) which may lead to significant impacts on solute fluxes out of the sediment (Huettel and Gust 1992) and affect the deposition of organic matter and the settlement of benthic invertebrate larvae (Dernie et al. 2003).

Along with seabed disturbance, large sediment clouds are produced in the water column as dredges are dragged along the seafloor. The residence time of dredging-induced plumes is governed by the hydrodynamism and sedimentary characteristics in the particular area, but plumes usually settle within a short-period (Medcof and Caddy 1971; Caddy 1973; Butcher et al. 1981; Meyer et al. 1981; Mayer et al. 1991; Black and Parry 1994; Gaspar 1996; Gaspar et al. 2003b; Pranovi et al. 2004). However, if they last for a long time, as discussed by Currie and Parry (1996), the increase in turbidity may result in increased mortality of invertebrates, especially suspension-feeding individuals. However, in shallow waters, turbidity also occurs during periods of natural disturbance such as during winter storms (Rees et al. 1977), so most of the species that live in such environments could be expected to have developed to be able to survive long periods of turbidity.

The horizontal dispersal of sediment particles after dredging will be accentuated if currents are above the critical threshold for deposition (Falcão et al. 2003). Dredging may result in the loss of fine materials from the area fished, since these particles will travel further than sand and coarse sediments (Mayer et al. 1991). Watling et al. (2001) noted that the immediate effect of dredging was the loss of the fine fraction of the top few centimetres of sediment, which was not restored six months later. Reduction in the percentage of silt in the fished areas immediately after fishing was also observed by Tuck et al. (2000), but after five days this difference was no longer significant. Notwithstanding, repeated sediment re-suspension due to continued dredging in an area may lead to a permanent change in the sediment's composition as fine materials are washed away by currents (Langton and Robinson 1990; Pranovi and Giovanardi 1994; Schwinghamer et al. 1996, 1998; Watling et al. 2001). In turn, changes in sediment grain size may influence the distribution of benthic species, although this may not be the primary determinant of infaunal species distribution, because the availability of food will play an important role (Hall et al. 1993). It is worth noting that, in some places, changes in sediment grain-size due to dredging are unlikely to occur. For example, on the south coast of Portugal, the bivalve fishery only occurs in very shallow waters (up to 15 m depth) on sandy bottoms that consist primarily of medium to coarse sediments. These areas are affected by tides, strong currents and wave action, so changes in the sediment grain size due to dredging are not expected (Gaspar et al. 2003b).

Effects of sediment re-suspension are very site-specific and can include an increase in the concentration of nutrients and contaminants in water column, an increase in turbidity (and, consequently, the reduction of available light for photosynthetic organisms), the burial of benthic biota, the smothering of spawning areas, and negative effects on feeding and metabolic rates of organisms. These effects depend on several factors such as sediment grain size and other sediment characteristics, water depth, hydrological conditions and the fauna present (Hayes et al. 1984; LaSalle 1990; Coen 1995).

Because sediment is a sink for nutrients (Henriksen et al. 1983; Sundby et al. 1992; Forja et al. 1994) re-suspension of sediment leads to the recycling of nutrients between the seabed and the water column (Fanning et al. 1982). Dredging enhances the upward flux of nutrients by releasing pore-water nutrients in a large pulse, rather than by slower and more stable mechanisms (Pilskaln et al. 1998). Therefore, immediately after dredging, it is expected that there would be an increase of nutrients in the water column near the bottom. Falcão et al. (2003) observed a decrease in ammonium, nitrates, organic nitrogen, phosphates and silicates dissolved in the pore water of dredge tracks immediately after dredging, suggesting their export to the water column. These findings are in accord with the work by Riemann and Hoffmann (1991), who showed that organic and inorganic nitrogen increased in the water column during fishing operations. The release of nutrients to the water column changes the chemical and biological stability of the sediment and may disrupt the biogeochemical processes along sediment profiles (Fanning et al. 1982). Mayer et al. (1991) investigated the immediate effects of scallop dredging off the coast of Maine in a shallow site with a substratum comprised of mixed mud, sand and shell. They found that organic matter profiles were strongly affected by dredging. The concentrations of total organic carbon and nitrogen at the new sediment-water interface were markedly reduced after dredging and carbon significantly increased at the 5–8 cm sediment depth interval.

The retention and release of phosphorus from sediments have been important foci for research due to their importance in the production and distribution of plankton in lakes, estuaries and coastal systems (Nixon 1981; Froelich et al. 1982; Benitez-Nelson 2000). The overturning of sediment during dredging probably allows oxygen to penetrate into the lower sediment layers. Therefore, when an oxidised surface layer is present, substantial amounts of phosphate can be retained in the sediment through adsorption to iron oxides, hindering its release into the water column (Krom and Berner 1981; Sundby et al. 1992; Anschutz et al. 1998;

Slomp et al. 1998; Falcão et al. 2003). The release of nutrients into the water column can accelerate the turnover of nutrients and so increase the overall productivity of the water column (Fanning et al. 1982). Thus, phytoplankton primary production (if it is controlled by nutrients) may increase as a result of dredging activities, but such an increase may be more significant during seasons when concentrations of nutrients in the water column are limiting. Pilskaln et al. (1998) demonstrated that the re-suspension of one millimetre of sediment could triple the nutrient fluxes to the water column, increasing the productivity by 100–200%. These authors also have observed changes in the composition of phytoplankton from picoplankton to diatom-dominated populations, in such situations, indicating that these populations take advantage of the nutrient pulse supply as nutrients are released from the bottom to the water column (Churchill 1989). On the other hand, such increases in primary production may not occur if turbidity reduces the light in the water column to levels that affect the growth of phytoplankton (Barnes et al. 1991). However, such an inhibition of primary production will depend on the residence time of sediment plumes in the water column.

Following dredging, a reduction in primary production of benthic microalgae might also be expected immediately after disturbance, because the concentration of nutrients in pore water decreases (Cahoon and Cooke 1992). Furthermore, the total microbial biomass may decrease in the top layers of the sediment. Mayer et al. (1991) observed a decrease in various classes of microbiota and an increase in anaerobic bacteria after dredging probably due to redox oscillations as a result of the exposure of anoxic sediments to an oxygenated water column (Davis 2003). This shift may have important implications in organically rich marine sediments, since sulfate reduction is the predominant pathway (Davis 2003). Moreover, the anaerobic bacterial oxidation converts the ammonium and nitrogen of organic matter to biologically unavailable nitrogen gas, stripping available nitrogen from the water column (Pilskaln et al. 1998; Sowles 2001).

5.5.2 Dredging Impacts on Benthic Communities

Impacts of dredging on species living in and on the benthos depend on the size of benthic animals (meiofauna or macrofauna), their life stage and phase of their reproductive cycle, the position of the individuals in the sediment (infauna or epifauna), the type of sediment (soft or hard), the fishing effort, the resilience and recovery of the ecosystem and the environmental characteristics of the area.

Benthic communities are structured, in part, by dynamically interacting factors that determine 'habitat quality'. Since benthic macrofauna demonstrate strong, but often narrow, affinities to certain conditions, forces altering (or disturbing) the environment will be of considerable importance to their distribution and abundance (Dayton 1971; Dayton and Hessler 1972; Thistle 1981; Lissner et al. 1991).

As we know, fishing affects the populations of target species, other species captured in the net (the by-catch), and potentially all the other species in the community with whom these species interact (Dayton et al. 1995). Moreover, bottom fishing gears affect the sea floor, causing mortality and injuries to surface-living and shallowly-buried fauna (Tuck et al. 1998), and altering features of the physical habitat (e.g., Auster et al. 1996), sedimentation (e.g., Churchill 1989) and nutrient cycling (e.g., Mayer et al. 1991). The effective level of fishing impacts is a function of all these factors and it is very difficult to predict the environmental consequences of all these factors in different areas and temporal periods. In fact, fishing disturbance promotes differing responses of *taxa* with different life-histories as demonstrated by, among others, Jennings et al. (1998, 1999), Frid et al. (1999) and Bradshaw et al. (2000). Therefore, the total responses to disturbance are not always predictable even though the intensity, severity and frequency of these events are known to be important (McConnaughey et al. 2000).

For most cases, the effective level of disturbance on benthic communities can be related to the duration of the fishing impact, so a cumulative impact is expected to be responsible for greater and longer-lasting changes in the benthic ecosystem than a short impact. However, several studies suggest that it is the first fishing event that may have the greatest impact in changing the benthos. The artificial selective pressure imposed by fisheries can affect community structure in the short-term and even perhaps drive changes in life-history traits in the long-term (Rodhouse et al. 1998).

Short-term impacts lead to modifications in the benthic ecosystem that are mainly caused by changes in the biota (e.g., Watling and Norse 1998). Despite the objective of the fishing activity to capture the target species, the major impacts tend to occur on the non-target species (e.g., Ardizzone et al. 2000; Ramsay et al. 2000). This is because the design of the fishing often attempts to be selective for the target species only. In fact, species' removal caused by fishing is the main factor responsible for ecosystem changes and therefore stopping the impact should allow the ecosystem to recover. After a short-term impact there is a strong possibility of recovery of the community due to recruitment events. In soft bottom areas, sediment structure and composition may rapidly return to pristine conditions, and so not causing a permanent change in the benthic community. On the other hand, after long-term impacts, both the biota and the physical and chemical characteristics of the ecosystem can be permanently affected. In such circumstances, a new equilibrium between biotic and non-biotic characteristics may results.

5.5.2.1 Short-term Impacts on the Benthos

The short-term environmental effects of dredging on the sea-bottom have received increased attention in recent years, and several studies have characterised changes in the ecosystem due to dredge-fishing (e.g., Hall et al. 1990; Michael et al. 1990; Eleftheriou and Robertson 1992; Kaiser and Spencer 1996; Lambert and Goudreau 1996; Bergman et al. 1998). Many experimental studies have shown that changes in the benthos are measurable after a short period of intense bottom fishing (e.g., de Groot 1984; Kaiser and Spencer 1996; Currie and Parry 1999; Jennings and Kaiser 1998 for a review). Dredging activities modify or destroy benthic habitats and reduce the biomass and diversity of invertebrate communities, affecting secondary production at large spatial scales (Kaiser 1998; Collie et al. 2000; Kaiser et al. 2000; Jennings et al. 2001). The most common approach used to study effects of fishing has been to perform random sampling of species along dredge-tracks (e.g., Medcof and Caddy 1971; Caddy 1973; Michael et al. 1990; Bergman et al. 1998) but this can lead to biased data being collected because impacts are not even throughout the track (Lambert and Goudreau 1996; Meyer et al. 1981; Chícharo et al. 2002a).

Studies have indicated that a variety of fishing gears, such as beam trawls (Bergman and Hup 1992; Kaiser and Spencer 1994), otter trawls (Van Dolah et al. 1987; Rumohr and Krost 1991) and dredges (van der Veer et al. 1985), can cause mortality of some epi- and infaunal benthic organisms. However, according to Kaiser et al. (1998), effective consequences of such effects on the benthic ecosystem may vary depending on whether or not the fishing impact coincides with peak periods of settlement of benthic invertebrate larvae. In fact, immediate changes in the abundance of most species can be undetectable after their next recruitment – usually within 6 months (Currie and Parry 1996; Kaiser et al. 1998). Therefore, although the direct effects of fishing impacts on benthic communities appear to be obvious, their magnitude and effective consequences are more difficult to evaluate and have often been considered equivocal (Thrush et al. 1998).

5.5.2.2 Long-term Impacts on the Benthos

Long-term changes to benthic communities due to dredging are more difficult to evaluate than short-term effects and, consequently, have been less analysed. Analyses of long-term effects are usually based on comparisons between areas that are fished and unfished, but the interpretation can be confounded, as unfished areas usually differ physically from fishing grounds.

Long-term impacts can affect biotic and abiotic components of the ecosystem. For example, the alteration of sediment characteristics may influence the colonisation and presence of benthic species and, therefore, in the long-term, the structure of benthic communities. Changes in abundance, biomass, diversity and dominant trophic groups within macro- and meiobenthic communities have been attributed to long-term fishing impacts. Decreases in these parameters in fished areas may be attributed to the continuous long-term effect of dredge passage along the bottom, damaging and killing most of the macrofauna species. Species like sea urchins, sea stars, gastropods or larger bivalves like *Acanthocardia tuberculata* (more than 60 mm) become rare or absent in areas under continuous fishing pressure (Chícharo et al. 2002b). Moreover, fishing impacts also cause the destruction of algal mats with the consequent decreases in abundances of herbivores and the removal of natural spawning areas for fishes and other species (Chícharo et al. 2002b).

While fishing impacts can be catastrophic for injured individuals, they may not be significant at a larger scale (Frid et al. 2000), as populations may be sustained by individuals living in patches not impacted by gears, even in heavily fished areas (Kaiser et al. 1997; Rijnsdorp et al. 1998). Nevertheless, as previously described, the dead and injured fauna left on the sea floor or exposed in trawl tracks, and also the addition to the benthos of offal and dead/dying by-catch, increases the opportunities for mobile scavengers/predators (Kaiser and Spencer 1994; Kaiser and Ramsay 1997). In addition, the continuous disturbance of the sea floor may benefit opportunistic infauna, as polychaetes, namely Nephtyidae, Nereididae and Cirratulidae, are reported as very abundant in impacted areas (Tuck et al. 1998; Rumohr et al. 1998).

Meiofauna is believed to be highly selective with distinct and often highly specialised food niches (Kennedy 1994). The analysis of meiofauna as a potential indicator of anthropogenic perturbation in aquatic ecosystems has often been limited to pollution monitoring surveys (Pranovi et al. 2000), however, meiofaunal analysis may also reveal long-term dredging disturbances. Chícharo et al. (2002b) found high abundances of meiofauna suspension feeders in soft bottom fishing areas off the south coast of Portugal, attributed to the continuous re-suspension of organic detritus during dredge trawls. However, despite the increase in scavengers' abundances, a similar increase was not observed in biomass. In fact, fishing impacts forced the selection towards the survival of small-sized animals, as gear selectivity hampered the survival of larger individuals. Therefore, long-term trends in benthic ecosystems impacted by fishing activities may reflect a shifting from a 'pristine' benthic community (dominated by long-lived, *k*-selected species) to an impacted community (dominated by *r*-selected, opportunistic, short-lived species, such as some polychaetes) as described by Commito (1982), Bemvenuti (1994), Mortimer et al. (1999), Christensen et al. (2000) and Chícharo (2002b).

One major limitation in identifying long-term changes on the benthic ecosystem due to dredging results from the difficulty in distinguishing between changes in the community caused by fisheries' disturbances and those caused by natural phenomena (Currie and Parry 1996). Kaiser et al. (1998) have noted the importance of evaluating the ecological relevance of fishing disturbance *versus* natural perturbations, which will vary between different habitats. Complex benthic communities fluctuate from year-to-year due to changes in local conditions differently affecting the reproduction, recruitment or survival of species. These variations could include temperature (Beaumont and Budd 1982; Southward 1995; Kroncke et al. 1998; Lindley 1998), seasonal currents (Pingree and Griffiths 1978; Hill et al. 1995, 1997), frequency and intensity of storms (Hall 1994; Kaiser et al. 1998), and abundance of phytoplankton (Frid et al. 1996). Climatic changes, such as global warming and other large-scale phenomena also affect ecosystems and benthic communities, and thus need to be considered.

So far, in this chapter, we have identified and discussed the problems associated with dredging and its impacts on the environment, both in term of the physical structure of affected habitats and the associated benthic communities. However, it is important to note that a great deal of the research effort that has examined effects of dredging has focussed on descriptions and quantifications of the impacts and efforts. In contrast, there have been relatively few studies that have addressed possible solutions to such problems.

In the rest of this chapter, we have elected to use a case study to describe how some of these problems can be resolved and, based on our experience, we identify how such solutions can be applied to other dredge fisheries in the world.

5.6 The Mitigation of Dredging Impacts on the Ecosystem: Case Study – The Portuguese Dredge Fishery

The evidence that dredges injure benthic organisms, and reduce habitat complexity and biodiversity has led environmental groups to question the use of these fishing gears. However, simply closing dredge fisheries would cause serious socio-economical problems, so the question arises: 'How can we reduce the environmental impacts of dredging whilst maintaining the fisheries that rely on it?' One of the solutions for mitigating the adverse ecological effects of dredging is to modify the dredge design. To achieve this goal, it is of utmost importance to assess the actual impacts caused by dredging and understand which parts of the fishing gear are responsible. Developing and introducing more 'environmentally friendly' dredges is a lengthy process that involves a large number of studies and underwater observations. The following case study off Portugal serves to demonstrate how ecosystem dredging impacts can be mitigated by developing more efficient and selective dredges.

Clam and razor clam dredges are extensively used along the Portuguese coast. At present, the dredge fleet comprises 172 boats, 4 – 15 m long, with engines of 17 – 150 Hp and a crew of 1 to 5 fishers. This fleet directs fishing effort towards the clams *Spisula solida*, *Donax* spp., *Chamelea gallina* and *Callista chione* and the razor clam *Ensis siliqua*. These species inhabit sandy bottoms, forming extensive and dense beds, sometimes several kilometres in length. In this fishery, only mechanical dredges composed of a rigid iron structure with a toothed lower bar and a collecting system are allowed. These dredges are dragged across the seafloor and are designed to dig clams and razor clams out of the sediment (up to 60 cm depth), impacting the benthic habitat, both in terms of its physical structure and its biological communities.

Because of the scarcity of information about the ecological impacts of dredging in Portugal, a new programme was implemented in 1990 to bring together researchers, managers and fishers, to identify and mitigate any adverse immediate effects of dredging on the ecosystem. With this purpose, several studies were conducted culminating in the development of a new dredge that was introduced in the Portuguese bivalve fishery in 2001. This programme focused primarily on the selectivity and efficiency of dredges, by-catches, and the immediate impact of dredging on the target species and benthic communities. This case study provides useful examples to illustrate how solutions to the adverse effects of dredges can be developed and implemented in the world's dredge fisheries. Below we describe the studies that contributed most to the development of the new dredge.

In the early nineties, two types of dredges were used along the Portuguese coast: the mesh dredge (MD) and the north dredge (ND) (Fig. 5.6). These dredges comprise a metallic frame where a net bag is attached and a tooth lower bar. The main difference between the dredges relate to their shape (semi-circular in the MD versus rectangular in the ND) and the width of the dredge mouth (64 cm in the MD versus 150 cm in the ND).

Fig. 5.6. Photographs of the mesh dredge (MD) and the north dredge (ND)

The first step in the programme was to adjust these fishing gears to account for the biology of the target species by introducing a minimum mesh size that would prevent the harvest of undersized individuals, allowing them to grow to a more valuable market size and to reach a size at which they can reproduce at least once before capture. Under this objective, selectivity experiments were done (Gaspar 1996; Gaspar et al. 1999, 2003c). It is well known that dredge selectivity is affected by several factors including the type of seabed, depth, tow duration and speed, the hanging coefficient of the net bag, the twine material and its diameter, tooth spacing and mesh size (e.g., Drinkwater 1974). It was impractical to assess the effects of all these factors on selectivity, so the investigations focused on tooth spacing and mesh size only. However, great care was taken to ensure that all the other factors remained constant during the experiments. The dredge selectivity experiments were performed using the cover method which involves fitting the dredge net with a cover made of smaller mesh netting that retains those individuals escaping from the dredge net. In our selectivity experiments a cover bag with a 20 mm mesh was attached to the gear. This bag was 1.6 times longer and wider than the primary net bag and did not impede the natural flow of water through the net (Gaspar et al. 1999). To assess the effect of tooth spacing and mesh size on the catch, two commercial dredges, equipped with different tooth spacings and mesh sizes, were towed side by side. Three tooth spacings and four mesh sizes were investigated. In one of these selectivity studies (Gaspar et al. 2003c), the effect of tooth length on the catches of *Spisula solida* was also assessed, comparing the structure of the catches from dredges with and without a tooth bar.

Surprisingly, the results showed that the space between the teeth does not have an effect on selectivity (Gaspar 1996; Gaspar et al. 1999, 2003c). Therefore, the only factor that contributed to dredge size selection in the experiments was mesh size, with the retained proportion of undersized clams decreasing with increasing mesh size. Although the effect of tooth spacing has been found to be significant in other studies (Baird and Gibson 1956; Drinkwater 1974; Nashimoto et al. 1983; Nashimoto 1984), these authors also recognized that the effect of tooth spacing on selectivity was of minor importance when compared to mesh size. Concerning the length of teeth, it was observed that this factor influenced the size composition of *Spisula solida* catches (Gaspar et al. 2003c). Catches from dredges without teeth consisted of a greater proportion of juveniles than catches from dredges equipped with teeth, indicating that larger clams burrow deeper in the sediment than smaller ones. Thus, it can be concluded that the capture efficiency of dredges is directly related to tooth length. Nevertheless, we believe that tooth spacing cannot be considered irrespective of tooth length. Hence, there should be a particular spacing between teeth after which the length of teeth will not have an effect on the catch. Bearing this in mind, we could hypothesise how the dredge would act as if a tooth bar was not present (Gaspar et al. 2003c).

In another study, Gaspar et al. (1998) demonstrated that tooth length had an effect on the proportion of damaged razor calms (*Ensis siliqua*) where increased tooth length resulted in lower proportions of damaged razor clams. This result is explained by the defensive behaviour of the species. Without perturbation, these animals are burrowed close to the surface, with the siphon protruding out of the sediment. When they feel any perturbation such as dredging, they quickly burrow into the sediment (up to 60 cm below the surface) in a defensive response. Thus, longer dredge teeth (penetrating deeper in the substrate) allow *Ensis siliqua* to be caught below the lower edge of the shell, lifting the razor clams into the dredge without causing damage. Conversely, when shorter teeth are used, the teeth hit a larger number of razor clams directly, leading to more damage to the shells. In this experiment, it was also observed that damage to razor clams was inversely proportional to catching efficiency. Therefore, in order to increase catching efficiency and to decrease the proportion of damaged individuals in the catch, the tooth length stipulated for a certain bivalve fishery should take into consideration certain ecological characteristics of the target species, especially its maximum burrowing depth.

Along with these selectivity experiments, a study to assess the effect of mesh size and tooth spacing on the proportion of damaged individuals was conducted. The main objectives of this study was to determine if changes in mesh size and tooth spacing could reduce the number of macrofaunal organisms damaged or killed by dredges (Gaspar et al. 2002). However, the results showed that the range of mesh sizes and tooth combinations tested had no effect on the numbers of damaged macrofauna caught. Such a result may be related to the way this gear is operated. Scuba divers observed that the tooth bar of the dredge penetrated 10 cm into the sediment, acting as a rake, pushing sand to the front of the mouth frame creating a 'sand wave', which precluded infauna from passing through the space between the teeth (Fig. 5.7). This result corroborates the one obtained in the selectivity study, where it was observed that tooth spacing did not have any effect on the catch. Regarding mesh size, underwater observations showed that the mesh of the net bag closed as it was stretched due to the weight of the material in the bag, preventing the escape of individuals from the bag. Therefore, most of the escapes through the mesh of the net bag occurred only when the dredge was hauled vertically and washed (to release the sand from the net bag). It is important to highlight, however, that most of the non-target individuals that entered the dredge did not escape, indicating that the selectivity of this gear is quite poor. Notwithstanding, independent of mesh size, the individuals that are retained in the net bag are susceptible to injuries due to the abrasion among animals and/or debris inside the bag. Therefore, it is expected that damage increases with tow duration. Gaspar et al. (1998) observed that the number of damaged razor clams in the catch was greater in longer tows, mainly due to the fragility of the *Ensis siliqua* shell. However, with respect to non-target species, it was found that tow duration did not have any effect on the overall proportion of damaged individuals, although some species were more vulnerable to dredging than others (Gaspar, unpublished data). Thus, one must conclude that the degree of damage inflicted by dredging on different macrobenthic species is related to their morphology and fragility.

During the experimental phase of these studies, scuba divers also observed that the 'sand wave' created in front of the gear mouth by the action of

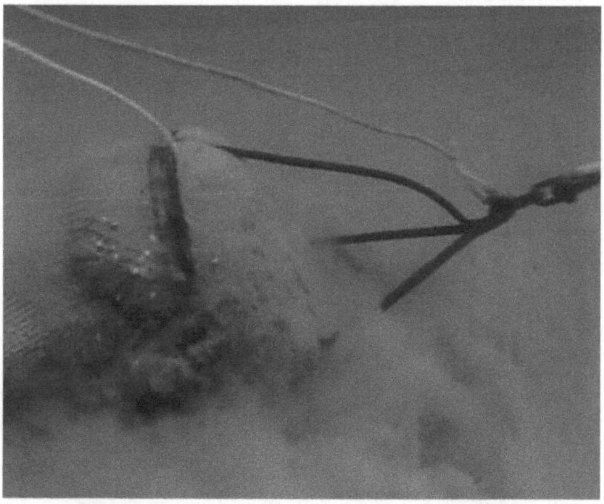

Fig. 5.7. Underwater photograph showing the clogging of the mesh dredge mouth by sand, during the deployment

teeth is formed shortly after the start of the tow, pushing sediment sideways and above the dredge, limiting the amount of material entering the net bag and, consequently, decreasing the efficiency of the dredge. The low efficiency of the dredge was also evident by the clogging of the dredge mouth by sand (Fig. 5.7) due to the closure of the mesh during the tow. *In situ* observation of the dredge tracks by divers showed that a high proportion of the animals that made contact with the dredge and escaped were injured.

Based on all these results and on underwater observations made on the performance of the dredge during the tow, it was concluded that: (i) the catch efficiency of dredges is drastically reduced during the tow; (ii) macrofauna mortality increases with this decrease in efficiency; and (iii) dredge selectivity is relatively low. These results reinforced the need to modify the design of the traditional dredge in order to enhance its efficiency and selectivity.

A small modification was introduced to the metallic structure of the dredge by welding a grid to its mouth (Fig. 5.8). This new dredge was named the net-grid traditional dredge. This modification aimed to prevent the clogging of the dredge mouth by sand during the tow. Although this alteration proved to be effective, only catching efficiency was enhanced. That is, for the same tow duration, the yield from this dredge was significantly greater than the one obtained with the traditional dredge. Gear selectivity remained unchanged, however, mainly because the retention method was not modified. As it was observed for the traditional dredge, during the tow, the meshes of the net bag closed due to the weight of the catch, water flow and tow speed, preventing the escape of individuals. Moreover, it was observed that, during dredging, the caught individuals remained inside the bag, near the bottom and rolling, increasing the probability of being damaged.

Fig. 5.8. Photograph of the net-grid traditional dredge (TD)

In disseminating the results of our research, several workshops were held with the fishing sector, scientists, administrators and other stakeholders. When the above benefits of using the net-grid traditional dredge were presented to fishers, they rapidly adopted this modified fishing gear.

It was next considered that the impact of this type of fishery upon the macrobenthic community could be further minimised by developing an even more efficient and, simultaneously, more selective dredge, in order both to reduce the number of non-target individuals in the catch and to allow the escape of by-catch during the tow. Therefore, our next goal was to modify/design a dredge that fulfilled the following requisites:

- maximal efficiency;
- low by-catch of non-target species;
- retention of very few undersized individuals;
- maintenance of selectivity throughout the tow; and
- low proportion of damaged individuals.

Taking these objectives into consideration, the design of the net-grid traditional dredge was modified with the collaboration of the fishers from the Setúbal region (south-western coast of Portugal) (Gaspar et al. 2001). The basic difference between the new dredge (named the grid dredge – GD) and traditional dredges is in the retention structure for the bivalves. In this new dredge, the net bag is replaced by a rectangular metallic grid, which is approximately 10 cm apart from the sediment (Fig. 5.9). In order to evaluate the possible introduction of this dredge in the bivalve fishery, a study was done to compare the efficiency and selectivity of the two

Fig. 5.9. Photograph of the grid dredge (GD)

dredges (net-grid traditional dredge and the new dredge design) and evaluate their impact on the benthic community associated with *Callista chione* (as estimated by the proportion of individuals that entered the dredges and were damaged) (Gaspar et al. 2001).

The results showed that catches from the traditional dredge (TD) comprised a great fraction of *Callista chione* juveniles, while in the new, grid dredge (GD), catches were almost entirely composed of individuals with a length above the minimum landing size. Additionally, the proportion of non-target species in the catches was significantly lower in the GD. The modified design significantly reduced the incidental capture of juveniles and sub-adults of commercially important bivalves by up to 95%. Divers also observed that small individuals (independent of the species) escaped immediately through the metallic bars of the grid while the dredge was being towed. This was due to the bottom of the metallic grid not directly contacting the seafloor, allowing individuals to escape during the tow and not only during the hauling of the dredge, as is the case for the traditional dredges. Divers also reported that undamaged individuals that escaped, burrowed immediately (in the case of the infauna) or recovered their activity (in the case of epifauna). With this rapid reburying response, dislodged organisms are less likely to be eaten by predators. Another result was that the mean fishing yield obtained with the GD was always greater than that of the TD. It was also shown that the TD caused mortalities to the target and by-catch species in the same order of magnitude as the GD (Gaspar et al. 2001). Nevertheless, despite the gear type used, there were significant direct effects of dredging on some benthic species, as certain groups of animals suffered heavy damage while others were less affected.

For the *Callista chione* fishery it was demonstrated that the new dredge was more efficient and selective than the net-grid traditional dredge. However, it was felt that this new dredge could prove its importance even more in other bivalve fisheries. Moreover, the total direct impact on the macrobenthic community needed to be investigated in order to determine if it was less for the new dredge. Therefore, a study was done to compare the total direct mortality on the macrobenthic community caused by three types of clam dredges (north dredge – ND, net-grid traditional dredge – TD, and the grid dredge – GD) used in the *Spisula solida* fishery (Gaspar et al. 2003a). Total direct mortality was assessed by considering the degree of damage sustained by individuals that entered the dredges, but also the damaged individuals that were left in the dredge tracks. The correlation between mortality and catching efficiency for each type of dredge was also assessed.

The results revealed that the ND and the TD retained almost all individuals that entered the dredge (94 and 97%, respectively), while the GD retained a smaller proportion of individuals (76.1%). The ANOSIM test that accounted for retention-type effects (grid *versus* mesh bag) showed significant differences between the GD and both the TD and ND, reflecting differences in the selectivity of these fishing gears. From these results, it can be concluded that rigid structures, such as metallic grid cages, are more selective than flexible collecting systems, such as net bags, in these dredges. Significant differences in total direct mortality between ND (18%) and both TD (10%) and GD (8%) were also observed. These differences were largely attributed to the animals in the dredge track that died as a direct result of the physical damage inflicted by the dredge. It was also found that the damage to uncaught individuals was inversely correlated to gear efficiency (Fig. 5.10). The lower catch efficiency of the ND (64%) led to a higher proportion of damaged individuals left in the dredge path, when compared with the more efficient TD (90%) and GD (98%) dredges. The differences in the efficiency of capture observed between dredges are related to the flow of the sediment that enters the dredge during the tow (Fig. 5.11). The greater efficiency of the GD results from the fact that the sand that enters the dredge is rapidly filtered by both grids. Therefore, all the clams that are found on the dredge path enter the dredge. The correlation between catch efficiency and damage has also been observed by other authors (Caddy 1973; Meyer et al. 1981; McLoughlin et al. 1991). Since yield is directly related to the efficiency of capture, significant differences in the mean yield were also observed between dredges. These results showed that the grid dredge was effective in reducing by-catch and direct mortality, while increasing catches of the target species. Indirect mortality (due to desiccation and predation) was also expected to decrease, since small individuals that entered the grid dredge escaped throughout the tow, burrowing almost immediately.

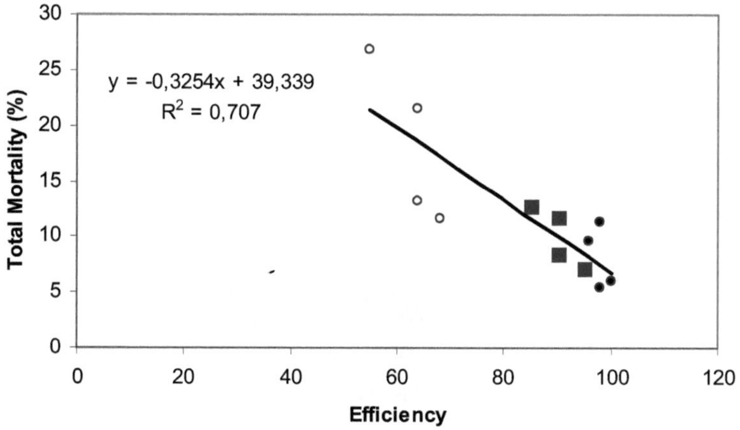

Fig. 5.10. Correlation between total direct mortality and catching efficiency. (O) north dredge; (■) traditional dredge; (●) grid dredge

Fig. 5.11. Underwater photographs of the north dredge (left), net-grid traditional dredge (centre) and grid dredge (right) during their deployment

Although the three dredge types used in the *Spisula solida* fishery were not compared for their impacts on the sediment, the above trends are expected to also apply because the capture method (toothed lower bar) is identical. It is worth noting, however, that the Portuguese bivalve fishery is managed using daily quotas per species and boat, meaning that these more efficient gears will result in the reduction of the dredged area for a given amount of harvested clams (Fig. 5.12). If these quotas were not in place, the more efficient gears could lead to greatly expanded fishing effort, increased pressure on stocks and the ecosystem.

The modification or introduction of a new gear in any fishery can only succeed if it accomplishes two objectives: first, it should reduce the impact on the ecosystem; and second, it should bring some benefits to the fishers, or else any gear modification is unlikely to be endorsed by fishers. With the development of the grid dredge in this case study, both objectives were achieved and, thus, many advantages for the ecosystem and fishers were

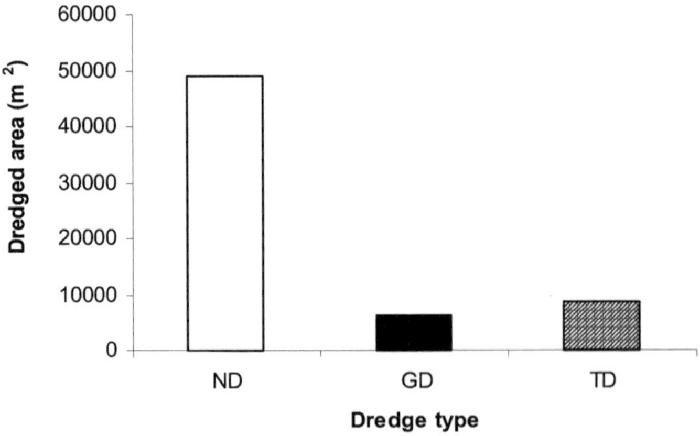

Fig. 5.12. Comparison of the daily dredged area estimated per boat to attain the daily quota, for the three dredges tested. ND, north dredge; GD, grid dredge; TD, net-grid traditional dredge

attained when this fishing gear was introduced. For the ecosystem, these were: the daily fished area is significantly reduced, lower amounts of by-catch, lower number of damaged individuals and consequently lower mortality. For the fishers, there was: less time spent sorting the catch (because of a lower amount of by-catch and debris), less time spent fishing, less working hours of the vessel engine, lower fuel consumptions, lower operational costs and higher incomes.

So, from fisheries management and ecological perspectives, the results showed that there are clear advantages in developing more efficient and selective dredges in order to reduce the number of damaged individuals and by-catch, and consequently decreasing the impact of dredging on macrobenthic communities. However, the mitigation of the impacts is only effective if other measures are also implemented in the fishery, such as daily quotas per boat and species. Otherwise, the development of more efficient fishing gears can contribute to increase fishing pressure on targeted stocks.

5.7 Discussion and Research Needs

Mobile fishing gears, such as clam and razor clams' dredges, have a deleterious effect on the ecosystem, impacting both the habitat and the benthic community. Aiming to reduce the effects resulting from such dredging, the IPIMAR implemented a research programme to enhance gear selectivity and efficiency, to reduce by-catch and, consequently, to mitigate fishing effects. In this chapter we have described the studies that were done towards this.

The improvements that were made to the dredges in this case study demonstrated that the impact of fishing gears on non-target species and habitats could be significantly reduced without negative effects on the profits of the fishing operation. In our opinion, two factors have contributed to the success of the implemented programme. First, was the close cooperation between the fishing industry and researchers, by involving fishers in the fieldwork. During experimental fishing, fishers shared and discussed their ideas and experiences, largely contributing to the improvements that were made. The second factor was the underwater observations that were done. These were crucial in the whole process, since they allowed us to perceive the performance of the fishing gear during towing and, therefore, to understand which gear modifications should be introduced in order to enhance selectivity and efficiency and to reduce unwanted mortality.

Our research showed the advantages of developing dredges equipped with a rigid collecting system, which proved to be much more selective than dredges that use flexible collecting systems (such as net bags) to retain the catch. In fact, it was observed that the grid dredge retained a significantly smaller proportion of individuals that entered the gear than traditional dredges. This result is related to differences in the geometry of the gear during dredging. When a net bag is used to retain individuals, the mesh stretches while the dredge is being towed, preventing the escape of organisms through the mesh. Thus, dredges equipped with net bags only become slightly selective during the hauling process. Furthermore, almost all the individuals that entered the dredge were retained. In contrast, when a metallic grid is used, selection of the captured individuals occurs throughout the tow, so the amount of by-catch was significantly reduced. Another important finding was that undamaged individuals, mostly juveniles, which passed through the parallel rods of the grid, burrowed or recovered their activity immediately. This rapid response decreased the probability of dislodged organisms being eaten and, therefore, their indirect mortality. Further, since by-catch from the grid dredge was significantly lower than that from traditional dredges, the mortality due to desiccation and catch handling on the boat deck is also expected to decrease as catches almost totally comprise large individuals, which have a greater resistance than juveniles to such factors.

Although we succeeded in developing a more efficient and selective dredge in this fishery, the amount of by-catch is still high in some bivalve fisheries. In some periods (late spring, early summer), we have observed that the quantity of by-catch could surpass the catch of the target species (Gaspar, unpublished data). Therefore, efforts to reduce even further the by-catch in Portuguese dredge fisheries must continue. This will certainly involve the development of modifications to the grid dredge to further improve selectivity and minimise by-catch. We believe that the amount of unwanted catch can be significantly reduced by the introduction of by-catch reduction devices (BRDs) in dredges, so studies to evaluate their effectiveness in dredge fisheries should be conducted. These devices have been successfully used in trawl fisheries and are divided into two categories: i) those that exploit behavioural differences between species; and ii) those that separate species by their size (see Broadhurst 2000 for a review). In dredge fisheries, the by-catch is mainly composed of bivalves and other invertebrate species, so the most correct approach would involve BRDs that select individuals by their size. BRDs in this category for trawls are designed to exclude those individuals that are larger than the openings in the separating grid (Broadhurst 2000). For dredges equipped with a metallic cage to retain the catch, a BRD could be introduced by incorporating, in the middle of the collecting system, an oblique metallic grid terminating at an escape exit at the top of the cage. Thus, individuals larger than the openings could be expected to be guided upwards to the escape exit, while smaller individuals should pass through the openings of the cage.

However, it is also important to emphasise that these excluder devices will only be accepted by fishers if the catch of the target species is not affected while reducing by-catch. Although we may succeed in reducing by-catch, BRDs should only be implemented if the mortality of escaping individuals is lower than if those individuals were captured and discarded.

Regarding efficiency, it was found during our case study that damage on uncaught individuals was directly related to gear efficiency. The lower the catching efficiency, the higher the proportion of damaged individuals left in the dredge tracks. Moreover, for the same tow duration, higher efficiency also led to higher fishing yields. These results proved the importance of developing more efficient dredges. It is important to note that for dredges that use long teeth to dig clams out of the sediment, the efficiency is related to the accumulation of sand in the dredge. In the case of traditional dredges, and since the meshes of the net bag closes during the tow, the amount of sand that was trapped inside the bag led to a sand wave in front of the dredge mouth, decreasing their efficiency. On the other hand, in the grid dredges the sand entering the dredge rapidly escaped through the metallic bars of the grid. This occurred because the dredge was equipped with two sledges that prevented the metallic grid contacting the seafloor, allowing the sand to escape through the sides and the bottom of the grid.

Despite the greater efficiency of the new dredge, significant direct effects of dredging were observed on some benthic species. Certain groups of animals suffer heavy damage while others are less affected, leading to immediate, short-term and long-term effects. It is important to remember, however, that the significance of dredging effects on benthic communities must take into account the magnitude and frequency of natural disturbances. Biological communities that occur in a particular habitat have adapted to their environment through natural selection and therefore any impacts of mobile fishing gears on the habitat structure and biological community should be considered against the impacts that natural disturbances have. Impacts from dredging are expected to be greater in low hydrodynamic areas, because in such areas, benthic communities may be less capable of sustaining and overcoming disturbance than those inhabiting more dynamic, coarser sediments in shallow waters (Jones 1992). However, even in shallow dynamic waters, chronic fishing disturbances may produce long-term changes to benthic communities (Sainsbury 1988; Collie et al. 1997; Jennings and Kaiser 1998; Bradshaw et al. 2000), depending on the scale and intensity of the activity. If the fished area is a large proportion of the habitat, a dilution effect of the impact cannot occur (Kaiser 1998) and, therefore, recovery will take longer (Hall 1994; Thrush et al. 1995). In this context, it is of utmost importance to understand if dredging has long-term effects. Therefore, one of the biggest current challenges that researchers are facing is to develop studies and/or methodologies to determine the magnitude of dredging impacts on the ecosystem. Benthic communities are very dynamic naturally. The abundance, biomass and diversity of benthic communities change locally by being influenced by seasonality, freshwater discharges from coastal rivers, recruitment events, storms, etc. This is one level of disturbance occurring to benthic communities. Another level of disturbances involves those due to natural phenomena at larger scales, such as global warming, ocean's pH changes caused by greenhouse effects or decreases in nutrient inputs in coastal waters. Furthermore, a third level of impact caused by anthropogenic activities includes pollution, modification of rivers discharge due to dams, and fishing activities that modify the structure and dynamics of benthic communities.

When we are examining a dataset searching for changes in the benthic community that can be attributable to effects of fishing, it is necessary to consider also the other levels of disturbance occurring to the communities. Only after accounting for the effects of these levels of natural disturbance is it possible to determine the real environmental impacts of a fishing activity. Long-term and exhaustive datasets are needed, and most of the information about fished areas was collected long after the start of the fishing activity, which, in some places, began centuries ago. Moreover, building new datasets for comparing natural and fishing impacts is difficult because fishing areas have particular characteristics, such as greater abundances of the target species (which play their vital role in the benthic community) that will not be similar in non-fished areas – otherwise that area would also be fished. One possibility to solve this problem is the creation of closure areas and to 'rewind' the effects of fishing by allowing the ecosystem to recover. However, after acquiring a certain level of impact, a new equilibrium may result, different from the 'pristine' case, and the recovery situation may not compare with the original ecosystem. Another limitation in detecting and separating natural changes from fishing impacts based on historical datasets is due to the gradual evolution and modification of fishing gears' designs, procedures and fishing efforts. In fact, two areas fished with the same gear but with different efforts present different levels of impact and, similarly, two areas fished with different gears but the same effort will present different levels of impact.

Notwithstanding, and regardless of the magnitude of the impact of dredging on the ecosystem, this chapter has demonstrated that dredging effects can be ameliorated through gear modifications without compromising the profits of fishers. In achieving this for a particular fishery, however, it is important to undertake a series of studies, aimed to:

- estimate the selectivity of the gear;
- estimate its efficiency;

- determine the composition and structure of by-catch;
- estimate the amount of discards;
- estimate total mortality;
- understand the capture process of fishing gear in various environments; and
- determine which parts of the gear inflict more damage to benthic organisms.

Then, armed with this information, one can begin to develop alternative gears that reduce the impacts of dredging.

Finally, during the process of developing and introducing environmentally friendly fishing technology it is of utmost importance to involve the fishing industry, scientists and other stakeholders. It is unlikely that gear modifications will eliminate all adverse effects and, therefore, realistic short- and long-term objectives are necessary when attempting to minimise ecosystem impacts of a fishery.

Acknowledgements

The authors would like to thank Dr Steve Kennelly for reviewing this book chapter and for all suggestions that greatly improved the manuscript and Dr Costa Monteiro who provided valuable guidance and insight. We also thank Dr João Cúrdia, Dr Paulo Vasconcelos, Dr Susana Carvalho and Dr Marta Rufino for many helpful comments on drafts of this chapter. The contribution of many staff members of the IPIMAR, CRIPSul, Delegação de Setúbal and University of Algarve is gratefully acknowledged. In particular, we would like to thank the following persons for their support in conducting the field work and/or laboratory analysis: Dr Miguel Neves dos Santos, Dr Alexandra Chícharo, Dr Manuel Sobral, Dr Francisco Leitão, Dr Manuela Falcão, Dr Aida Campos, Dr Damásia Dias, Dr Paulo Fonseca, Dr João Tata Regala, Dr Filipe Alves, José Luís Sofia, Tibério Simões, Eugénio Brás and António Laranjo. We also would like to thank the fishery organisations BIVALPESCA and OLHÃOPESCA for their collaboration. We thank the skippers and crews of the RV 'Donax' and of the fishing boats 'Surpresa' and 'Família Santa' for their expertise and valuable participation in some of the experiments reported.

References

Aller RC (1988) Benthic fauna and biogeochemical processes in marine sediments: the role of burrow structures. In: Blackburn TH, Sørensen J (eds) Nitrogen Cycling in Coastal Marine Environments. Wiley, New York, pp 301–338

Anschutz A, Zhong S, Sundby B, Mucci A, Gobeil C (1998) Burial efficiency of phosphorus and the geochemistry of iron in continental margin sediments. Limnol Oceanogr 43(1):53–64

Ardizzone GD, Tucci P, Somaschini A, Belluscio A (2000) Is bottom trawling partly responsible for the regression of *Posidonia oceanica* meadows in the Mediterranean sea? In: Kaiser MJ, de Groot SJ (eds) Effects of Fishing on Non-target Species and Habitats. Blackwell Science, Oxford, pp 37–46

Arntz WE, Weber W (1970) *Cyprina islandica* L. (Mollusca, Bivalvia) als Nahrung von Dorschg und Kliesche in der Kieler ucht. Ber Deut Wissen Komm Meeresforschung 21:193–205

Auster PJ, Malatesta RJ, Langton RW, Watling L, Valentine PC, Donaldson CLS, Langton EW, Shepard AN, Babb IG (1996) The impacts of mobile fishing gear on seafloor habitats in the Gulf of Maine (northwest Atlantic): implications for conservation of fish populations. Rev Fish Sci 4(2):185–202

Baird RH, Gibson FA (1956) Underwater observations on scallop (*Pecten maximus* L.) beds. J Mar Biol Assoc UK 35:555–562

Barnes D, Chytalo K, Hendrickson S (1991) Final Policy and Generic Environmental Impact Statement on Management of Shellfish in Uncertified Areas Program. NY Department of Environment and Conservation, 79 pp

Beaumont AR, Budd MD (1982) Delayed growth of mussel (*Mytilus edulis*) and scallop (*Pecten maximus*) veligers at low temperatures. Mar Biol 71:97–100

Bemvenuti CE (1994) O poliqueta *Nephtys fluviatilis* Monro, 1937, como predador da infauna na comunidade de fundos moles. Atlântica 16:87–98

Benitez-Nelson CR (2000) The biogeochemical cycling of phosphorus in marine systems. Earth-Sci Rev 51:109–135

Bergman MJN, Hup M (1992) Direct effects of beam trawling on macrofauna in sandy bottom sediment in the southern North Sea. ICES J Mar Sci 49:5–11

Bergman MJN, van Santbrink JW (2000) Mortality in megafaunal benthic populations caused by trawl fisheries on the Dutch continental shelf in the North Sea in 1994. ICES J Mar Sci 57(5):1321–1331

Bergman MJN, Ball B, Bijleveld C, Craeymeersch JA, Munday BW, Rumohr H, van Santbrink JW (1998) Direct mortality due to trawling. In: Lindeboom HJ, de Groot SJ (eds) The Effects of Different Types of Fisheries on the North Sea and Irish Sea Benthic Ecosystems. Nioz-Rapport 1998-1/Rivo-Dlo Report C003/98, pp 167–184

Bergman MJN, Fonds M, Hup M, Lewis M, van der Puyl P, Stam A, Uyl D (1990) Direct effects of beamtrawl fishing on benthic fauna in the North Sea – a pilot study. In: Effects of Beamtrawl Fishery on the Bottom Fauna of the North Sea. BEON-rapport 8:33–57

Black KP, Parry GD (1994) Sediment transport rates and sediment disturbance due to scallop dredging in Port Phillip Bay. Mem Queensl Mus 36:327–341

Bradshaw C, Veale LO, Hill AS, Brand AR (2000) The effects of scallop dredging on gravelly sea-bed communities. In: Kaiser MJ, de Groot SJ (eds) Effects of Fishing on Non-target Species and Habitats. Blackwell Science, Oxford, pp 83–104

Broadhurst MK (2000) Modifications to reduce bycatch in prawn trawls: A review and framework for development. Rev Fish Biol Fish 10:27–60

Broadhurst MK, Suuronen P, Hulme A (2006) Estimating collateral mortality from towed fishing gear. Fish Fish 7:180–218

Butcher T, Matthews J, Glaister J, Hamer G (1981) Study suggests scallop dredges causing few problems in Jervis Bay. Aust Fish 40:9–12

Caddy JF (1968) Underwater observations on scallop (*Plactopecten magellanicus* L.) behaviour and drag efficiency. J Fish Res Bd Can 25:2123–2124

Caddy JF (1973) Underwater observations on tracks of dredges and trawls and some effects of dredging on scallop ground. J Fish Res Bd Can 30:173–180

Cahoon LB, Cooke JF (1992) Benthic microalgal production in Onslow Bay, North Carolina. Mar Ecol Prog Ser 84:185–196

Chícharo L, Regala J, Gaspar M, Alves F, Chícharo MA (2002a) Macrofauna spatial differences within clam dredge-tracks and their implications for short-term fishing effects studies. Fish Res 54(3):349–353

Chícharo L, Chícharo MA, Gaspar M, Alves F, Regala J (2002b) Ecological characterisation of dredged and recently dredge-free areas of a bivalve fishing ground, off South Portugal. J Mar Biol Assoc UK 82:41–50

Chícharo L, Chícharo A, Gaspar M, Regala J, Alves F (2002c) Reburial time and indirect mortality of *Spisula solida* clams caused by dredging. Fish Res 59 (1–2):247–257

Chícharo MA, Amaral A, Condinho S, Alves F, Regala J, Gaspar M, Chícharo L (2003a) Adenylic-derived indices and reburying time as indicators of the effects of dredging-induced stress on the clam *Spisula solida*. Mar Biol 142:1113–1117

Chícharo MA, Chícharo L, Amaral A, Condinho S, Gaspar M (2003b) Chronic effects of dredging-induced stress on the clam (*Spisula solida*): nucleic acid and lipid composition. Fish Res 63:447–452

Christensen B, Vedel A, Kristensen E (2000) Carbon and nitrogen fluxes in sediment inhabited by suspension-feeding (*Nereis diversicolor*) and non-suspension-feeding (*N. virens*) polychaetes. Mar Ecol Prog Ser 192:203–217

Churchill JH (1989) The effect of commercial trawling on sediment ressuspension and transport over the Middle Atlantic Bight continental shelf. Cont Shelf Sci 9(9):841–864

Coen LD (1995) A review of the potential impacts of mechanical harvesting on subtidal and intertidal shellfish resources. Unpublished Report. South Carolina Department of Natural Resources, Marine Resources Research Institute

Coleman FC, Williams SL (2002) Overexploiting marine ecosystem engineers: potential consequences for biodiversity. Trends Ecol Evol 17:40–44
Collie JS, Escanero GA, Valentine PC (1997) Effects of bottom fishing on the benthic megafauna of Georges Bank. Mar Ecol Prog Ser 155:159–172
Collie JS, Hall SJ, Kaiser MJ, Poiner IR (2000) A quantitative analysis of fishing impacts on shelf-sea benthos. J Anim Ecol 69:785–798
Commito JA (1982) Importance of predation by infaunal polychaetes in controlling the structure of a soft-bottom community in Maine, USA. Mar Biol 68:77–81
Conner WG, Simon JL (1979) The effects of oyster shell dredging on an estuarine benthic community. Estuarine Coastal Mar Sci 9:749–758
Currie DR, Parry GD (1996) Effects of scallop dredging on a soft sediment community: a large-scale experimental study. Mar Ecol Prog Ser134:131–150
Currie DR, Parry GD (1999) Impacts and efficiency of scallop dredging on different soft substrates. Can J Fish Aquat Sci 56:539–550
Dare PJ, Key D, Connor PM (1993) The efficiency of springloaded dredges used in the Western English Channel fishery for scallops, *Pecten maximus* (L.). *ICES Fish Capture Committee* CM1993/B:15, p 8
Davis G (2003) Physical disturbance and biological implications of sediment resuspension from commercial fishing gear in the Gulf of Maine. Unpublished report
Dayton PK (1971) Competition, disturbances and community organization: the provision and subsequent utilization of space in a rocky intertidal community. Ecol Monogr 41:351–389
Dayton PK, Hessler RR (1972) Role of biological disturbance in maintaining diversity in the deep sea. Deep-Sea Res 19:199–208
Dayton PK, Thrush SF, Agardy MT, Hofman RJ (1995) Environmental effects of marine fishing. Aquat Conserv Mar Freshwater Ecosys 5:205–232
DeAlteris J, Skrobe L, Lipsky C (1999) The significance of seabed disturbance by mobile fishing gear relative to natural processes: a case study in Narragansett Bay, Rhode Island. In: Benaka L (ed) Fish habitat: essential fish habitat and rehabilitation. American Fisheries Society, Symposium 22, Bethesda, Maryland, pp 224–237
De Groot SJ (1984) The impact of bottom trawling on benthic fauna of North Sea. Ocean Manag 9:177–190
Demestre M, Sánchez P, Kaiser M (2000) The behavioural response of benthic scavengers to otter trawling disturbance in the Mediterranean. In: Kaiser MJ, de Groot SJ (eds) Effects of fishing on non-target species and habitats. Blackwell Science, Oxford, United Kingdom, pp 121–129
Dernie KM, Kaiser MJ, Richardson EA, Warwick RM (2003) Recovery of soft sediment communities and habitats following physical disturbance. J Exp Mar Biol Ecol 285–286:415–434
Dias J (2004) Contribuição para o estudo da pescaria de *Solen marginatus* na Ria Formosa. Relatório de estágio da licenciatura em Biologia Marinha e Pescas da Universidade do Algarve

Drinkwater J (1974) Scallop dredge selectivity experiments. ICES C.M. 1974/K:25, 5 pp

Eleftheriou A, Robertson MR (1992) The effects of experimental scallop dredging on the fauna and physical environment of a shallow sandy community. Neth J Sea Res 30:289–299

Falcão M, Gaspar MB, Caetano M, Santos MN, Vale C (2003) Short-term environmental impact of clam dredging in coastal waters (south of Portugal): chemical disturbance and subsequent recovery of seabed. Mar Environ Res 56:649–664

Fanning KA, Carder KL, Betzer PR (1982) Sediment resuspension by coastal waters: a potential mechanism for nutrient re-cycling on the ocean's margins. Deep-Sea Res 29(8A):953–965

Fonds, M, 1994 Mortality of fish and invertebrates in beam trawl catches and the survival chances of discards. In: de Groot SJ, Lindeboom HJ (eds) Environmental Impact of Bottom Gears on Benthic Fauna in Relation to Natural Resources Management and Protection of the North Sea. NIOZ Report 12:131–146

Fonds M, Groenewold S, Hoppe I, Kaiser MJ, Munday BW, Ramsay K (1998) Scavenger responses to trawling. In: Lindeboom HJ, de Groot SJ (eds) The effects of different types of fisheries on the North Sea and Irish Sea benthic ecosystems. NIOZ Report 1998–1, RIVO-DLO Report C003/98, pp 85–244

Forja JM, Blasco J, Gómez-Parra A (1994) Spatial and seasonal variation of in situ benthic fluxes in the Bay of Cadiz (south-west Spain). Estuarine Coastal Shelf Sci 39:127–141

Franceschini G, Pranovi F, Raicevich S, Farrace MG, Giovanardi O (1999) 'Rapido' trawl fishing in the Northern Adriactic Sea: direct impact on epifauna. In: Giovanardi O (ed) Impact of Trawl Fishing on Benthic Communities. ICRAM Workshop Proceedings, pp 49–60

Francois F, Poggiale J-C, Durbec J-P, Stora G (2001) A new model of bioturbation for functional approach to sediment reworking resulting from macrobenthic communities. In: Aller JY, Woodin SA, Aller RC (eds) Organism-Sediment Interactions. University of South Carolina Press, Columbia, pp 73–86

Frid CLJ, Buchanan JB, Garwood PR (1996) Variability and stability in benthos: twenty-two years of monitoring off Northumberland. ICES J Mar Sci 53: 978–980

Frid CLJ, Clark RA, Hall JA (1999) Long-term changes in the benthos on a heavily fished ground off the NE coast of England. Mar Ecol Prog Ser 188:13–20

Frid CLJ, Harwood KG, Hall SJ, Hall JA (2000) Long-term changes in the benthic communities on North Sea fishing grounds. ICES J Mar Sci 57:1303–1309

Froelich, PN, Bender ML, Luedtke NA, Heath GR, DeVries T (1982) The marine phosphorus cycle. Am J Sci 282:474–511

Gaspar MB, Leitão F, Santos MN, Chícharo L, Damásia D, Chícharo A, Monteiro CC (2003a) A comparison of direct macrofaunal mortality using three types of Portuguese clam dredge. ICES J Mar Sci 60(4):733–742

Gaspar MB, Santos MN, Leitão F, Chícharo L, Chícharo A, Monteiro CC (2003b) Recovery of substrates and macro-benthos after fishing trials with a new Portuguese clam dredge. J Mar Biol Assoc UK 83:713–717

Gaspar MB, Leitão F, Santos MN, Sobral M, Chícharo L, Chícharo A, Monteiro CC (2003c) Size selectivity of the *Spisula solida* dredge in relation to tooth spacing and mesh size. Fish Res 60(2–3):561–568

Gaspar MB (1996) Bivalves do litoral oceânico algarvio. Aspectos da biologia, ecologia e das pescarias dos mananciais de interesse económico: aplicação à gestão dos recursos. PhD thesis, Universidade do Algarve, Faro, 282 pp

Gaspar MB, Monteiro CC (1998) Indirect mortality caused by dredges on the survival of juveniles of *Spisula solida* and *Donax trunculus*. Proceedings of the ICES Symposium – marine benthos dynamics: environmental and fisheries impacts. Crete, Greece

Gaspar MB, Monteiro CC (1999) Indirect mortality caused to juveniles of *Spisula solida* due to deck exposure. J Mar Biol Assoc UK 79(3):566–568

Gaspar MB, Castro M, Monteiro CC (1998) The influence of tow duration and tooth length on the number of damaged razor clams, *Ensis siliqua*. Mar Ecol Prog Ser 169:303–305

Gaspar MB, Castro M, Monteiro CC (1999) Effect of tooth spacing and mesh size on the catch of the Portuguese clam and razor clam dredge. ICES J Mar Sci 56:103–110

Gaspar MB, Dias MD, Campos A, Monteiro CC, Santos MN, Chícharo A, Chícharo L (2001) The influence of dredge design on the catch of *Callista chione* (L. 1758). Hydrobiologia 465:153–167

Gaspar MB, Leitão F, Santos MN, Sobral M, Chícharo L, Chícharo A, Monteiro CC (2002) Influence of mesh size and tooth spacing on the proportion of damaged organisms in the catches of the Portuguese clam dredge fishery. ICES J Mar Sci 59(6):1228–1236

Gilkinson KD, Fader GBJ, Gordon Jr DC, Charron R, McKeown D, Roddick D, Kenchington ELR, MacIsaac K, Bourbonnais C, Vass P, Liu Q (2003) Immediate and longer-term impacts of hydraulic clam dredging on an offshore sandy seabed: effects on physical habitat and processes of recovery. Cont Shelf Sci 23:1315–1336

Hall SJ (1994) Physical disturbance and marine benthic communities: life in unconsolidated sediments. Oceanogr Mar Biol Annu Rev 32:179–239

Hall SJ, Basford DJ, Robertson MR (1990) The impact of hydraulic dredging for razor clams *Ensis* sp. on an infaunal community. Neth J Sea Res 27(1): 119–125

Hall SJ, Robertson MR, Basford DJ, Heaney SD (1993) The possible effects of fishing disturbance in the northern North Sea: An analysis of spatial patterns in community structure around a wreck. Neth J Sea Res 31:201–208

Hall-Spencer JM, Moore PG (2000) Impact of scallop dredging on maerl grounds. In: Kaiser MJ, de Groot SJ (eds) Effects of Fishing on Non-Target Species and Habitats. Blackwell Science, Oxford, pp 105–107

Hauton C, Atkinson RJA, Moore PG (2003) The impact of hydraulic blade dredging on a benthic megafaunal community in the Clyde Sea area, Scotland. Journal of Sea Research 50:45–56

Hayes DF, Raymond GL, McLellan TN (1984) Sediment resuspension from dredging activities. Proceedings of the ASCE Specialty Conference, Dredging '84. Clearwater, Florida, pp 72–82

Henriksen K, Rasmussen MB, Jensen A (1983) Effect of bioturbation on microbial nitrogen transformations in the sediment and fluxes of ammonium and nitrate to the overlying water. Ecol Bull 35:193–205

Hill AE, Brown J, Fernand L (1995) The western Irish Sea gyre: A retention system for Norway lobster (*Nephrops norvegicus*)? Oceanol Acta 19:357–368

Hill AE, Brown J, Fernand L (1997) The summer gyre in the western Irish Sea: Shelf sea paradigms and management implications. Estuarine Coastal Shelf Sci 44:83–95

Houghton RG, Williams T, Blacker RW (1971) Some effects of double beam trawling. ICES CM 1971/B:5, 12 pp

Huettel M, Gust G (1992) Impact of bioroughness on interfacial solute exchange in permeable sediments. Mar Ecol Prog Ser 89:253–267

Jenkins SR, Brand AR (2001) The effect of dredge capture on the escape response of the great scallop, *Pecten maximus* (L.): implications for the survival of undersized discards. J Exp Mar Biol Ecol 266:33–50

Jenkins SR, Mullen C, Brand AR (2004) Predator and scavenger aggregation to discarded by-catch from dredge fisheries: importance of damage level. Journal of Sea Research 51:69–76

Jennings S, Kaiser MJ (1998) The effect of fishing on marine ecosystems. Adv Mar Biol 34:201–252

Jennings S, Greenstreet SPR, Reynolds JD (1999) Structural change in an exploited fish community: a consequence of differential fishing effects on species with contrasting life histories. J Anim Ecol 68:617–627

Jennings S, Reynolds JD, Mills SC (1998) Life history correlates of responses to fisheries exploitation. Proc R Soc Lond Ser B Biol Sci 265:333–339

Jennings S, Dinmore TA, Duplisea DE, Warr KJ, Lancaster JE (2001) Trawling disturbance can modify benthic production processes. J Anim Ecol 70: 459–475

Jones JB (1992) Environmental impact of trawling on the seabed: a review. NZ J Mar Freshw Res 26: 59–67

Kaiser MJ (1998) Significance of bottom-fishing disturbance. Conserv Biol 12(6):1230–1235

Kaiser MJ, Ramsay K (1997) Opportunistic feeding by dabs within areas of trawl disturbance: possible implications for increased survival. Mar Ecol Prog Ser 152:307–310

Kaiser MJ, Spencer BE (1994) Fish scavenging behavior in recently trawled areas. Mar Ecol Prog Ser 112:41–49

Kaiser MJ, Spencer BE (1995) Survival of by-catch from a beam trawl. Mar Ecol Prog Ser 126:31–38

Kaiser MJ, Spencer BE (1996) The effects of beam-trawl disturbance on infaunal communities in different habitats. J Anim Ecol 65:348–358

Kaiser MJ, Ramsay K, Richardson CA, Spence FE, Brand AR (2000) Chronic fishing disturbance has changed shelf sea benthic community structure. J Anim Ecol 69:494–503

Kaiser MJ, Edwards DB, Armstrong PJ, Radford K, Lough NEL, Flat RP, Jones HD (1998) Changes in megafauna benthic communities in different habitats after trawling disturbance. ICES J Mar Sci 55:353–361

Kaiser MJ, Hill AS, Ramsay K, Spencer BE, Brand AR, Veale LO, Prudden K, Rees EIS, Munday BW, Ball B, Hawkins SJ (1997) Benthic disturbance by fishing gear in the Irish Sea: a comparison of beam trawling and scallop dredging. Aquat Conserv Mar Freshwater Ecosys 7:269–285

Kennedy AD (1994) Predation within meiofaunal communities: description and results of a rapid-freezing method of investigation. Mar Ecol Prog Ser 114:71–79

Kennelly SJ (1999) Areas, depths, and times of high discard rates of scup, *Stenotomus chrysops*, during demersal fish trawling off the northeastern United States. Fish Bull 97:185–192

Kristensen E (1988) Benthic fauna and biogeochemical processes in marine sediments: microbial activities and fluxes. In: Blackburn TH, Sørensen J (eds) Nitrogen Cycling in Coastal Marine Environments. Wiley, New York, pp 275–299

Krom MD, Berner RA (1981) The diagenesis of phosphorus in a near-shore marine sediment. Geochim Cosmochim Acta 45:207–216

Kroncke I, Dippner JW, Heyen H, Zeiss B (1998) Long-term changes in macrofaunal communities off Norderney (East Frisia, Germany) in relation to climate variability. Mar Ecol Prog Ser 167:25–36

Lambert J, Goudreau P (1996) Performance of the New England Hydraulic dredge for the harvest of Stimpson's surf Clams (*Mactromeris polynyma*). Canadian Industry Report of Fisheries and Aquatic Science 235, vii + 28 pp

Langton RW, Robinson WE (1990) Faunal associations on scallop grounds in the western Gulf of Maine. J Exp Mar Biol Ecol 144:157–171

LaSalle MW 1990 Physical and chemical alterations associated with dredging: An overview. In: Effects of Dredging on anadromous Pacific Coast Fishes. Workshop Proceedings, Seattle, September 8–9, 1988. Report of Washington Sea Grant Program, pp 112

Lindley JA (1998) Diversity, biomass and production of decapod crustacean larvae in a changing environment. Invertebr Reprod Dev 33:209–219

Lissner AL, Taghon GL, Diener DR, Schroeter SC, Dixon JD (1991) Recolonization of deep-water hard substrate communities: potential impacts from oil and gas development. Ecol Appl 1:258–267

Maguire JA, Coleman A, Jenkins S, Burnell GM (2002) Effects of dredging on undersized scallops. Fish Res 56:155–165

Mayer LM, Schick DF, Findlay RH, Rice DL (1991) Effects of Commercial Dragging on Sedimentary Organic Matter. Mar Environ Res 31 (4):249–261

McConnaughey RA, Mier KL, Dew CB (2000) An examination of chronic trawling effects on soft-bottom benthos of the eastern Bering Sea. ICES J Mar Sci 57:1377–1388

McLoughlin RJ, Young P, Martin R, Parslow J (1991) The Australian scallop dredge: estimates of catching efficiency and associated indirect fishing mortality. Fish Res 11:1–24

Medcof J, Bourne N (1964) Causes of mortality of the sea scallop *Placopecten magellanicus*. Proceedings of the National Shellfisheries Association 53: 33–50

Medcof JC, Caddy JF (1971) Underwater observations on the performance of clam dredges of three types. ICES CM 1971/B:10, 7pp

Meyer TL, Cooper RA, Pecci KJ (1981) The performance and environmental effects of a hydraulic clam dredge. Marine Fisheries Review 43:14–22

Michael KP, Olsen GP, Hvid BT, Cranfield HJ (1990) Design and performance of two hydraulic subtidal clam dredges in the New Zealand. New Zealand Fisheries Technical Reports 21, 16pp

Minchin D, Skjaeggestad H, Haugum AG, Strand O (2000) Righting and recessing ability of wild and native cultivated scallops. Aquac Res 31:473–474

Morgan L, Chuenpagdee R (2003) Shifting Gears: Addressing the Collateral Impacts of Fishing Methods in USA Waters. Island Press Publication Services, Washington DC, 42 pp

Mortimer RJG, Davey JT, Krom MD, Watson PG, Frickers PE, Clifton RJ (1999) The effect of macrofauna on porewater profiles and nutrient fluxes in the intertidal zone of the Humber Estuary. Estuarine Coastal Shelf Sci 48:683–699

Nashimoto K (1984) The selectivity of sunary surf clam dredge. Bull Jpn Soc Sci Fish 50:1145–1155

Nashimoto K, Miyzawa H, Hiraishi T (1983) The tooth selectivity of the Japanese surf clam dredge. Bull Jpn Soc Sci Fish 49:379–385

Nixon SW (1981) Remineralization and nutrient cycling in coastal marine ecosystems. In: Neilson BJ, Cronin LE (eds) Estuaries and nutrients. Humana Press, Clifton, New Jersey, pp 151–162

Persson L, Eklov P (1995) Prey refuges affecting interactions between piscivorous perch and juvenile perch and roach. Ecology 76:70–81

Phelps HL, Hardy JT, Pearson WH, Apts CW (1983) Clam burrowing behaviour: Inhibition by copper-enriched sediment. Mar Pollut Bull 14(12):452–455

Pickett G (1973) The impact of mechanical harvesting on the Thames Estuary cockle fishery. MAFF Laboratory Leaflet (New Series) no. 29, Lowestoft Suffolk, pp 1–22

Pilskaln CH, Churchill JH, Mayer LM (1998) Resuspension of Sediment by Bottom Trawling in the Gulf of Maine and Potential Geochemical Consequences. Conserv Biol 12(6):1223–1229

Pingree RD, Griffiths KD (1978) Tidal fronts on shelf seas around the British Isles. J Geophys Res 83:4615–4622

Pranovi F, Giovanardi O (1994) The impact of hydraulic dredging for short-necked clams, *Tapes* spp., on an infaunal community in the lagoon of Venice. Sci Mar 58:345–353

Pranovi F, Da Ponte F, Raicevich S, Giovanardi O (2004) A multidisciplinary study of the immediate effects of mechanical clam harvesting in the Venice Lagoon. ICES J Mar Sci 61:43–52

Pranovi F, Raicevich S, Franceschini G, Farrace MG, Giovanardi O (2000) Rapido trawling in the northern Adriatic Sea: effects on benthic communities in an experimental area. ICES J Mar Sci 57:1–8

Ramsay K, Kaiser MJ (1998) Demersal fishing disturbance increases predation risk for whelks (*Buccinum undatum* L.). Journal of Sea Research 39:299–304

Ramsay K, Kaiser M, Hughes R (1996) Changes in hermit crab feeding patterns in response to trawling disturbance. Mar Ecol Prog Ser 144(1–3):63–72

Ramsay K, Kaiser MJ, Hughes RN (1998) Responses of benthic scavengers to fishing disturbance by towed gears in different habitats. J Exp Mar Biol Ecol 224:73–89

Ramsay K, Kaiser MJ, Moore PG, Hughes RN (1997) Consumption of fisheries discards by benthic scavengers: utilization of energy subsidies in different marine habitats. J Anim Ecol 66:884–896

Ramsay K, Kaiser MJ, Rijnsdorp AD, Craeymeersch JA, Ellis J (2000) Impact of trawling on populations of the invertebrate scavenger *Asteria rubens*. In: Kaiser MJ, de Groot SJ (eds) Effects of Fishing on Non-target Species and Habitats. Blackwell Science, Oxford, pp 151–162

Rees EIS, Nicholaidou A, Laskaridou P (1977) The effects of storms on the dynamics of shallow water benthic associations. Pages 465–474 In: Keegan BF, Ceidigh PO, Boaden PJS (eds) (1977) Biology of Benthic Organisms. Pergamon, Oxford, United Kingdom

Reeves SA, Armstrong DW, Fryer RJ, Coull KA (1992) The effects of mesh size, cod-end extension and cod-end diameter on the selectivity of Scottish trawls and seines. ICES J Mar Sci 49:279–288

Riemann B, Hoffmann E 1991 Ecological consequences of dredging and bottom trawling in the Limfjord, Denmark. Mar Ecol Prog Ser 69:171–178

Rijnsdorp AD, Buys AM, Storbeck F, Visser EG (1998) Micro-scale distribution of beam trawl effort in the southern North Sea between 1993 and 1996 in relation to the trawling frequency of the sea bed and the impact on benthic organisms. ICES J Mar Sci 55:403–419

Robinson RF, Richardson CA (1998) The direct and indirect effects of suction dredging on a razor clam (*Ensis arcuatus*) population. ICES J Mar Sci 55(5):970–977

Rodhouse PG, Dawe EG, O'Dor RK (1998) Squid recruitment dynamics. The genus *Illex* as a model. The commercial *Illex* species. Influences on variability. FAO Fish Tech Pap, Rome, no. 376, 273 pp

Rooker JR, Holt GJ, Holt SA (1998) Vulnerability of newly settled red drum (*Scianops ocellatus*) to predatory fish: Is early-life survival enhanced by seagrass meadows? Mar Biol 131:145–51

Rosenberg R, Nilsson HC, Grémare A, Amouroux J-M (2003) Effects of demersal trawling on marine sedimentary habitats analysed by sediment profile imagery. J Exp Mar Biol Ecol 285–286:465–477

Rumohr H, Krost P (1991) Experimental evidence of damage to benthos by bottom trawling with special reference to *Arctica islandica*. Meeresforschung 33:340–345

Rumohr H, Ehrich S, Knust R, Kujawski T, Philippart CJM, Schroeder A (1998) Long-term trends in demersal fish and benthic invertebrates. In: Lindeboom HJ, de Groot SJ (eds) The effects of different types of fisheries on the North Sea and Irish Sea benthic ecosystems. NIOZ-Rapport 1998-1/Rivo-Dlo Report C003/98, pp 280–352

Sainsbury KJ (1988) The ecological basis of multispecies fisheries and management of a demersal fishery in tropical Australia. In: Gulland JA (ed) Fish population dynamics, 2nd edition. Wiley, New York, pp 349–382

Schwinghamer P, Guigne JY, Siu WC (1996) Quantifying the impact of trawling on benthic habitat using high resolution acoustics and chaos theory. Can J Fish Aquat Sci 53:288–296

Schwinghamer P, Gordon DC, Rowell TW, Prena J, McKeown DL, Sonnichsen G, Guigne JY (1998) Effects of experimental otter trawling on surficial sediment properties of a sandy-bottom ecosystem on the Grand Banks of Newfoundland. Conserv Biol 12:1215–1222

Slomp CP, Malschaert JFP, van Raaphorst W (1998) The role of absorption in sediment-water exchange of phosphate in North Sea continental margin sediments. Limnol Oceanogr 43(5):832–846

Southward AJ (1995) The importance of long time-series in understanding the variability of natural systems. Helgol Meersunters 49:329–333

Sowles J (2001) Nitrogen in the Gulf of Maine: Sources, Susceptibility and Trends. In: A Workshop on Nutrient Management in the Gulf of Maine, Portsmouth, NH

Sundby B, Gobeil C, Siverberg N, Mucci A (1992) The phosphorus cycle in coastal marine sediments. Limnol Oceanogr 37:1129–1145

Thistle D (1981) Natural physical disturbances and communities of soft bottoms. Mar Ecol Prog Ser 6:223–228

Thrush SF, Hewitt JE, Cummings VJ, Dayton PK (1995) The impact of habitat disturbance by scallop dredging on marine benthic communities – What can be predicted from the results of experiments. Mar Ecol Prog Ser 129:141–150

Thrush SF, Pridmore RD, Hewitt JE, Cummings VJ (1992) Adult infauna as facilitators of colonization on intertidal sandflats. J Exp Mar Biol Ecol 159:253–265

Thrush SF, Hewitt JE, Cummings VJ, Dayton PK, Cryer M, Turner SJ, Funnell GA, Budd RG, Milburn CJ, Wilkinson MR (1998) Disturbance of the marine benthic habitat by commercial fishing: impacts at the scale of the fishery. Ecol Appl 8:866–879

Tuck ID, Hall SJ, Robertson MR, Armstrong E, Basford DJ (1998) Effects of physical trawling disturbance in a previously unfished sheltered Scottish sea loch. Mar Ecol Prog Ser 162:227–242

Tuck ID, Bailey N, Harding M, Sangster G, Howell T, Graham N, Breen M (2000) The impact of water jet dredging for razor clams, *Ensis* spp, in a shallow sandy subtidal environment. Journal of Sea Research 43:65–81

van der Veer HW, Bergman MJN, Beukema JJ (1985) Dredging activities in the Dutch Wadden Sea: effects on macrobenthic infauna. Neth J Sea Res 19: 183–190

van Dolah RF, Wendt PH, Nicholson N (1987) Effects of a research trawl on a hard bottom assemblage of sponges and corals. Fish Res 5:39–54

Wassenberg TJ, Hill BJ (1987) Feeding by the sand crab, *Portunus pelagicus*, on material discarded from prawn trawlers in Moreton Bay. Mar Biol 95: 387–393

Watling L, Norse EA (1998) Disturbance of the seabed by mobile fishing gears: a comparison to forest clearcutting. Conserv Biol 12(6):1180–1197

Watling L, Findlay RH, Mayer LM, Schick DF (2001) Impact of a scallop drag on the sediment chemistry, microbiota, and faunal assemblages of a shallow subtidal marine benthic community. Journal of Sea Research 46(3–4):309–324

6 Technical Measures to Reduce Seabed Impact of Mobile Fishing Gears

PINGGUO HE

6.1 Introduction

Mobile fishing gears such as demersal otter trawls leave behind tracks and other physical changes as they are towed over the seabed. The effects of fishing by these gears have been debated since the otter trawl was invented in the 1880s when the Royal Commission on Trawling (the Dalhousie Commission) was charged to examine whether and how the newly introduced otter trawl technology affected fish stocks and the seabed (Wardle 1986). Many public debates and scientific investigations have continued to this day to answer the same question posed over a century ago.

In the last ten years, there has been more intensive questioning about the impact of mobile gears on the seabed after declines of major commercial stocks in the early 1990s. As a result, several major reviews, books, symposia and comprehensive studies have been completed, and many are ongoing, concerning the impacts of mobile gears such as demersal otter trawls, beam trawls and shellfish dredges on the seabed and marine habitat (see Dorsey and Pederson 1998; Hall 1999; Kaiser and de Groot; 2000; Linnane et al. 2000; Anon. 2001; NRC 2002; Sinclair and Valdimarsson 2003; Lart et al. 2003; Barnes and Thomas 2005). These are in addition to several working groups and focused topic groups of the International Council for the Exploration of the Sea (ICES), especially the Working Group on Ecosystem Effects of Fishing Activities (WGECO) and the Working Group on Fishing Technology and Fish Behaviour (WGFTFB) (ICES 1988, 1999, 2000a, 2000b, 2004). The special topic group of WGFTFB on 'Mitigation measures against seabed impact of fishing operations' concentrated on measures to reduce seabed impact through gear designs and operations (ICES 2004) and this chapter evolved from that meeting.

Mobile gears alter physical features of the seabed and may cause direct and indirect mortalities of bottom-dwelling species. Direct mortalities can occur to organisms crushed by gear components that contact the seabed

S. J. Kennelly (ed.), By-catch Reduction in the World's Fisheries, 141–179.

(Lindeboom and de Groot 1998). Gilkenson et al. (1998) suggested that smaller (and therefore lighter) animals might be pushed aside by the pressure wave in front of mobile fishing gears, resulting in less direct mortality for such organisms. Large bivalves, gastropods, and attached epifauna may suffer more direct mortality than small animals and infauna. Consequently, more heavily trawled or dredged areas may have fewer larger organisms (Jennings et al. 2001). Bottom fishing gears may also flatten the seabed through removing, modifying or redistributing physical features (NRC 2002; Steele et al. 2005). Reductions in habitat complexity, either biological or physical, through reduced numbers of species or reduced sizes of particular species (especially large macrofauna), can have negative consequences on fish populations because such locations provide important habitats for bottom-dwelling fishes (Collie et al. 1997; 2000).

In general, initial trawling and dredging in an unfished area have greater impacts on benthic habitats than repeated fishing in previously fished areas (Jennings et al. 2005). Patchy fishing over the same, few grounds is thus more desirable than fishing over many grounds in terms of habitat conservation. Commercial fishing is often patchy because of the uneven distribution of target species, unsuitable seabed for bottom fishing (e.g., too rocky, or the existence of wrecks), distance from port and/or fisheries regulations (Jennings et al. 2005). Recovery of habitats from bottom fishing operations may depend on the intensity and frequency of operations, physical and biological characteristics of the area and local oceanographic conditions (NRC 2002).

However, natural variation, the innate complexity of benthic communities, variations in gear designs and differences in the methodologies used to evaluate impacts have resulted in very few general conclusions regarding the impact of towed bottom fishing gears on seabeds and marine ecosystems (Løkkeborg 2005). Despite these uncertainties over the actual impacts of mobile gears, several papers have described, discussed and proposed technical modifications that aim to lessen seabed impacts of fishing activities (Carr and Milliken 1998; Goudey 1999; Rose et al. 2000; Ball et al. 2003; He and Foster 2000; He 2001; Matsushita et al. 2001; van Marlen 2000; Pol and Carr 2000; Fonteyne and Polet 2002; NRC 2002; Valdemarsen and Suuronen 2003; Lart et al. 2003; He and DeLouche 2004; Polet et al. 2005a, 2005b; van Marlen 2005). In accordance with the precautionary approach to ecosystem and fisheries management, negative impacts of fishing on the seabed should, ideally, be reduced whenever feasible. This chapter summarises recent advances and work in progress on measures to reduce impacts on the seabed and benthic communities of mobile fishing gears including otter trawls, beam trawls and shellfish dredges. These measures include increasing the fishing efficiency to reduce fishing time and seabed impacts, and gear modifications to reduce weight and intrusiveness of gear components that contact the bottom. Some new methods which have potential to lessen seabed impacts are also discussed.

6.2 Review of Mobile Fishing Gears and their Operations

6.2.1 Otter Trawls

The otter trawl is the most important fishing gear in the world for harvesting groundfish and many shellfish species. The otter trawl was developed from the beam trawl in the 1880s and was patented by Scott of Granton in 1894 (Wardle 1986; Gabriel et al. 2005). An otter trawl consists of a pair of trawl doors, a set of cable assemblies connecting the doors and the trawl, and the trawl net (Fig. 6.1). The doors are also called otter boards. The trawl doors expand the trawl horizontally through hydrodynamic and ground sheer forces. Sand or mud clouds stirred up by the doors and sweeps towed over soft bottom help herd fish into the mouth of the trawl (Wardle 1983). The trawl net is typically composed of wings, the square, the body of the net (belly and back), the extension piece and the codend. Floats are attached to the headline to open the trawl vertically, while the groundgear attaches to the fishing line through a series of toggle chains. The middle part of the groundgear is called the bosom, while the quarter and the wing sections continue towards the wingend. Depending on bottom and sea conditions, the groundgear includes light wires, chains, small rubber discs (sometimes called cookies), large size rubber discs, wheels or spherical bobbins (Fig. 6.1). Larger size discs and bobbins have been used when fishing is expanded to less desirable grounds with rough seabed and strong currents. The rockhopper gear has wire threaded through the side of the discs, preventing the discs from rolling. But even in roller and bobbin gears, only a few rollers and bobbins in the bosom roll freely – discs and rollers on the quarters and wings do not roll as their axes are not perpendicular to the towing direction. The body of the net tapers so that the belly part of the net is kept away from the seabed to avoid damage. The extension piece is added to many trawl designs to aid stability of the codend immediately following it. The codend usually does not contact the bottom, but large catches may bring the codend down. Therefore, chafing gear is typically used to protect the codend from damage from chafing with the bottom.

Bottom trawls such as typical groundfish and shrimp trawls are towed with both doors and trawl groundgears on the seabed, while pelagic or midwater trawls fish completely off the bottom (Fig. 6.2A). Between the

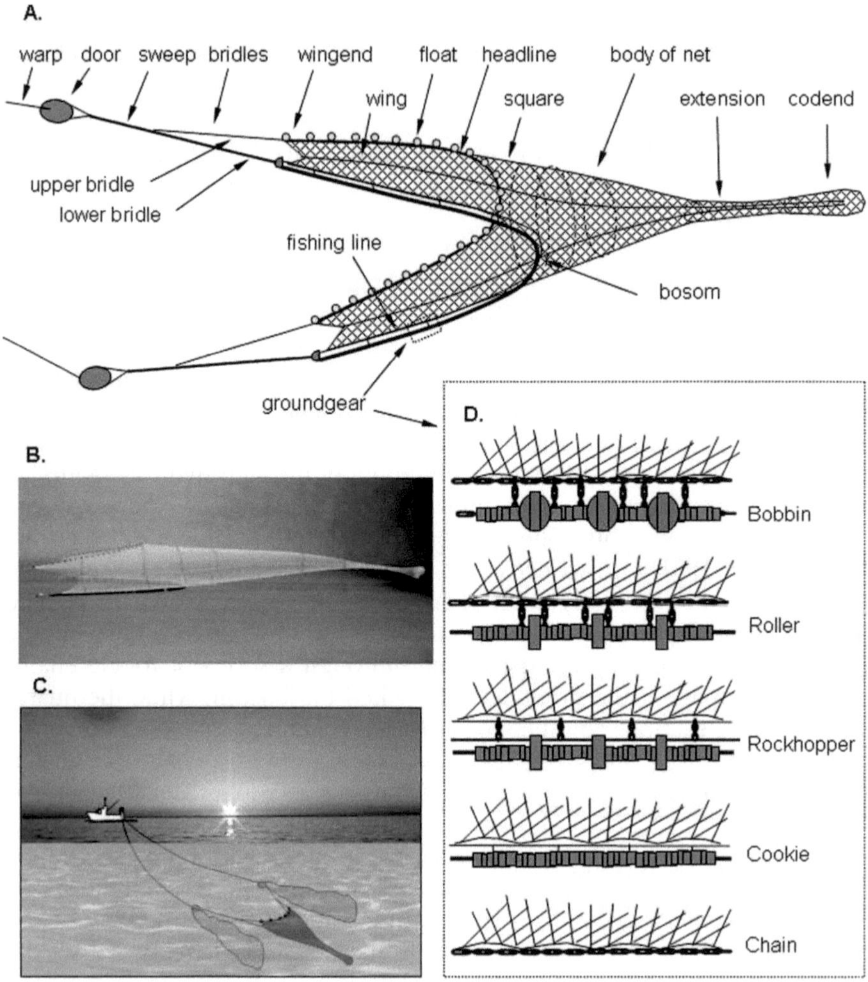

Fig. 6.1. A. Parts and names of a trawl, B. a model trawl in a flume tank, C. schematic illustration of an otter trawl in action, and D. various types of groundgears

bottom trawl and the pelagic trawl, there are some intermediate trawling styles: the 'semi-pelagic' trawl which has either the door or the trawl off the bottom, the 'quasi-pelagic' trawl which is a pelagic trawl with only a few chains or similar trailing objects lightly in contact with the bottom, and the 'quasi-bottom' trawl which is a bottom trawl with heavy groundgears replaced by lighter drop chains or other similar trailing objects. Fig. 6.2 illustrates the trawling styles mentioned in this chapter.

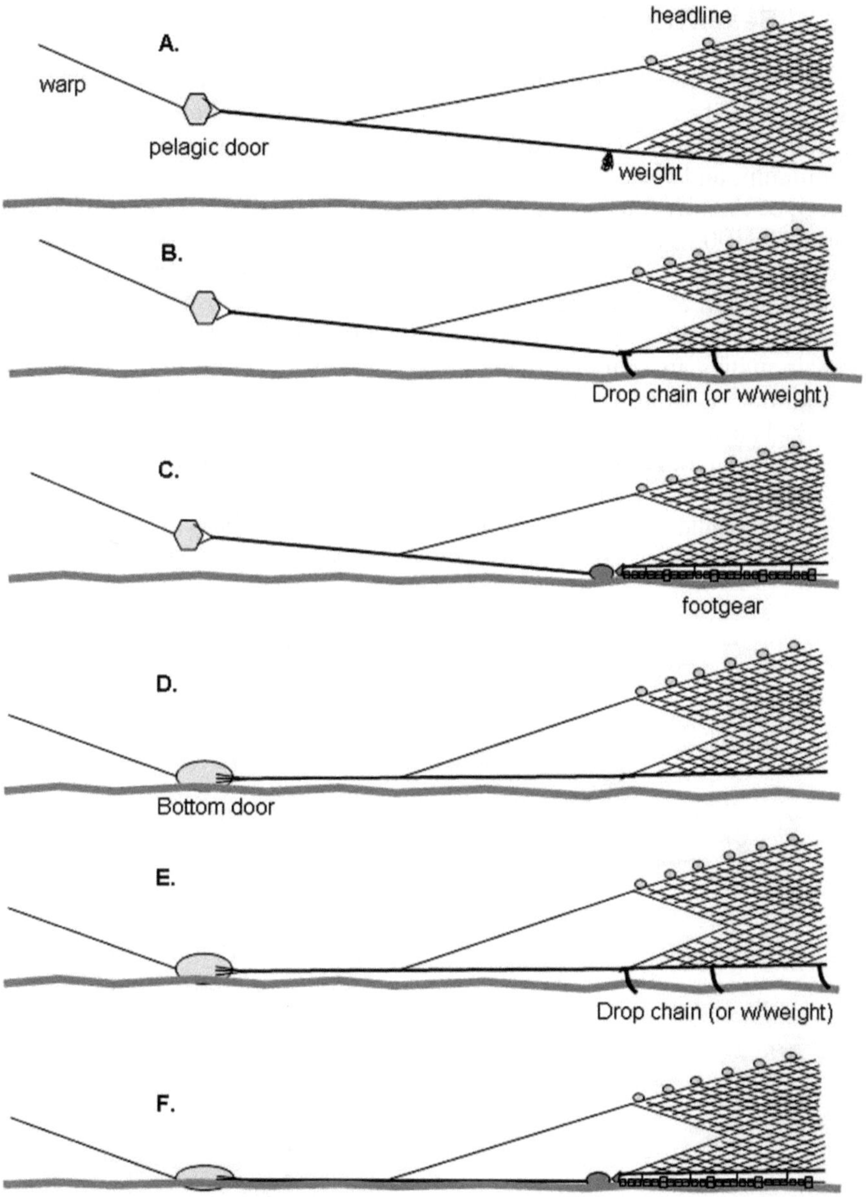

Fig. 6.2. Different otter trawls in relation to potential seabed impacts of their operations: A. pelagic (midwater) trawl – no part of the gear contacts the bottom; B. quasi-pelagic trawl – only drop chains contact the bottom; C and D. semi-pelagic trawls – either doors are on the bottom with the trawl off the bottom, or doors are off the bottom with the trawl on the bottom; E. quasi-bottom trawl – doors and drop chains on the bottom; F. bottom trawl – doors and groundgear heavily on the bottom

6.2.2 Beam Trawls

Beam trawls are used for harvesting flatfishes and crustaceans, especially shrimps. The beam trawl uses a beam to spread the trawl horizontally instead of doors (Fig. 6.3). The beam is usually made of wood, bamboo, steel or aluminium, and varies in length depending on the size of the vessel and the design of the trawl, usually ranging between 4 and 12 m (Valdemarsen and Suuronen 2003). A pair of shoes (also called heads) supports the beam and keeps it off the bottom to avoid damage. The shoes connect to the wingends through a set of short bridles. Large wheels are used in some beam trawls to help the gear roll over the seabed (Gabriel et al. 2005). Chains or a chain matrix are commonly used in flatfish beam trawls in the North Sea to stimulate bottom-dwelling species such as soles (Fig. 6.3B, van Marlen 2000). The headline is usually fastened to the shoes. The net otherwise resembles an otter trawl net but, because of the species targeted, large groundgears are seldom used. Beam trawls weigh from a few hundred kg to several tonnes. In the North Sea flatfish fisheries, beam trawls are usually towed at high speeds of up to 7 knots (Valdemarsen and Suuronen 2003).

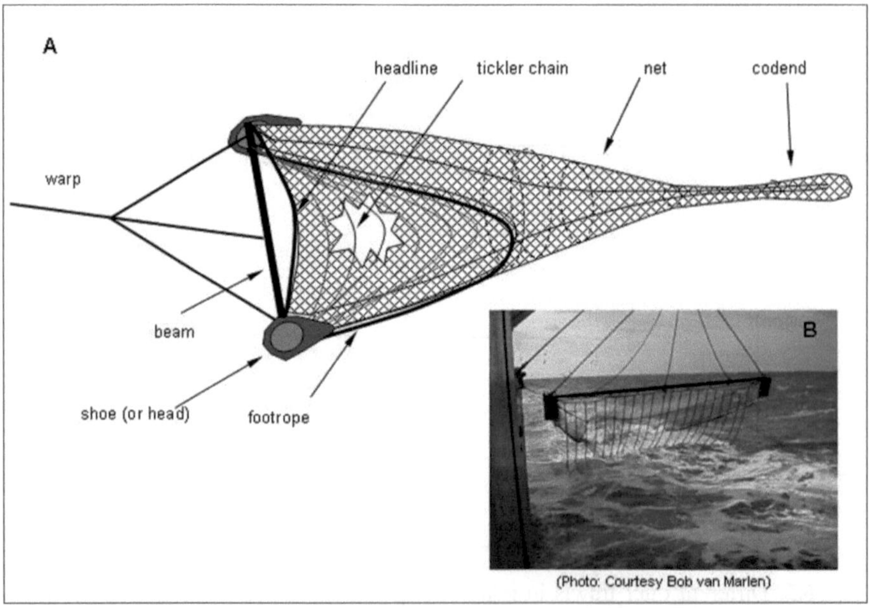

Fig. 6.3. A. Schematic illustration of a beam trawl, and B. a beam trawl with tickler chains being hauled from the side of a Netherlands beam trawler (Photo: Bob van Marlen)

6.2.3 Scallop Dredges

Dredges are mainly used to catch bivalves such as scallops and clams which live on or in the substratum. A typical New Bedford style scallop dredge used on Georges Bank is illustrated in Fig. 6.4. Dredges are primarily made of steel frames and cutting bars, and the bags are made of steel rings. In some dredges, the covering on the top has been replaced by synthetic materials, and the use of large mesh in the top netting has aided the release of by-catch fish species (Carr and Milliken 1998). The leading edge of many dredges has tooth bars of 8–15 cm in length, although the New Bedford style dredges are without teeth (Pol and Carr 2003). Another type of dredge is the hydraulic dredge which uses compressed air to 'fluidise' the sediment in front of the dredge to expose the target species for easier harvesting (Valdemarsen and Suuronen 2003). These are best used on sandy or fine bottoms. In some dredges, a suction hose is used to fetch the catch (including trash and substratum) continuously to the vessel without the need to lift the dredge. Figure 6.5 shows a hydraulic dredge for surf

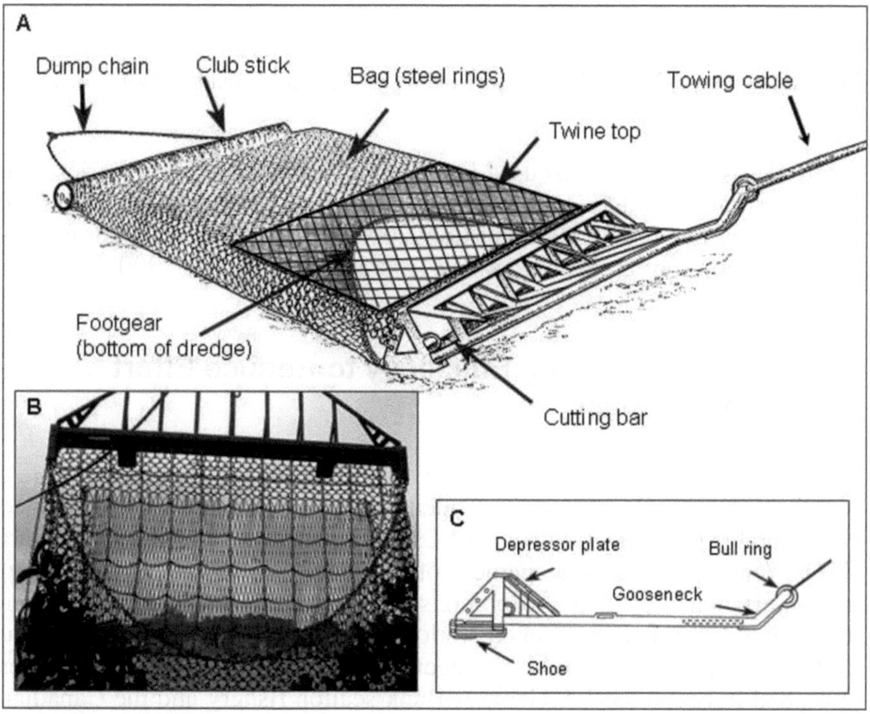

Fig. 6.4. A. Schematic illustration of a New Bedford style scallop dredge, B. view of a New Bedford style scallop dredge from underneath, and C. details of the towing frame (Courtesy of Ron Smolowitz)

Fig. 6.5. A hydraulic dredge for surf clams in Atlantic Canada (Photo: Fisheries and Marine Institute)

calms (*Spisula solidissima*) used in eastern Canada. In many fisheries, two or more smaller dredges are towed behind a single vessel for easy handling of the heavy metal gear (Pol and Carr 2000). In the United Kingdom, it is quite common for vessels to tow 4 to 16 dredges (Rose et al. 2000).

6.3 Improving Fishing Efficiency to Reduce Effort and Seabed Impact

Seabed impacts and fishing efficiency are closely linked because both are a function of time. In output-controlled fisheries, such as those managed with Total Allowable Catches (TACs) or Individual Transferable Quota (ITQs) regimes, if the management of total mortality is effective, improving the efficiency of fishing operations can reduce the contact time between the fishing gear and the seabed. Below are examples of two output-controlled scallop fisheries whose management has lead to significant decreases in seabed impacts: the US Georges Bank scallop fishery and the Canadian Browns Bank scallop fishery.

6.3.1 Effort and Bottom Time Reduction in the US Scallop Fishery on Georges Bank

The New Bedford dredge (Fig. 6.4) is the primary gear used to harvest sea scallops (*Placopecten magellanicus*) on Georges Bank, the world's largest single resource for the species. The dredges are typically 4.3 m wide and two dredges are usually towed by a single vessel at a speed of 4 to 5 knots. Unlike dredges used in Europe and in the Pacific, this dredge is toothless. The front edge of the New Bedford dredge includes the cutting bar, which rides above the surface of the substratum and creates turbulence that stirs up the substratum and animals, resulting in scallops, by-catch species and other organisms and debris (trash) accumulating in the bag of the dredge. The shoes and the bottom of the bag are the primary contact area of the dredge with the seabed. The turbulence behind the cutting bar also results in suspension of sediment and the smoothing of irregularities of the seabed. Other physical impacts can occur during setting and hauling but they are minor in comparison to those caused by the shoes, the bag and the cutting bar (Pol and Carr 2002).

Large areas on the US side of Georges Bank (approximately 50% of the overall area) were closed in 1994 to protect fish populations (Hart 2001). Scallop fishing was diverted to other areas where fishing efficiency was reduced and the scallops were, on average, of a smaller size. By 2000, survey indices for scallop biomass in the closed areas had increased by twenty times (Hart 2001). Stokesbury (2002) developed a video camera apparatus to survey scallops in these areas and measured scallop densities among the highest reported in any Georges Bank survey. Sea scallops were found to be highly concentrated in beds several square nautical miles in area. Moreover, depletion experiments and underwater photography permitted improved estimates of the efficiency and selectivity of commercial dredges.

A fully developed rotational fishery was then established on these high-density scallop beds in the closed areas. Compared to other grounds, catching efficiency was very high in the closed areas, resulting in less bottom time of the dredges for the same quantity of catch. In addition, landings were limited to 4,540 kg per trip, and vessels were deducted 10 fishing days from their annual allotment for each trip into these areas, although actual trip durations averaged 6 days. As a result, this program reduced the annual available days-at-sea (the number of days a vessel is allowed to fish) in this fishery by 2,576 days or about 10%. The total annual fishing time (the time towing) was reduced from 408,000 hrs in 1999 to 384,000 hrs in 2000, a reduction of 6%. The consequent reduced bottom time of the dredge resulted in reduced habitat impacts (Howard 2004).

6.3.2 Seabed Mapping and Effort Reduction in the Canadian Offshore Scallop Fishery

The Canadian scallop fishery off Nova Scotia is managed under ITQs that limit the annual amount of harvest. The fishery is dominated by several large corporations with a few large vessels. In the early 1990s, these corporations formed the Canadian Offshore Scallop Industry Mapping Group and worked with government agencies and universities to map seabed characteristics using multibeam technology. The three dimensional topographic charts were overlaid by geological features and sediment characteristics of commercial scallop areas. The resultant three dimensional maps of bathymetry, sediment and benthic habitats were used by fishers to identify productive scallop beds consisting of light gravels – called 'pea' gravels. The locations of the beds were then made available to vessel captains, who optimised dredge efficiencies by directly fishing on those areas (Fig. 6.6). Follow-up research determined that the maps were 94% accurate in identifying the presence of scallops. Because the total catch was limited by quotas, total fishing time (i.e., with dredges on the bottom) for harvesting a quota of 13.46 tonnes of scallop was reduced from 162 hours to 43 hours, a reduction of 73%. Likewise, fuel consumption and the area towed (and therefore disturbed) were reduced by 36 and 74% respectively (Table 6.1) (Robert et al. 2002).

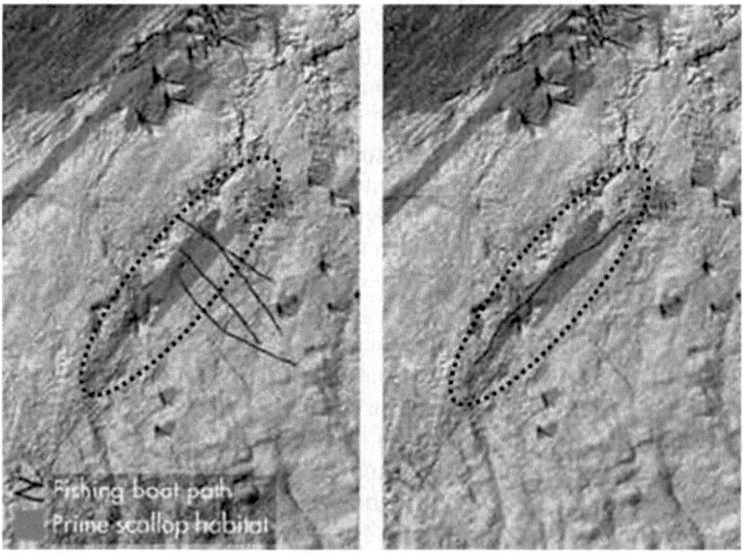

Fig. 6.6. Towing tracks (solid lines) of the scallop dredge without multibeam mapping in 1999 (left) and with multibeam mapping in 2000 (right). Dark area in the middle of the image encircled by the oval is prime scallop beds made up of 'pea' gravel. (From Robert et al. 2002)

Table 6.1 Operational characteristics of harvesting a quota of 13.64 tones scallop meat with or without multibeam maps in the Canadian offshore scallop fishery on Browns Bank off Nova Scotia (from Robert et al. 2002)

	Without map	With map	Reduction
Bottom Time (hr)	162	43	73%
Area towed (km^2)	1.176	0.311	74%
Fuel usage (liters)	27697	17545	36%

The offshore scallop industry was especially well-suited to this technology. Scallops can be mapped reliably because they are relatively sedentary and closely associated with a particular substratum. The industry's quota management system under the Canadian Department of Fisheries and Oceans assures the industry a secure level of access, thereby providing a strong incentive for the industry to invest in science and technology such as mapping, and other activities with long-term benefits.

The success of this project has inspired others. A similar project in Ireland is currently underway, with multibeam imaging of important scallop grounds already completed (D. Rihan, pers. comm.).

The above mitigation measures for seabed impacts of scallop dredges resulted indirectly from increases in efficiencies that arose differently in each case. In US waters, dredging was allowed in areas previously closed to fishing where very high scallop densities were determined by video imaging. Habitat impacts were reduced by the reduced bottom time of dredges due to high densities of scallops, trip limits and effort reduction measures. In Canadian waters, the increased efficiency was the result of the interpretation of comprehensive multibeam imaging leading to accurate mapping of scallop beds. In combination with individual transferable quotas, bottom time of dredges was reduced by more efficient fishing.

It should be noted, however, that new technologies similar to multibeam seabed mapping could only contribute to ecosystem conservation when accurate stock assessment and strict output controls can be assured. Overestimations of stocks, excessive TACs and/or the inability to control outputs in combination with high-resolution seabed maps could result in significant depletion of resources and lead to serious ecological impacts on the benthos.

It is worth noting that attempts to limit fishing efficiency can prevent the use of technical innovations that lead to less seabed impacts. For example, pair trawling for groundfish was banned in 1993 in multispecies fisheries in the northeastern United States because of its high fishing efficiency (Corey and Williams 1995), even though a well-adjusted pair trawl would have less seabed impact than an equivalent otter trawl because the former does not use trawl doors. Similarly, multi-rig trawl systems are illegal for British-registered vessels (A. Revill, pers. comm.), though they use smaller and lighter nets to cover the same ground area compared to single trawls, and could reduce overall fishing impacts.

6.4 Pelagic and Semi-pelagic Trawls

6.4.1 Alaskan Pollock Trawls

Alaskan pollock were fished using bottom trawls prior to 1990, but concerns over the by-catch of shellfish (mainly crabs) and other groundfish (mainly Pacific halibut) prompted the North Pacific Fisheries Management Council (NPFMC) to allocate a large proportion of the pollock TAC to the pelagic trawl sector. In an attempt to discourage by-catch, the NPFMC set up a performance standard for pelagic trawls by regulating the maximum number of tanner crabs caught by the trawl (Pereyra 1995). Onboard observers determined whether the trawl was operating in pelagic or non-pelagic mode and accordingly assign the catch against the appropriate TAC. Because the non-pelagic trawl was assigned only a small portion of the TAC, industry soon adopted the pelagic method to harvest pollock. Ultimately, with industry support, NPFMC banned bottom trawling in the Bering Sea pollock fishery in 1999 (NRC 2002).

While the original concern in this example was the by-catch of shellfish and groundfish, the resultant pelagic trawls for pollock have probably benefited the seabed and the benthic ecosystem due to the huge reduction in contact between the trawl components and the seabed. Although the pelagic trawls used in the pollock fishery may still make bottom contact when pursuing fish near the bottom, they are generally much lighter and likely less intrusive on the seabed.

6.4.2 Semi-pelagic Trawls for Pink Shrimps in the Northwest Atlantic

Shrimp trawls are generally similar to groundfish designs except that mesh sizes are much smaller. In groundfish trawls, sand clouds stirred up by the doors, sweeps and bridles are known to herd fish toward the mouth of the trawl (Wardle 1986). Shrimps (such as the pink shrimp *Pandalus borealis*), on the other hand, are not herded by sand clouds and bridles due to their poor swimming ability and inability to react to fast-moving trawl components. Therefore, a semi-pelagic trawl with the trawl doors off the bottom (Fig. 6.2C) and therefore no sand clouds should not reduce the capture efficiency of the gear for pink shrimp, but would reduce the disturbance of the seabed by the doors and bridles.

A project to test the feasibility of a semi-pelagic shrimp trawling system was done in the Gulf of Maine (He et al. 2002; He and DeLouche 2004; He et al. 2006). In the experiment, the primary control of the door height off the seabed was achieved through the shortening of warps and monitored in real time through the use of door height monitoring devices of the NetMind system (Northstar Technologies, St. John's, Canada). High lift-coefficient and high lift-to-drag ratio Poly-Ice® El Cazador doors (Hampidjan, Iceland) were selected for the project (Fig. 6.7).

Fig. 6.7. The Poly-Ice© Al Cazador semi-pelagic trawl door

When the trawl was towed on a straight track, the doors could be kept off the bottom with shorter warps. However, during the fishing trials, the trawl often had to tow in a curved track because of the small fishing areas. When towing along a curved track, the door on the inside of the curve would become closer to the bottom, while the other would be lifted to a higher point in the water column (Fig. 6.8). After 38 tows in the western Gulf of Maine in 2003, only about one-third of the door shoes were polished, indicating very light and intermittent bottom contact during turning and changes in depth. In that experiment, the amount of shrimp caught by the experimental trawl operating in semi-pelagic mode was comparable to catches by similar vessels fishing commercially with regular shrimp trawls on the same grounds (Fig. 6.9), suggesting the possibility of using such a trawling system in that fishery.

Although the results are preliminary, this example demonstrated the potential of semi-pelagic trawling for shrimps if the door height and the groundgear bottom contact can be better controlled. Semi-pelagic trawling with doors, sweeps and bridles off the bottom may also reduce the herding of fish by these trawl components, resulting in reduced fish by-catch.

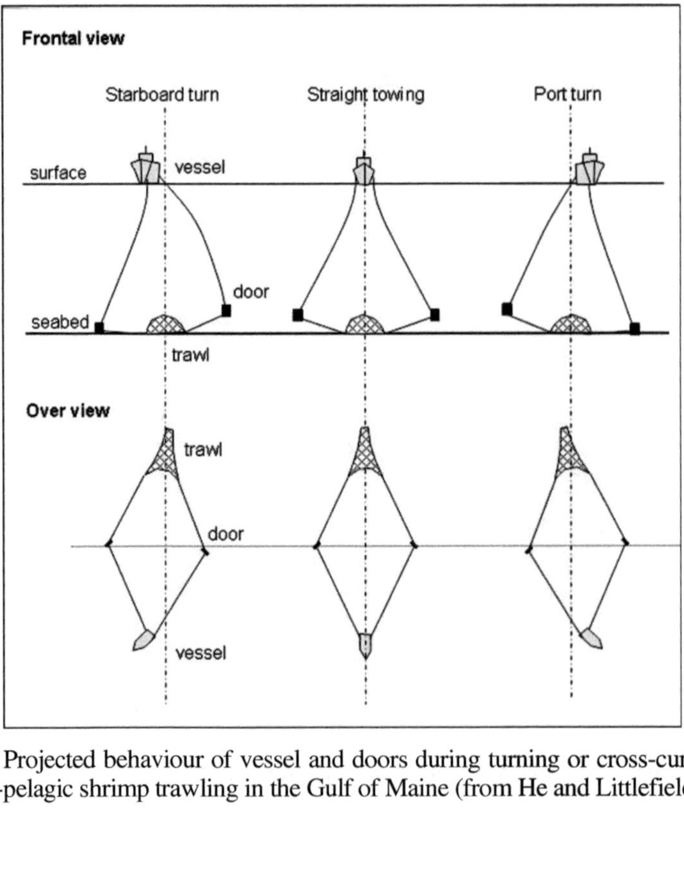

Fig. 6.8. Projected behaviour of vessel and doors during turning or cross-current during semi-pelagic shrimp trawling in the Gulf of Maine (from He and Littlefield 2003)

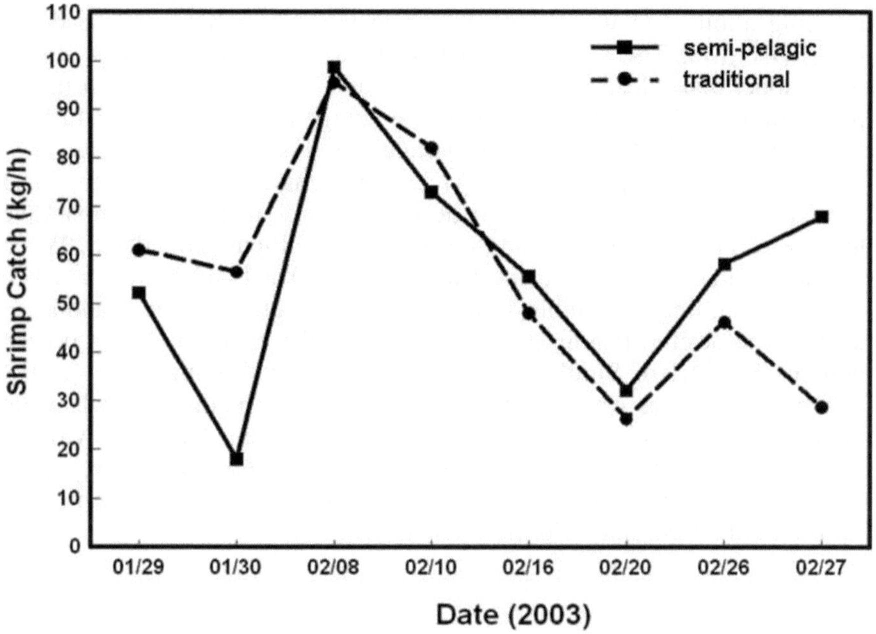

Fig. 6.9. The average daily catch rate of the pink shrimp (*Pendalus borealis*) by the semi-pelagic trawl and other commercial trawls during sea trials (from He and Littlefield 2003)

6.4.3 Quasi-bottom Trawls for Red Snappers in Australia

Red snappers (*Lutjanus malabaricus* and *L. erythropterus*) are harvested by bottom trawls in northern Australia. Concerns with high discards of unwanted fish and invertebrates, and the bottom impact of these trawls, resulted in tests and subsequent adoption of a quasi-bottom trawling system (Fig. 6.2E, Fig. 6.10) in Australia's northern trawl fishery (Ramm et al. 1993; Brewer et al. 1996). The doors in the trawling system of Brewer et al. (1996) were left on the bottom while the groundgear was replaced by several drop chains and weights.

Initial tests by Ramm et al. (1993) used a trawl named the 'Julie Ann trawl' with seven drop chains (0.5 m long, 10 kg each) on the bosom and one heavier weight (60 kg) on each of the wingends, and with the doors on bottom. This rig kept the fishing line 0.3 m off the bottom. The catch rates for commercial species were similar as those for traditional demersal trawls, while the catch of non-target species was reduced by 57%, and benthos by 97%. The trawling system left only nine furrows 0.1 to 0.3 m wide, totalling about 2 m width out of a 65 m trawl path between the doors (i.e., only 3%). The traditional demersal trawl's groundgear was rigged with rubber disks, lead weights and steel cables meaning that a substantial proportion of the bottom between the wingends would be disturbed. Because the 'Julie Ann' trawl was particularly sensitive to changes in fishing and operating conditions, the 'McKenna' wing trawl was designed and subsequently tested in the same fishery (Brewer et al. 1996). Comparative tests were made using the same net with and without groundgear. When operating without groundgear, steel weights (10, 20 and 40 kg) were attached to the fishing line, in combination with added flotation on the headline, to achieve the desired fishing line heights of 0.4–0.5 m or 0.8–0.9 m off the bottom (Fig. 6.10). Similar to earlier findings by Ramm et al. (1993), commercial species were not reduced by using this modified trawl, while discard species and benthos were substantially reduced. The trawl with the fishing line 0.4–0.5 m off bottom had the best performance in retaining commercial species and reducing discard species, and was recommended for use in the Australia's Northern fish trawl fishery.

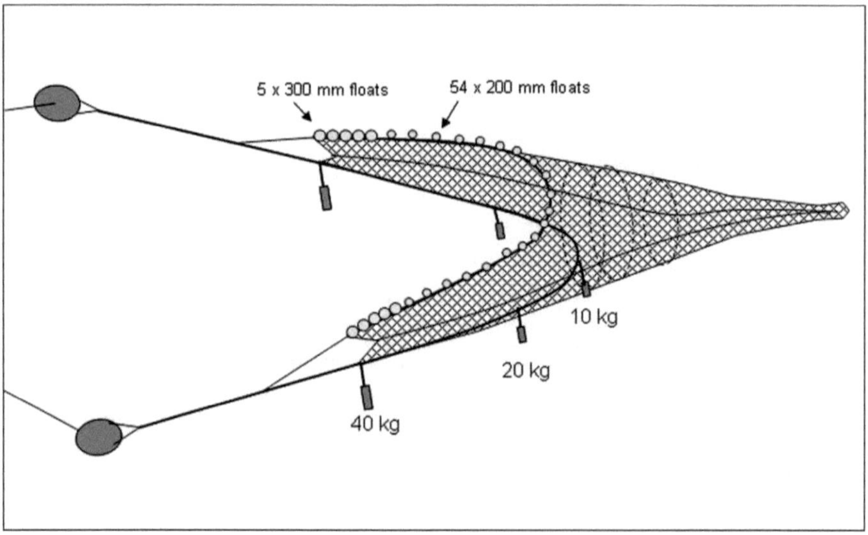

Fig. 6.10. A quasi-bottom trawl design tested in the northern Australian red snapper fishery (redrawn after Brewer et al. 1996)

6.4.4 Quasi-bottom 'Sweepless' Trawl for Whiting in the Gulf of Maine

The trawl groundgear is called the 'sweep' in the northeastern United States. A 'sweepless' trawl is therefore a trawl without groundgear. Drop chains are used to keep the trawl at suitable distances from the bottom. The sweepless trawl is thus a 'quasi-bottom' trawl according to our definition in Fig. 6.2E.

The 'sweepless' trawl was modified from the 'raised footrope' trawl which was developed for the Gulf of Maine silver hake (whiting, *Merluccius bilinearis*) with minimal catch of controlled groundfish species (Pol 2003) (Fig. 6.11). In the raised footrope trawl (A), the fishing line was

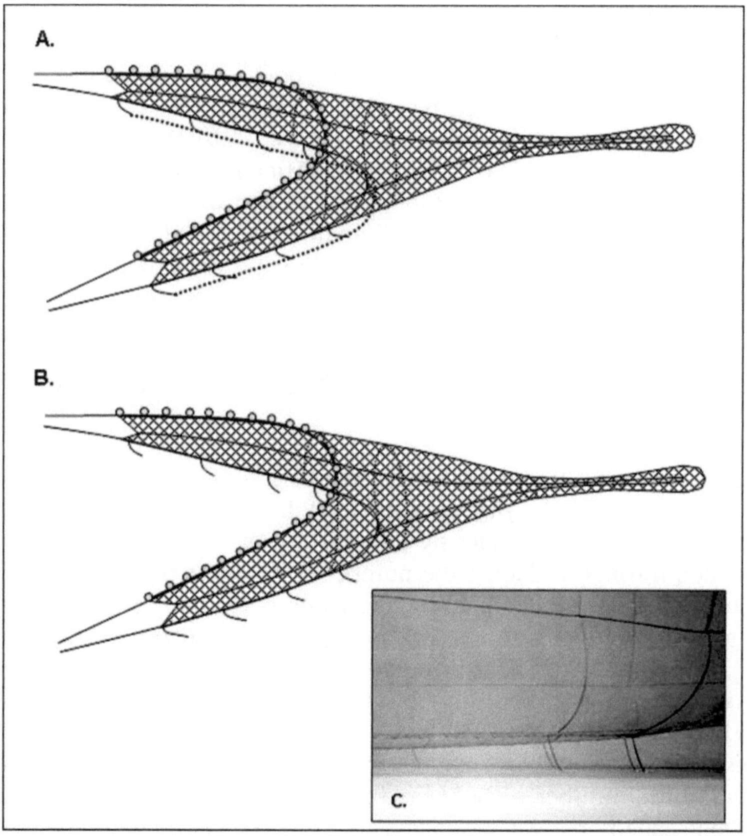

Fig. 6.11. Raised footrope (A) and 'sweepless' trawl (B) for whiting (*Merluccius bilinearis*) in the Gulf of Maine (redrawn after Pol 2003), and a similar 'sweepless' trawl as seen in a flume tank (C)

raised by the attachment of a sweep chain by a number of toggle chains 1 m long. The sweepless trawl (B and C) has no chain sweep or other groundgear assembly. Additional weight to replace the weight of the chain sweep was provided either by increasing the link size of drop chains, or by hanging two chains at each attachment point.

The sweepless trawl represents several improvements over the raised footrope trawl. It is easier to rig and less likely to become entangled with debris. The sweepless trawl also has less impact on the sea floor, because contact is reduced to a limited number of points, instead of the whole width between the wingends (Pol 2003).

6.5 Other Trawl Modifications to Reduce Effects on the Seabed

Gear modifications to achieve ecosystem objectives including the reduction of impacts on the seabed have been discussed by Carr and Milliken (1998), Rose et al. (2000), Ball et al. (2003), CEFAS (2003), and Valdemarsen and Suuronen (2003). These modifications include reducing the weight of groundgear, reducing the contact of trawl doors, the use of rolling components such as wheels, rollerballs, and bobbins instead of fixed, skidding and plowing components. The following are examples of developments in fishing gears to decrease impacts on seabeds.

6.5.1 Lighter Groundgear for the Offshore Shrimp Fishery off Labrador

He and Foster (2000) and He (2001) reported a project to reduce the seabed impact of offshore shrimp trawls off Labrador. The project investigated whether seabed contact by the existing offshore shrimp trawl could be reduced through reducing the number of footgear bobbins without significantly altering the engineering and catch performance of the gear. The fishing gear tested was a three-bridle Skjervoy 3600 shrimp trawl with 31 bobbins (Fig. 6.12A). The full footgear weighed 5,698 kg in air and 2,984 kg in water. The modified 9-bobbin footgear weighed 2,187 kg in air and 1,306 kg in water (Fig. 6.12B).

The total area of seabed contact by the trawl was calculated from the width and the number of bobbins in the groundgear. The percentage of impacted seabed area was defined as the ratio of the total contact width to the swept width between the wings (wingend spread). Analysis and visual

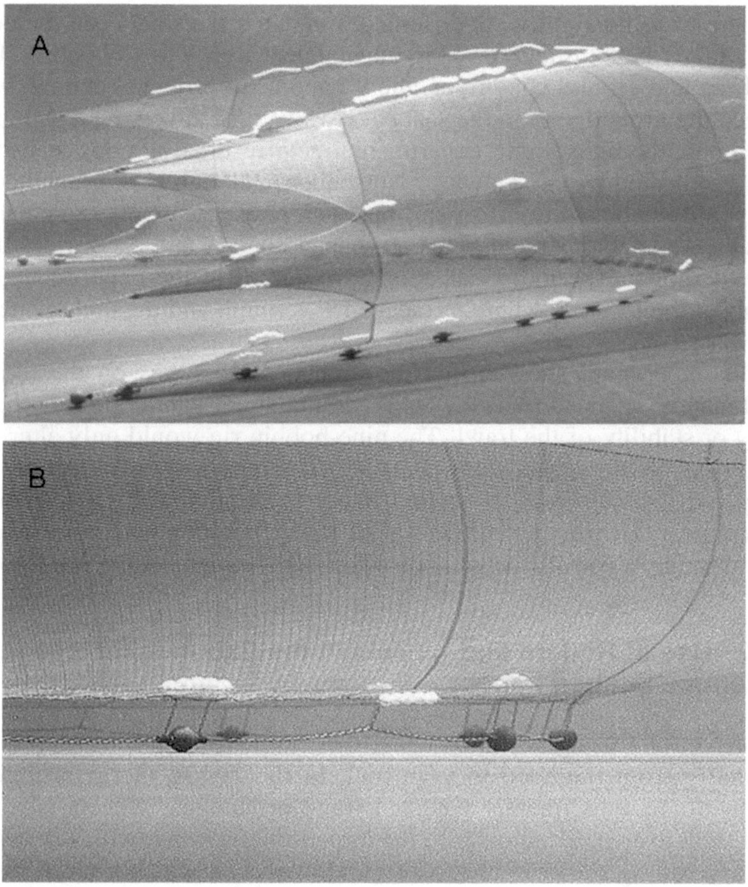

Fig. 6.12. The commercial offshore shrimp trawl Skjervoy 3600 with 31 bobbins (A) and experimental 9-bobbin trawl (B, only boson shown) of the same design, as seen in the flume tank in St. John's, Newfoundland

observation in the flume tank showed no measurable changes in the geometry or stability of the trawl when the number of bobbins was reduced from 31 to 9. The total bottom affected by the bobbins was reduced by 69% when their number was reduced from 31 to 9.

Sea trials were conducted onboard M/V 'Newfoundland Otter', a 60 m shrimp factory freezer trawler with a full set of Scanmar trawl monitoring devices (Scanmar A/S, Norway) including a Trawleye Netsonde. The fishing trials were off the Labrador coast in the northwest Atlantic at depths between 270 and 367 m. A total of 17 tows was completed targeting the pink shrimp.

There were no significant differences in catch rates between the control gear with 31 bobbins and the experimental gear with 9 to 19 bobbins. Under good sea and ground conditions, the 9 bobbin rig provided enough weight to keep the groundgear on the seabed. However, in adverse sea and ground conditions, the lightweight experimental gear resulted in poor seabed contact as indicated by the Trawleye Netsonde. Although poor seabed contact is generally believed to affect shrimp catch rates negatively, no clear relationship was demonstrated between seabed contact and catch rates during the experimental period. Gear damage occurred in the experimental gear mainly at the location where the seabed was rough and the underwater current was strong.

This experiment showed that the number of bobbins on the Skjervoy trawl may be reduced to as few as 9 without significantly altering the geometry or stability of the trawl. The nine-bobbin rig would only alter as little as 4% of the seabed between the wingends, a 69% reduction when compared with the area of seabed likely to be altered by the 31-bobbin control gear. However, the trawl with fewer bobbins was more likely to incur damage, especially on grounds with rough sea and bottom conditions.

6.5.2 Use of Rollers and Wheels in the Groundgear to Reduce Seabed Impact

The drive for a design of groundgear that can wheel over the seabed came originally from the need to save fuel. In the 1940s, German engineers designed and subsequently patented trawl groundgear wheels which had all of their axes perpendicular to the direction of towing, which is essential for easy rolling (Gabriel et al. 2005). However, it was not until 50 years after the invention that a full-scale groundgear was constructed and tested at sea (Fig. 6.13). In 1993, a German shrimp fisher constructed such a roller groundgear which collected much less sand and shells in his catch and no reduction in commercial shrimp species (Gabriel et al. 2005). It was conceivable that the roller gear would have less drag and bottom impact as indicated by the smaller quantity of substratum caught in his net.

Ball et al. (1999) and Ball et al. (2003) also tested a prawn trawl using rollers on its groundgear, as well as a dropout panel behind the groundgear to reduce seabed impact and the catch of benthos. Fourteen rollers of 4 kg each were used on each wing of the trawl and six smaller rollers (2 kg each) were used around the mid-section of the footgear. The dropout panel was 6 x 3 m of 90 mm mesh rigged as square-mesh. The rest of the net, including the codend, was made of 80 mm mesh. Sea trials compared this

Fig. 6.13. Bobbin groundgear with all axis of the bobbins perpendicular to the towing direction (modified from Gabriel et al. 2005)

rollerball net with a commercial shrimp net on the west coast of Ireland. Catches of commercial species were similar, but reductions of 32 and 66% in unwanted invertebrates and debris were reported. The actions of the rollers appeared to stimulate fish to rise off the seabed, eliminating the need for tickler chains. Because of reduced ground friction, the roller gear was easier to tow. It was estimated that the experimental net required 12% less power than the commercial net when towed at the same speed of 2.6 knots. The catch from the rollerball net was also cleaner, with less silt around fish's gills, indicating that the experimental net 'was not penetrating into the seabed to the extent of the standard design' (Ball et al. 2003).

Researchers in Faeroe Island and Norway are also testing swivelled rollers and wheels to replace rockhoppers typically used in their trawls (K. Zachariassen, pers. comm.; E. Grimaldo, pers. comm.). Among several configurations tested by Zachariassen, the most successful rolling gear consisted of 0.22 m wide rubber disks with steel axles (Fig. 6.14). Between the wheels, there was a combination of small discs and rollers. Each wheel can rotate independently, and maintain orientation in the towing direction. This rolling gear reduced catches of the target species compared with the rockhopper gear, but it clearly reduced sand clouds behind the roller gear. The design seemed to be workable and practical, and further tests are planned.

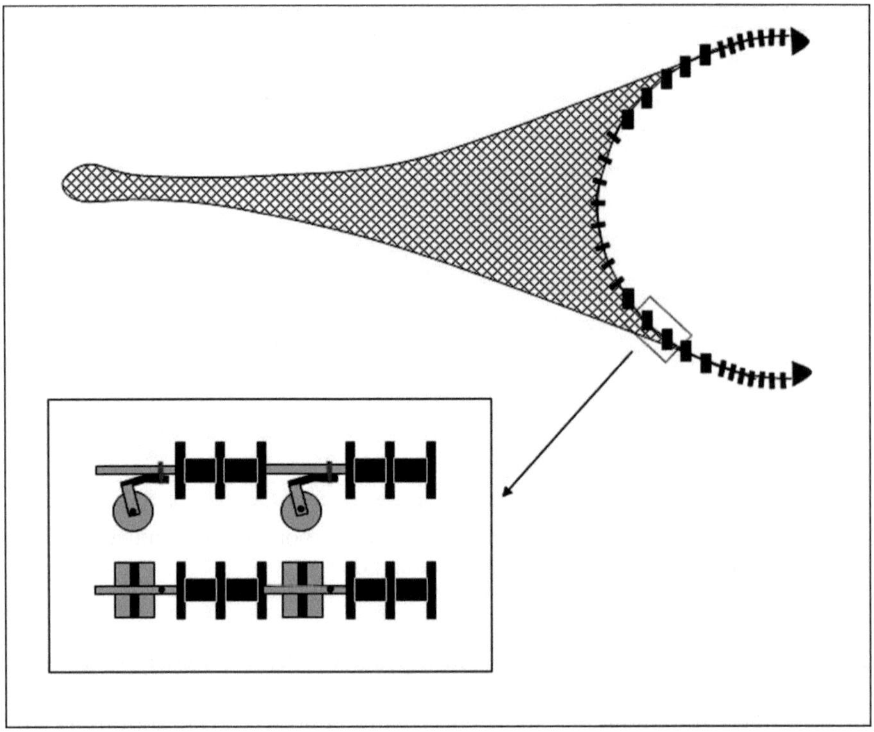

Fig. 6.14. Groundgear swiveled wheels being tested in Faeroe Islands (courtesy: K. Zachariassen)

6.5.3 Tickler Brushes

Tickler chains are commonly used in flatfish fisheries in the North Sea. Tickler brushes were tested by Faeroe researchers as a replacement for tickler chains to reduce seabed impact of trawling (K. Zachariassen, pers. comm.) and these brushes were made of nylon and cylindrical in shape. Alternate tows indicated that the catches of target species were not affected using the experimental gear. Underwater observations showed a great reduction in suspended sediments, indicating less seabed disturbance by the brushes. However, a similar groundgear device, called the 'Bristal sweep' or the 'street sweeper' (Fig. 6.15), was used and subsequently banned in the northeastern United States in the mid-1990s due to its high fishing efficiency (Pol and Carr 2000).

Fig. 6.15. 'Street sweeper' groundgear used and subsequently banned in the 1990s in the Gulf of Maine

6.5.4 The Use of Different Trawl Doors

Trawl doors vary in size and design. In addition to their function of horizontal spread, sand clouds stirred up by the doors are known to herd demersal fish species (Main and Sangster 1981; Wardle 1983, 1986). Early trawl doors had simple designs and relied heavily on ground sheer to spread the trawl and would not function when they were off the bottom. Newer trawl doors are more complicated in design and rely primarily on hydrodynamic forces to spread the trawl. They usually have a higher aspect ratio (the ratio of height to width) than older designs and some remain stable both on and off the bottom.

Main and Sangster (1981) measured the geometry of sand clouds stirred up by four types of trawl doors used in Scotland (Fig. 6.16). The rectangular flat and Vee doors were found to have wider sand clouds than polyvalent (cambered slotted oval) and rectangular cambered doors, indicating more disturbance of the seabed from the former two types.

The use of high-aspect pelagic trawl doors may also reduce seabed impacts even if they are on the bottom. Hydrodynamically efficient doors typically have a narrow width and operate at a lower angle of attack, leaving a narrower 'footprint' compared with traditional bottom doors (Goudey and Loverich 1987; McCallum 2001). For example, for a 2 m^2 door, the low aspect ratio door may be 2 m long by 1 m high with an aspect ratio of 0.5. A high aspect door of the same area (2 m^2) could be 1 m wide by 2 m

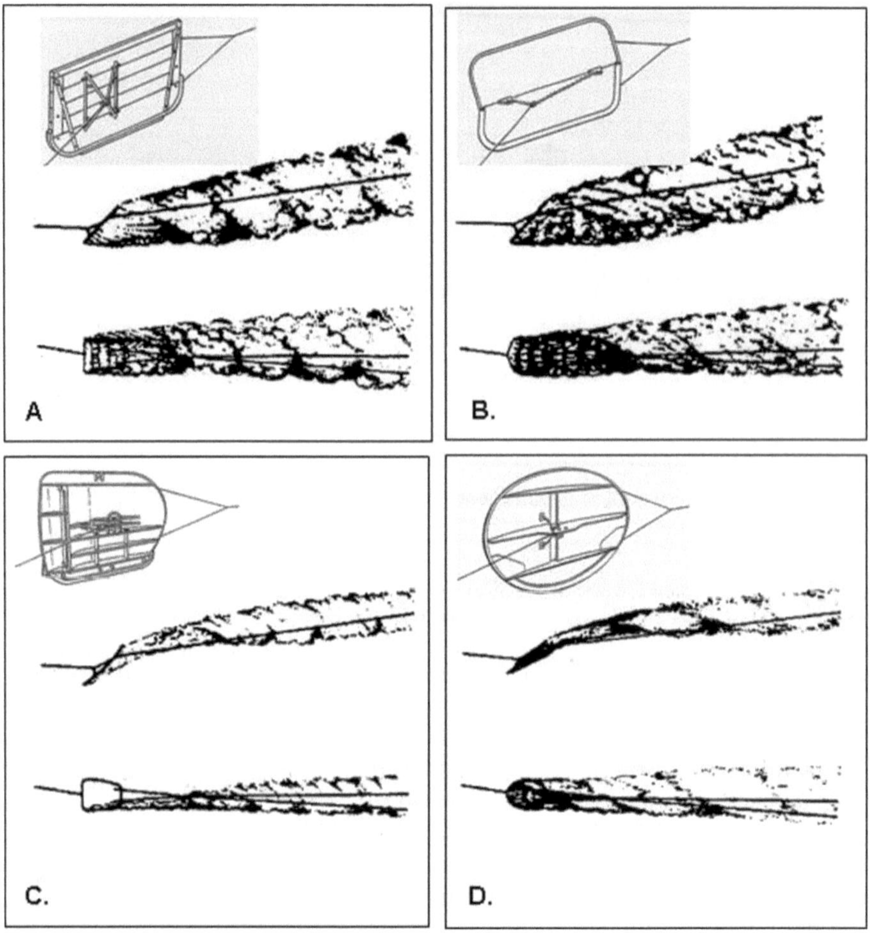

Fig. 6.16. Sand clouds generated by the four types of doors as measured by divers during towing at sea. A. Rectangular flat door, B. Vee type door, C. Rectangular cambered door, D. Oval cambered slotted door (modified from Main and Sangster 1981)

high with an aspect ratio of 2. The low aspect ratio door often operates at a large attack angle (about 43°), while the high aspect ratio door often operates at about 30°. As seen in Fig. 6.17, the area of seabed affected by the high aspect ratio door would only be 40% of that affected by the low aspect door. High aspect trawl doors also reduce the bottom contact of ground wires behind the door by keeping a large proportion of them off the seabed. For fisheries where herding by sand clouds is not critical, such as those for shrimps and small fish species, the use of high aspect trawl doors may be feasible, and help reduce seabed disturbance.

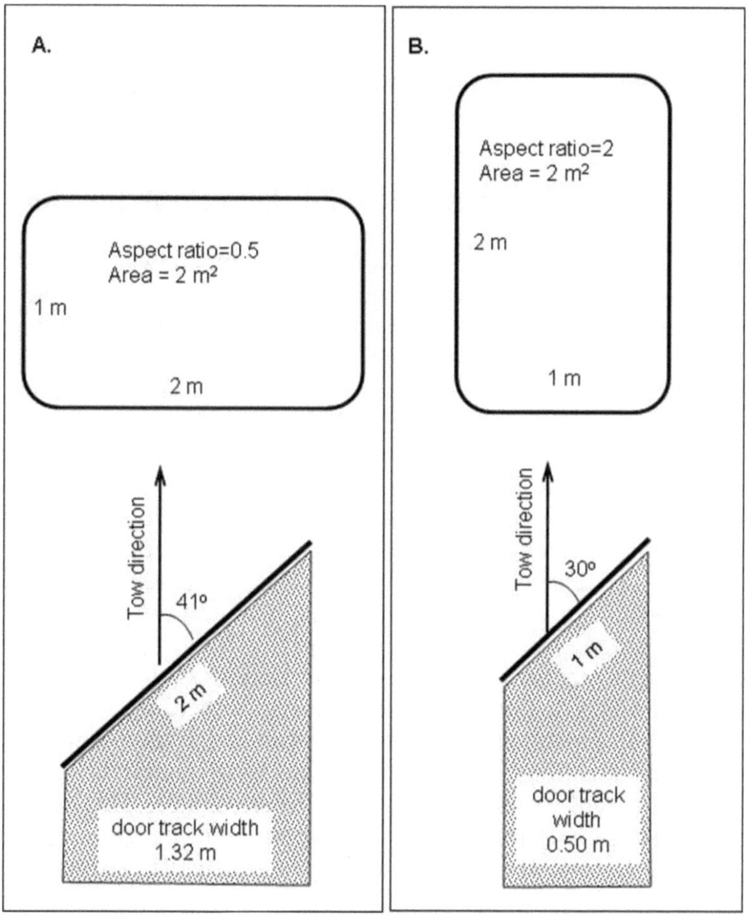

Fig. 6.17. Comparison between high and low aspect trawl doors and their width of track

6.5.5 Short Bridles and Sweeps

Bridles and sweeps connect the wingend and the door. Heavy sweeps and lower bridles strung with rubber discs, called 'cookies', are used in flounder trawls in New England to stir up the sediment to improve herding efficiency. In fisheries where bridle herding is not important or undesirable, such as shrimp trawl fisheries, shorter and lighter bridles are used. In the Gulf of Maine pink shrimp fishery, regulations require that the wire between the wingend and the door does not exceed 27.5 m, and only bare wires are allowed. The primary intent of this regulation was to reduce herding and the catch of finfish, especially flounder, as catch rates of flounders are strongly correlated to bridle length (Somerton and Munro 2001). Shorter and lighter wires may help reduce seabed impact of the trawl as they are less likely to be in constant contact with the seabed.

6.5.6 The 'Active Trawl' System and 'Auto-trawl' System

Fohl (1967) investigated how the fishing depth of midwater trawls might be controlled. This concept was further pursued by Shenker (1995; 1996) who developed the Active Trawl System to improve the performance of trawl doors and the active control of the height of doors. The Active Trawl System expands the trawl by using 'variable thrust vector devices' (VTVDs) which are powered from the ship using cables. VTVDs are based on the 'Magnus Effect' – where towed rotating cylinders generate side forces perpendicular to the axis of the cylinder and the towing direction. The system was reported to have a 'bottom-contour' mode in which the VTVDs would maintain light contact with the bottom or operate at a set height above the seabed (Shenker 2005). This has potential for reducing seabed impact if applied to bottom trawls, but the product is still in the development stage, and no commercial use has been reported. Successful application of this technology may result in a doorless 'otter' trawl for certain fisheries which do not involve sand clouds for herding.

Similar developments using acoustic control of the trawl doors' vertical and horizontal positions have been pursued by Scanmar, a fishing gear equipment company based in Norway (CEFAS 2003). This is part of a more comprehensive research and development program called the 'Auto-trawl' system. It is reported that the vertical positions of doors can be controlled by acoustic manipulators fitted onto the doors, but detailed information is not yet available. Successful devices like this would allow for semi-pelagic trawling without any bottom impact by the doors, as well as near-bottom pelagic trawling for certain species with the elimination of any gear effects on the seabed.

6.5.7 Use of Kites, Depressors and Other Flexible Devices in Trawls

Water-borne kites and other flexible devices have been investigated to reduce the weight of fishing gear, and eliminate the use of trawl doors or reduce their size (Goudey 1999). Instead of heavy weight rollers and chains, Goudey tested a band of fabric panels between the fishing line and the footgear designed to generate downward forces to keep the gear on the bottom. However, soft materials used for the device could suffer significant damage if they contact rough seabed. The 'self-spreading' groundgear developed by SINTEF of Norway uses a series of rubber plates (instead of soft canvas) hanging under the fishing line (Fig. 6.18) (SINTEF 2004). In flume tank and field tests, increased wingspread (15–20%) was observed with this arrangement in comparison to typical rockhopper gear. This increased spread from the groundgear suggests that the door size might be reduced, which in return would reduce seabed impact of the doors. In addition, because the individual plates can flip horizontally when encountering rocks and other obstructions, this gear appears to be less disruptive to the bottom and may produce lower quantities of suspended sediments. The gear also seems more efficient in catching some species of groundfish. Underwater observations indicated very few or no fish escaping under the self-spreading groundgear, while it is quite common for fish to escape under the rockhopper gear (Engas and Godo 1989).

Kites were also installed at various locations in the trawl including the square, and the adjacent side panels, mid belly, and in the rear part of the trawl (Goudey 1999). 'Parafoil' trawl doors were also tested to replace traditional doors (Fig. 6.19). Comprehensive flume tank tests were performed for various designs, but no sea trials have yet been done on these gears.

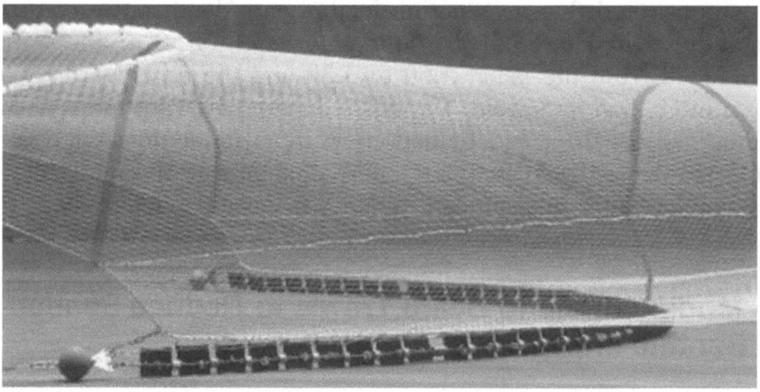

Fig. 6.18. The self-spreading groundgear using footgear plate (plate gear) as developed by SINTEF of Norway. (Courtesy of SINTEF, Denmark)

Fig. 6.19. The 'soft' trawl door install at the wingend (from Goudey, 1999)

6.5.8 Small Warp/Depth Ratio to Reduce Door Pressure on the Seabed

Vincent (2001) discussed an option to reduce the weight or pressure of the trawl door on the seabed by changing the warp/depth ratio (also called the shooting ratio, or scope). With the French survey gear 25/47 GOV trawl, it was demonstrated that a reduction in the warp/depth ratio resulted in a reduced downward force of the door. In a depth of 50 m, the warp ratio was reduced from the usual 5.60 to 3.30 and the downward force was reduced more than three-fold. Various reductions in warp ratios and corresponding downward force reductions were also demonstrated at other depths. However, trawl geometry may change due to changes in warp ratios, and its effect on catching efficiency needs to be determined.

6.6 Modifications to Beam Trawls to Reduce Seabed and Benthic Impact

Modifications to beam trawls to reduce seabed contact include the use of electric stimuli to replace chain mats in the North Sea flatfish beam trawls (van Marlen, et al. 2001b) and in shrimp beam trawls in Belgium (Polet et al. 2005a, 2005b). Dropout panels have also been tested in beam trawls to reduce the catch of benthos in the Belgian flatfish beam trawl.

6.6.1 Experiments on Electrical Stimuli in Netherlands' Flatfish Beam Trawls

Early tests on the use of alternative stimuli, such as electricity, in fish and shrimp trawls were aimed to increase fishing efficiency and to reduce drag (Stewart 1975; Watson 1976). Practical problems and high power requirements when using electricity in sea water prevented its use in marine fisheries. Heavy chain mats are thus still used in many fisheries, such as the North Sea flatfish beam trawls, to drive buried species into the mouth of the trawl, resulting in biological and physical damage to the seabed community (Valdemarsen and Suuronen 2003).

Considerable efforts have been made to evaluate and reduce the impact of flatfish beam trawls in the Netherlands and Belgium. Catch comparisons and tests of discard mortality were done with electric pulse trawls in the Netherlands. Initial tests were conducted on a 7 m prototype trawl developed by Verburg-Holland Ltd. (van Marlen et al. 2005). This experiment is continuing using the full-scale 12 m trawl with promising results. Catches of benthos were about 60% of that caught by the conventional beam trawl and fewer species of infauna were caught (van Marlen et al.; 2001; 2005), indicating the potential of using electrical pulses as stimuli in the flatfish beamtrawl fishery. For 15 species in the catch, significantly lower direct mortality was found using the 7 m pulse beam trawl compared with the conventional beam trawl. The short-term discard mortality of hermit crabs (*Pagurus bernharus*) was also reduced from 64 to 38%.

6.6.2 Drop-out Panels to Reduce Benthos Catch and Dislocation in Beam Trawls

Belgium, Dutch and British researchers have been working on drop-out zones and escape panels in the belly of beam trawls for flatfish and shrimps to reduce catch of benthos and other seabed materials (van Marlen 2000; Fonteyne and Polet 2002; Polet 2003).

The Belgium tests included escape zones (large mesh panel or openings) just behind the fishing line and square-mesh panels just ahead of the codend (Fig. 6.20) (Fonteyne and Polet 2002). Sea trials showed that escape openings just behind the fishing line were not effective in releasing benthos and had an unacceptable loss of commercial catch. Similar results were obtained in the Dutch experiment on similar designs (van Marlen 2000). However, square-mesh panels just ahead of the codend (Fig. 6.20B) significantly reduced benthos caught in the Belgium flatfish beam trawls. For commercial species, the square-mesh panels produced a mixed result

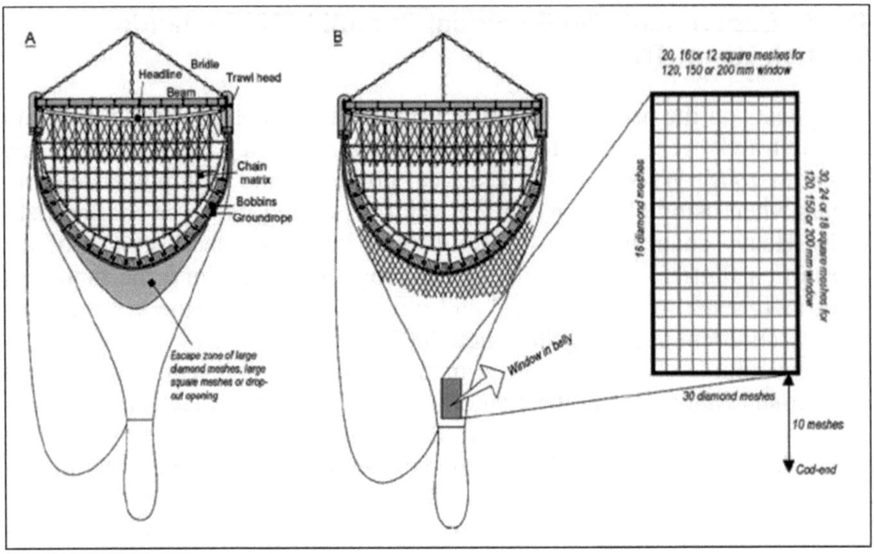

Fig. 6.20. Drop-out panel in flatfish beam trawls as tested by Fontyne and Polet (2002)

with some showing reduced catches while others had increased catches (Fonteyne and Polet 2002).

British researchers reported a study done in the English Channel beam trawl fishery to evaluate a variety of square-mesh drop-out panels (A. Revill, pers. comm.). Similar to the Belgian rigging mentioned above, panels of 140–150 mm full mesh fixed into the belly of a beam trawl a few meshes in front of the codend proved most effective. Around 80% of unwanted benthic invertebrates were released from the beam trawls with drop-out panels and escapees exhibited a high survival rate. No loss of target species was observed with this simple technology.

6.6.3 Electric Pulses as a Stimulus in Brown Shrimp Beam Trawls

Belgian researchers tested the feasibility of electric pulses as an alternative stimulus to the traditional heavy groundgear for brown shrimps (*Crangon crangon*) to improve the species selectivity of the shrimp beam trawl and to reduce bottom contact of the groundgear and thus seafloor disturbance (Polet et al. 2005a, 200b). The study found that the shrimps were very responsive to electrical stimuli while fish (with the exception of dabs and sole) and other invertebrates showed weak responses to electric pulses. When a reaction was observed, the animal kept close to the bottom, indicating that species-selective fishing for brown shrimps may be possible with electric pulses as an alternative stimulus.

Subsequent sea trials also demonstrated the potential for a species-selective and benthos-friendly electro-trawl without loss of targeted brown shrimps (Polet et al. 2005b). The raised groundrope design (Fig. 6.21) of the electro-trawl created an escape opening for most of the discard species and benthos caught in the shrimp trawls. An alternative electrode arrangement (parallel rather than perpendicular to the towing direction) may also reduce seabed impacts of the gear.

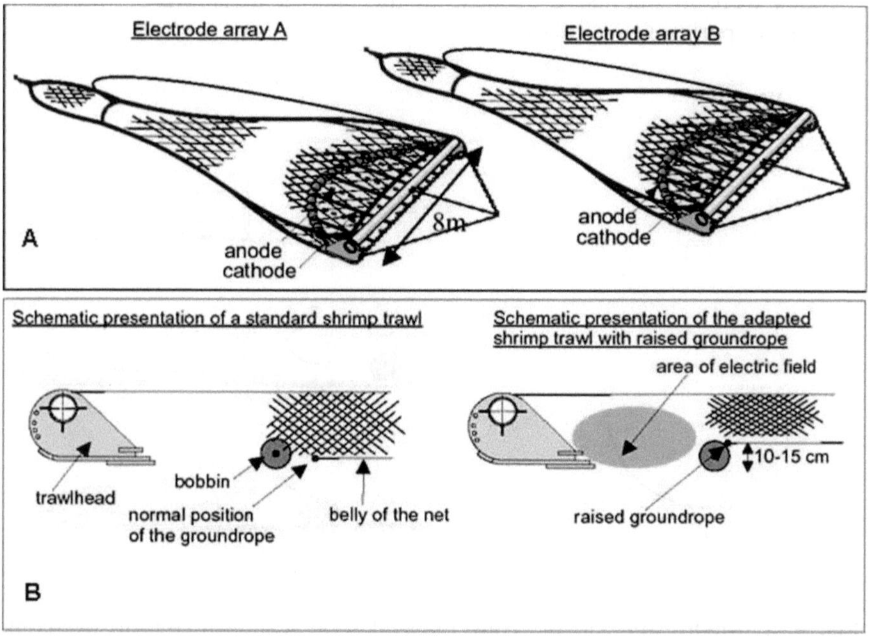

Fig. 6.21. Different arrangement of electrodes (A) and raised footrope beam trawl (B) as tested in Belgium (Polet et al. 2005b)

6.7 Scallop Dredge Modifications to Reduce Seabed Impact

6.7.1 Design and Test of a Low Impact Scallop Dredge in New England

New England scallop dredges are heavy (2500 kg in air) and towed at a high speed (4 to 5 knots). The weight is to ensure good bottom contacts over a wide range of bottom conditions at high speeds in order to cover large fishing areas with limited fishing days (which are regulated) (Goudey 1999). Goudey (1999) evaluated the traditional dredge design, and explored the use of hydrodynamic forces to keep the dredge on the bottom rather than heavy dredge weight. The modified dredge was able to roll over rocks while keeping good contact with the bottom (Fig. 6.22). While the new dredge showed promise, comparisons of catch rates with traditional dredges were inconclusive because of the lack of a fishing permit at the time the experiment was done.

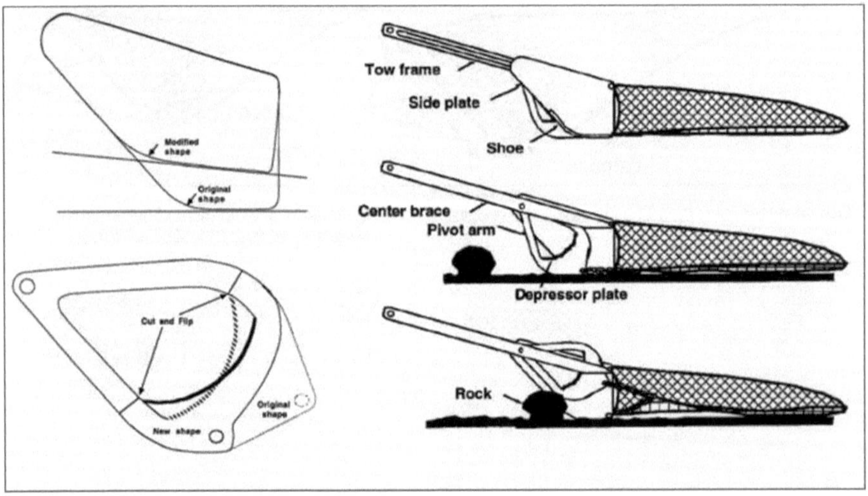

Fig. 6.22. Modification to the New England style scallop dredge to overcome boulders on the seabed whilst reducing the weight of the gear (Goudey 1999)

6.7.2 Acoustic and Electric Stimuli for Scallop Dredges

Pol and Carr (2002) tested acoustic and electric stimuli for sea scallop and bay scallop (*Argopecten irradians*) in the US Georges Bank scallop fishery in order to reduce dredge weight and penetration into the seabed. Bay scallops were observed to swim up into the water column following the passage of a boat with an outboard engine. Based on multiple observations of this phenomenon, bay scallops and sea scallops were exposed to selected acoustic frequencies, recordings of sounds from engines, and the original engine-type that produced the observed reaction, both in the field and in captivity. However, testing of the reactions of scallops to acoustic stimuli did not result in expected reactions which might be used for the capture of the species.

Research efforts with electric pulses resulted in a more promising and safe dredge design, and these experiments on the use of electric pulses to stimulate scallops are continuing.

6.8 General Discussion

Demersal mobile fishing gears such as otter trawls, beam trawls and shellfish dredges alter physical and biological structures of the seabed, although their impact on benthic communities and ecosystems may vary with the diversity, sensitivity and natural disturbances of the seabed. While research is continuing to quantify the various impacts of different fishing gears used under various fishing conditions, technical measures to reduce seabed impacts are occurring and should be encouraged.

Measures that improve fishing efficiencies in strictly-enforced output-controlled fisheries can reduce fishing time and consequently seabed impacts. Alternative gears that have less seabed contact, such as pelagic or semi-pelagic trawls, may be used instead of traditional bottom-tending gear in some fisheries where herding of the target species by sand clouds is less critical. Other gear modifications that have less seabed impact include those that: reduce contact area/points of trawl groundgear; reduce the weights of groundgear and doors; use more efficient and high-aspect ratio trawl doors; provide dropout openings in beam trawls; involve 'sweepless' trawls; and 'wheeled' or 'rollerball' groundgear replacing rockhoppers. Electrical stimuli may be employed in beam trawls to replace traditional heavy tickler chains or at least reduce their number in some fisheries. Some novel gears which have potential for reducing seabed impact include the 'Active Trawl' system, the 'Auto-trawl' system, and the use of 'soft' door, and footgear depressors in trawls, though they are in very early stages of development.

The various approaches to reduce seabed impact discussed in this chapter are summarised in Fig. 6.23. However, it should be noted that some technical measures described here may have other negative or positive outcomes in addition to reducing seabed impacts. Caution should therefore be exercised when recommending or implementing their use in specific fisheries.

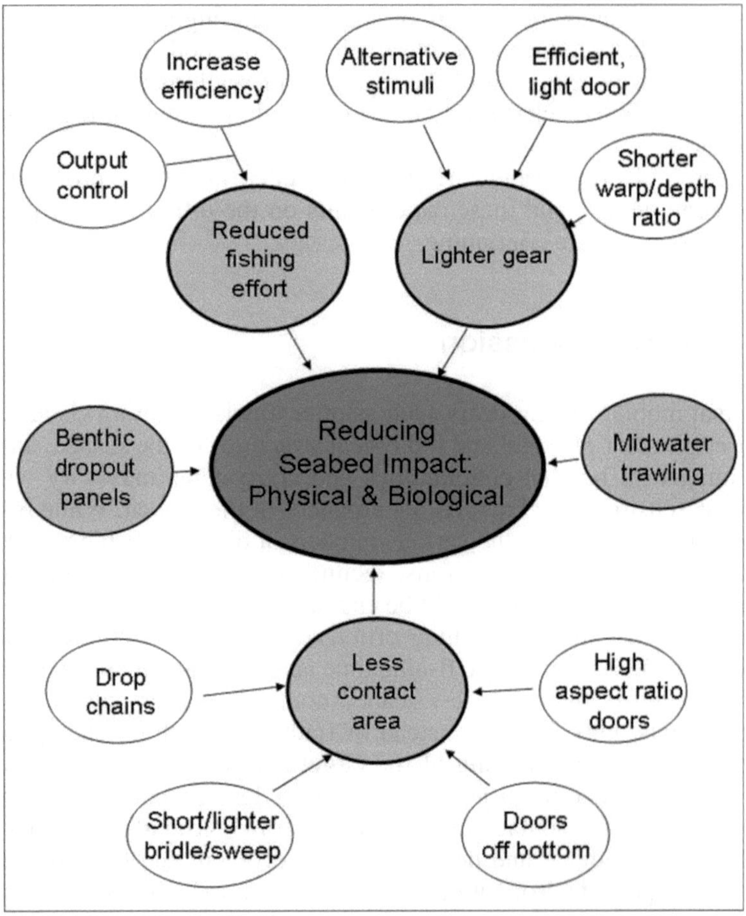

Fig. 6.23. Schematic summary of the approaches used in gear designs and operations that have potential benefits to reduce impacts on the seabed and benthic organisms

Acknowledgement

This chapter was based on a manuscript discussed at the ICES/FAO Fishing Technology and Fish Behaviour Working Group meeting in April 2004. The author would like to thank the following for input and comment: P. Winger, R. Fonteyne, M. Pol, P. MacMullen, S. Løkkeborg, B. van Marlen, T. Moth-Poulsen, K. Zachariassen, A. Sala, W. Thiele, U. Hansen, E. Grimaldo, A. Revill, H Polet, G. Loverich, C. Goudey and R. Smolowitz.

References

Anon (2001) Proceedings of the 'Satellite Workshop on Fishing Impacts – Evaluation, Solution and Policy'. October 2001, Tokyo, Japan, 63 pp

Ball B, Munday B, Fox G (1999) The impact of a *Nephrops* otter trawl fishery on the benthos of the Irish Sea. J Shellfish Res 18(2):708

Ball B, Linnane A, Munday B, Davis R, McDonnell J (2003) The rollerball net: A new approach to environmentally friendly ottertrawl design. Archive Fish Mar Res 50(2):193–203

Barnes PW, Thomas JP (eds) (2005) Benthic habitats and the effects of fishing. Am Fish Soc Symp 41 Bethesda, Maryland, 890 pp

Brewer D, Eayrs S, Mounsey R, Wang YG (1996) Assessment of an environmentally friendly, semi-pelagic fish trawl. Fish Res 26:225–237

Carr HA, Milliken H (1998) Conservation engineering: options to minimize fishing's impacts to the seafloor. In: Dorsey EM, Pederson J (eds) Effects of fishing gear on the sea floor of New England. Conservation Law Foundation, Boston, USA, pp 100–103

CEFAS (2003) A study on the consequences of technological innovation in the capture fishing industry and the likely effects upon environmental impacts. Submitted to the Royal Commission on Environmental Pollution, London, UK. Centre for Environment, Fisheries and Aquaculture, Lowestoft, United Kingdom, 181 pp

Corey T, Williams E (1995) Bycatch: Whose issue is it, anyway? Nor'easter – Magazine of Northeast Sea Grant Programs, Spring/Summer 1995. Available online at http://seagrant.gso.uri.edu/noreaster/noreasterSS95/Titlepage_ss95.html

Collie JS, Escanero GA, Valentine PC (1997) Effects of bottom fishing on the benthic megafauna of Georges Bank. Mar Ecol Prog Ser 155:159–172

Collie JS, Hall SJ, Kaiser MJ, Pointer IR (2000) A quantitative analysis of fishing impacts on shelf sea benthos. J Anim Ecol 69:785–798

Dorsey EM, Pederson J (eds) (1998) Effects of fishing gear on the sea floor of New England. Conservation Law Foundation, Boston, USA, 160 pp

Fohl T (1967) Development of a practical depth control system for midwater trawls. A report submitted to the US Department of the Interior, Bureau of Commercial Fisheries, under contract #14-17-007-365. Washington DC, USA, 19 pp

Fonteyne R, Polet H (2002) Reducing the benthos bycatch in flatfish beam trawling by means of technical modifications. Fish Res 55:219–230

Gabriel O, Lange K, Dahm E, Wendt T (2005) Fish catch methods of the world. Fourth edition. Blackwell, Oxford, United Kingdom, 523 pp

Gilkenson K, Paulin M, Hurley S, Schwinghammer P (1998) Impact of trawl door souring on infaunal bivalves: Results of a physical trawl door model/dense sand interaction. J Exp Mar Biol Ecol 224:291–312

Engas A, Godo OR (1989) Escape of fish under the fishing line of a Norwegian sampling trawl and its influence on survey results. J Cons Cons Int Explor Mer 45:269–276

Goudey C (1999) Progress in reducing the habitat impact of trawls and dredges. MIT Sea Grant College Program. MITSG 99-8, Cambridge, MA, USA, 35 pp

Goudey C, Loverich G (1987) Reducing the bottom impact of Alaskan groundfish trawls. In: Oceans 87 Proceedings – Vol. 2. Marine Engineering, Policy, Education and Technology Transfer. Marine Technology Society, Washington, DC., USA, pp 632–637

Hall SJ (1999) The effect of fishing on marine ecosystems and communities. Blackwell, Oxford, United Kingdom, 274 pp

Hart D (2001) Status of fishery resources off the Northeastern United States – Sea scallops. Available online at http://www.nefsc.noaa.gov/sos/spsyn/iv/scallop (accessed 12 April, 2004)

He P (2001) Reducing seabed contact of bottom trawls. In: Proceedings of the 'Satellite Workshop on Fishing Impacts – Evaluation, Solution and Policy' October 2001, Tokyo, Japan, pp 27–35

He P, DeLouche H (2004) Reducing seabed contact of trawling: Semi-pelagic shrimp trawling experiment in the Gulf of Maine and off Newfoundland ICES-FAO Working Group on Fish Behavior and Fishing Technology, 22–23 April 2004. Gdynia, Poland, 10 pp

He P, Foster D (2000) Reducing seabed contact of shrimp trawls. ICES Working Group on Fishing Technology and Fish Behavior, 10–14 April 2000, Haarlem, The Netherlands, 6 pp

He P, Hamilton R, Littlefield G, Sypgers R (2006) Design and test of a semi-pelagic shrimp trawl to reduce seabed impact. Final report submitted to the Northeast Consortium. University of New Hampshire, Durham, New Hampshire, USA, 24 pp

He P, McNeel B, Littlefield G (2002) Reducing seabed contact of trawling: design and test of a semi-pelagic shrimp trawl for the pink shrimp fishery. Final report submitted to the Northeast Consortium. University of New Hampshire, Durham, New Hampshire, USA, 7 pp

Howard PJ (2004) Scallop vessel access to groundfish closed areas: a management success. Available online at http://www.nfcc-fisheries.org/cs_pov_2e.html (accessed April 11, 2004)

ICES (1988) Report on the study group on the effects of bottom trawling. ICES C.M. 1988/B:56. ICES, Copenhagen, Denmark, 30 pp

ICES (1999) Report of the ICES Working Group on Fishing Technology and Fish Behavior. April 1999. ICES CM 1999/B:1. ICES, Copenhagen, Denmark, 37 pp

ICES (2000a) Report of the ICES Working Group on Fishing Technology and Fish Behavior. ICES CM 2000/B:03. ICES, Copenhagan, Denmark. 99 pp

ICES (2000b) Report of the ICES Working Group on Ecosystem Effects on Fishing Activities. ICES CM 2000/ACME:02. ICES, Copenhagan, Denmark. 93 pp

ICES (2004) Report of the ICES Working Group on Fishing Technology and Fish Behavior. ICES CM 2004/B:03. ICES, Copenhagan, Denmark. 189 pp

Jennings S, Dinmore TA, Duplisea DE, Warr KJ, Lancaster JE (2001) Trawling disturbance can modify benthic production processes. J Anim Ecol 70: 459–475

Jennings S, Freeman S, Parker R, Duplisea D, Dinmore T (2005) Ecosystem consequences of bottom fishing disturbance. In: Barnes PW, Thomas JP (eds) Benthic habitats and the effects of fishing. Am Fish Soc Symp 41. Bethesda, Maryland, USA, pp 73–90

Kaiser MJ, Spencer BE (1996) The effects of beam-trawl disturbance on infaunal communities in different habitats. J Anim Ecol 65:348–358

Kaiser MJ, de Groot SJ (2000) Effect of fishing on non-target species and habitats. Blackwell, Osney Mead, Oxford, United Kingdom, 398 pp

Lart W et al. (2003) Evaluation and improvement of shellfish dredge design and fishing effort in relation to technical conservation measures and environmental impact. [ECODREDGE FAIR CT98-4465]. Seafish Report CR 198-200. ISBN 0 903941 46 5. Hull, United Kingdom

Lindeboom HJ, de Groot SJ (eds) (1998) The effects of different types of fisheries on the North Sea and Irish Sea benthic eco-systems. EU-project AIR2-CT94-1664 (IMPACT-II), Final Report ISSN 0923-3210. RIVO-DLO, Ijmuden, The Netherlands. 404 pp

Linnane A, Ball B, Munday B, van Marlen B, Bergman M, Fonteyne R (2000) A review of potential techniques to reduce the environmental impact of demersal trawls. Ir Fish Invest 7:1–39

Løkkeborg S (2005) Impact of trawling and scallop dredging on benthic habitats and communities. FAO Fish Tech Pap 472. Food and Agricultural Organization, Rome, Italy, 58 pp

Main J, Sangster GI (1981) A study of the sand clouds produced by trawl boards and their possible effect on fish capture. Scott Fish Res Rep 20:1–20

Matsushita Y, Fujita K, Honda N (2001) Seabed disturbance by mobile gears. Proceedings of the 'Satellite Workshop on Fishing Impacts – Evaluation, Solution and Policy'. October 2001, Tokyo, Japan, pp 23–26

McCallum B (2001) Impact of mobile fishing gear on benthic habitat and the implication for fisheries management. Master of Marine Study Thesis. Memorial University of Newfoundland, St John's, Newfoundland, Canada, 82 pp

NRC (2002) Effect of trawling and dredging on seafloor habitat. National Research Council (US). National Academy Press, Washington DC, USA, 126 pp

Pereyra WT (1995) Midwater trawls and the Alaska pollock fishery: a management perspective. In: Solving Bycatch: Consideration for today and tomorrow. University of Alaska Sea Grant College Program Report 96-03, pp 91–94

Pol M (2003) Tuning gear research into effective management: a case study. Presented at the 'Managing Our Fisheries' conference. November 2003, Washington DC, USA

Pol M, Carr AH (2000) Overview of gear development and trends in the New England commercial fishing industry. Northeast Nat 7(4):329–336

Pol M, Carr AH (2002) Developing a low impact sea scallop dredge. Final Report NOAA/NMFS Saltonstall-Kennedy Program NA96FD0072. Massachusetts Division of Marine Fisheries, Pocasset, MA, USA. 12 pp

Polet H (2003) Evaluation of by-catch in the Belgian brown shrimp (*Crangon crangon* L.) fishery and technical means to reduce discarding. PhD thesis. University of Ghent. Belgium, 212 pp

Polet H, Delanghe F, Verschoore R (2005a) On electrical fishing for brown shrimp (Crangon crangon): I. Laboratory experiments. Fish Res 72:1–12

Polet H, Delanghe F, Verschoore R (2005b) On electrical fishing for brown shrimp (*Crangon crangon*): II. Sea trials. Fish Res 72:13–27

Ramm DC, Mounsey RP, Xiao Y, Poole SE (1993) Use of a semi-pelagic trawl in a tropical demersal trawl fishery. Fish Res 15:301–313

Robert G (2002) Impact of multibeam survey on scallop fishing efficiency and stock assessment. In: 'Bedford Institute of Oceanography 2001 in Review' Canadian Offshore Scallop Industry Mapping Group, Bedford Institute of Oceanography, Dartmouth, Nova Scotia, Canada, pp 40–41

Rose C, Carr A, Ferro D, Fonteyne R, MacMullen P (2000) Using gear technology to understand and reduce unintended effects of fishing on the seabed and associated communities: background and potential directions. ICES CM 2000/13:03. ICES, Copenhagen, Denmark, pp 106–122

Shenker MI (1995) Active trawl system – a revolution in trawling technology. INFOFISH International 2/95:64–67

Shenker MI (1996) The active trawl system – from concept to reality. INFOFISH International 6/1996:56-58

Shenker MI (2005) Active Trawl System – 'Four Wheel Drive' Trawl Technology'. Active Fishing Systems, Cape Town, South Africa. Available online at http://users.iafrica.com/m/ms/mshenker/html/active_trawl_system.html (accessed October 18, 2005)

SINTEFF (2004) Spreading ground gear. SINTEF Fisheries and Aquaculture. Hirtshals, Denmark, 1 pp

Sinclair M, Valdimarsson G (eds) (2003) Responsible Fisheries in the Marine Ecosystems. Food and Agricultural Organization. Rome, 426 pp

Somerton DA, Munro P (2001) Bridle efficiency of a survey trawl for flatfish. Fish Bull 99:641–652

Steele J, Roberts S, Alverson D, Auster P, Collie J, DeAlteris J, Deegan L, Briones E, Hall S, Kruse G, Pomeroy C, Scanlon K, Weeks P (2005) National Research Council study on the effects of trawling and dredging on seafloor habitat. In: Barnes PW, Thomas JP (eds) Benthic habitats and the effects of fishing. Am Fish Soc Symp 41, Bethesda, Maryland, pp 91–99

Stokesbury KDE (2002) Estimation of sea scallop abundance in closed areas of Georges Bank, USA. Trans Am Fish Soc 131:1081–1092

Sutton GD, Tully O, Hervas A, Hickey J (2003) Multibeam sonar mapping and scallop stock assessment: GIS data integration in support of sustainable fisheries management. Presented at the Fifth International Symposium on GIS and

Computer Cartography for Coastal Zone Management, 16–18 October 2003, Genova, Italy
Valdemarsen JW, Suuronen P (2003) Modifying fishing gear to achieve ecosystem objectives. In: Sinclair M, Valdemarsen G (eds) Responsible fisheries in the marine ecosystems. Food and Agricultural Organization. Rome, pp 321–341
van Marlen B (2000) Technical modifications to reduce the by-catches and impacts of bottom gears on non-target species and habitats. In: Kaiser MJ, de Groot SJ (eds) Effects of fishing on non-target species and habitats: Biological, conservation and socio-economic issues. Blackwell, Oxford, United Kingdom, pp 253–268
van Marlen B, Bergman MJN, Groenewold S, Fonds M (2005) New approaches to the reduction of non-target mortality in beam trawling. Fish Res 72:333–345
van Marlen B, Boon AR, Oschatz LG, van Duyn JB, Fonds M (2001) Experiments in 1999 on a beam trawl with electrical stimulation. RIVO Report C028/01. RIVO-DLO, Ijmuiden, The Netherlands, 49 pp
Vincent B (2001) A way to reduce impact of trawl door. Presented at the ICES WGFTFB meeting, April 2001. Seattle, Washington DC, USA
Wardle CS (1983) Fish reactions to towed fishing gears. In: MacDonald A, Priede IG (eds) Experimental biology at sea. Academic Press, New York, USA, pp 167–195
Wardle CS (1986) Fish behavior and fishing gear. In: Pitcher TJ (ed) The behavior of teleost fishes. Croom Helm, London, pp 463–494

7 The Fate of Fish Released by Recreational Anglers

STEVEN J. COOKE AND GENE R. WILDE

7.1 Introduction

Discussions of by-catch are usually restricted to the commercial fishing sector. Several papers and reviews (Dayton et al. 1995; Hall 1996; Greenstreet and Rogers 2000; Bache 2003; Cheunpagdee et al. 2003; Lewison et al. 2004) and even entire books have been devoted to this topic. Due to the well-documented role of commercial fishing practices in generating by-catch, by-catch reduction has arguably become one of the most pressing conservation issues today (Hall et al. 2000; Cheunpagdee et al. 2003). Indeed, commercial by-catch has been implicated in the global decline of marine fish resources (Hilborn et al. 2003). The realisation that the fishing mortality of large numbers of non-target organisms is a problem in the marine commercial fishing sector has led to much research and innovation into ways of mitigating by-catches (Hall et al. 2000; Bache 2003). New approaches, techniques and gear modifications are needed to reduce the mortality and sublethal disturbances that can arise from by-catch. However, efforts focusing on reducing by-catch in the commercial sector alone fail to recognise the potential contribution that the recreational fishing sector may have on by-catch-related issues and global fish declines. Before discussing the latter, however, it is useful to provide a summary of the terminology relevant to a discussion of recreational by-catch and discard issues (see Box 1).

Only recently has the recreational fisheries sector been considered as a potential threat to fisheries stocks, especially on a global scale (e.g., for freshwater, see Post et al. 2002, Arlinghaus and Cooke 2005; for marine, see McPhee et al. 2002, Coleman et al. 2004). For example, Cooke and Cowx (2004) argued that commercial and recreational fisheries share much in common, including the potential to degrade environments, alter ecosystems,

S. J. Kennelly (ed.), By-catch Reduction in the World's Fisheries, 181–234.

Box 1: Terminology. Definitions of terms associated with by-catch and discards originally proposed by McCaughran (1992). We have modified and expanded the terminology to place it in a more recreational context

Recreational fishing – Conducted by individuals for sport and leisure, with a possible secondary objective of catching fish for personal consumption (FAO 1997; Pitcher and Hollingworth 2002). Sometimes this definition is expanded to include selling surplus catch to offset costs (Cowx 2002).

Commercial fishing – The act of fishing with the intent to make a profit from selling the harvested fish to consumers.

Angler – One who fishes for recreation, generally with a rod and line.

Catch-and-release – The act of releasing fish that were caught by a recreational angler. The phrase "catch-and-release" was believed to develop from the various conservation organisations in the United States (e.g., Trout Unlimited) reflecting a strong conservation ethic (See Policansky 2002 for detailed discussion of the history of catch-and-release).

Discarded catch – The proportion of the total catch that is returned to the water – may be either target species or non-target species.

Fishing mortality – Death of fishes that can be directly or indirectly attributed to fishing activities.

Incidental catch – Catch of non-target species.

Target catch – Individuals that are primarily sought by the fishery. In recreational fisheries, many anglers fish opportunistically and thus it is sometimes difficult to determine the target catch.

By-catch – Discarded catch plus the incidental catch.

Hooking mortality – Death of fishes attributable to capture with standard fishing gears (baited hooks, artificial baits with various hook types and arrays). Mortality may result from fatal wounds or the accumulation of sublethal wounds and physiological disturbances.

Sublethal disturbances – The suite of non-lethal effects imparted by recreational angling to fish that are released. These can include physiological, behavioural and fitness impacts as well as physical injuries. In this context, fitness includes all metrics that can affect life-time reproductive success including factors that reduce growth (and thus fecundity or ability to compete for mates) or directly affect reproductive success or the quality or quantity of progeny (See Cooke et al. 2002a for a more detailed discussion of sublethal disturbances in catch-and-release).

Unaccounted fishing mortality – Death resulting from fishing that cannot be easily quantified. In commercial fishing this would result from fish passing through net webbing, freeing themselves from hooks, ghost fishing, etc. In recreational fishing, unaccounted fishing mortality has never been considered to our knowledge but may include fish that are snagged or hooked legally, but either break the line or get off the hook(s). Another example would include fish that are preyed on during the angling event.

Box 1. (cont.)

Immediate mortality – Immediate (or initial) mortality is defined as capture-related death that occurs during and following capture, up to the time the fish is released.

Post-release (delayed) mortality – Represents death from catch-and-release angling at some point after the released fish swims away. This mortality is usually determined by holding fish in cages, pens or hatchery ponds, or by affixing transmitters or tags to them prior to release in the wild.

impart evolutionary effects through selective fishing, collapse stocks and generate by-catch. But, as noted above, the current debate on by-catch has largely failed to include the recreational fishing sector. Because impacts from recreational fisheries are believed to be diffuse, there is the perception that any impacts on global fish capture are negligible. However, recent estimates suggest that the annual contribution of recreational fisheries to the global fish harvest may be quite high. Specifically, Cooke and Cowx (2004) estimated that recreational harvest may exceed 10 million t, compared to over 80 million t in the commercial sector. Furthermore, nearly 12% of the world's population engages in recreational fishing on a regular basis. With recreational discard/release rates believed to be at or near 60%, more than 30 billion angler-caught fish may be released annually (Cooke and Cowx 2004). This level of release warrants examination as a potential conservation concern (Box 2).

In this chapter, we discuss by-catch in the recreational fishing sector, alongside other contributions in this book on by-catch in commercial fisheries. We begin by providing an overview of recreational fishing, including the main reasons why large numbers of fish are released following their capture. We then review the developments in fishing gears and practices that have the potential to mitigate by-catch and provide several relevant species-specific case-studies. We outline a conceptual model of release and by-catch in recreational fisheries and synthesise existing knowledge to assess recreational fishing by-catch in a conservation and management context. Overall, we contend that this synthesis will help to evaluate and illuminate the issue of by-catch in recreational fishing. Furthermore, this will promote future developments in both gear and angling practices and ultimately minimise the injury, mortality or sublethal disturbances to fish released by recreational anglers. However, unlike commercial fishing, it is less likely that widescale reductions in by-catch or actual discards can be achieved in recreational fishing because in many cases, anglers have not targeted their effort solely on one species and because discards are really those fish that have been subject to catch-and-release angling. Thus, our

Box 2: Why has recreational by-catch been largely ignored by scientists, managers and conservationists? Modified from Hall et al. (2000).

Lack of visibility – Any discard mortality arising from recreational fishing will tend to be diffuse (both temporally and spatially) compared to that which occurs to commercial by-catch. Furthermore, mortality in recreational fishing can be delayed rather than immediate as is often seen in fish that are captured in commercial gears. An exception would be the visibility of moribund fish after mass release at competitive angling events. Such events have helped to drive change in this sector (e.g., Wilde et al. 2002).

Disbelief that there was a problem – Many anglers, product manufacturers and special interest groups do not want to publicise the negative aspects of recreational fishing. The onus has been placed on governments to document problems. The limited number of examples of recreational fisheries collapses (due in part to complex angler behaviour and stock supplementation) add to this problem.

Assumption that mortality following release is negligible – There are relatively few mortality studies for recreational fisheries, although this field is rapidly expanding. Earlier work tended to be short-term and failed to consider delayed mortality (which can be significant). Further, there is a significant difference between a dead fish and a fish that has negligible effects arising from the angling experience. Many fish experience sublethal disturbances that could affect fitness.

Assumption that the overall magnitude is small – There is a tendency to consider the effects of recreational fishing in the context of individual anglers as compared to a large commercial fishing fleet. Recent estimates place global angling participation rates and capture rates much higher than that which was previously thought.

Fisheries management versus fisheries conservation – There is a pervasive belief that recreational fishing is simply a resource management problem. By elevating recreational fisheries to a conservation issue through recognition that angling can affect fish populations (e.g., Coleman et al. 2004; Cooke and Cowx 2006), it will help to generate public interest and drive future improvements in angling gear and practices.

objective is to discuss efforts to ensure that released fish are minimally impacted by capture and handling processes.

7.1.1 What is Recreational Fishing?

Commercial and recreational fishing are both important sources of protein, and contribute substantially to local and national economies (e.g., Arlinghaus et al. 2002; Cowx 2002; Hilborn et al. 2003; Pitcher and Hollingworth 2002). Whilst commercial fishing is conducted specifically to catch fish for sale, recreational fisheries usually involve participants that fish for sport and leisure, with a secondary objective of catching fish for personal consumption (FAO 1997; Pitcher and Hollingworth 2002; note that in some jurisdictions, catch-and-release is dissuaded or illegal, with the primary purpose being for food consumption). Sometimes this definition is expanded to include the sale of surplus catch to offset costs (Cowx 2002). Cowx (2002) refined the FAO (1997) definition to categorise anglers into four main types: those who participate in leisure, competitive, game, and specimen or specialist fishing (when anglers focus all of their efforts on a specific type of fish and fishing activity). It must be noted, however, that many anglers participate in more than one type of recreational fishing activity. Although recreational fishing is most often perceived as involving anglers using hook-and-line fishing, recreational fishers also employ other gears and techniques. For example, in some countries, recreational fishers use spears (e.g., Nevill 2005), bows and arrows (i.e., bowfishing), rifles and even explosives (Cowx 2002). For these fisheries, few if any fish are released. Sometimes conventional hooks are used to snag fish in locations other than the mouth. Gillnets, cast nets, trawls and traps also are considered appropriate gear for recreational fishing in some places, but the delineation between artisanal, commercial and recreational fisheries can become difficult (See Cowx 2002 for discussion). In addition to these techniques, there are a number of regionally-specific gears and tactics (e.g., 'noodling' for large Ictalurids – this involves placing ones fist and forearm into underwater cavities that contain these fish and when they bite down the 'angler' attempts to pull the fish aboard the boat). Nonetheless, despite all the above variations, this chapter is restricted to recreational fisheries that use hook and line, generally with a rod, because these recreational fisheries are by far the largest and are most often characterised as having potentially problematic by-catch.

7.1.2 Why are Fish Released in Recreational Fisheries?

Although there are many fish released each year under the classification of discarded by-catch, the reasons for releasing fish vary significantly among different fishing sectors. In the commercial sector, most highly-regulated fisheries are managed using total allowable catch, quota systems and, for net-based gears, minimum mesh sizes. These strategies result in excessive catch with under-sized individuals, many of which do not survive, being dumped. In other instances, non-target species or undesirable-sized fish of target species are also discarded. In the recreational sector, while some anglers do harvest a portion of the fish they catch, many fish are immediately released. Among the reasons why anglers release fish is that they are undesirable (wrong gender, questionable food value), not the targeted species, or of an illegal size. In an attempt to conserve fisheries resources in some countries, regulations mandate release of some or all fish (Quinn 1996).

However, compared to commercial fisheries, recreational fisheries also include a significant voluntary catch-and-release component, in which anglers release fish for ethical, conservation or sporting reasons (e.g., the assumption that the released fish will survive to be caught again in the future; Quinn 1996; Aas et al. 2002; Policansky 2002). There is a growing debate concerning the ethics of recreational fishing, and in particular catch-and-release fishing (e.g., de Leeuw 1996; Balon 2000) although thorough discussion is beyond the scope of this paper. Irrespective of the reasons for releasing fish, however, mortality and sublethal effects can arise from capture and handling, which can lead to uncertainty in estimating fishing mortality.

7.1.3 How Many Fish are Released?

Alverson et al., (1994) estimated that between 17.9 and 39.5 million tonnes of fish are discarded each year in commercial fisheries. This compares with an estimated 19 million tonnes of fish, representing over 30 billion individuals, released globally in recreational fisheries (Cooke and Cowx 2004). Although there is some uncertainty in these estimates, it is clear that many fish in the recreational fishery are discarded, with discard rates varying considerably among species and countries. In some specialised recreational fisheries, such as in the coarse (i.e., a terminology used to describe 'non-game' fish that typically are benthivorous) fisheries of Western Europe or in elitist fisheries such as that for bonefish (*Albula* spp), voluntary release rates approach 100% (Policansky 2002). In other fisheries few, if any, fish are released (e.g., 8% of dorado [*Coryphaena hippurus*] and 9% of king mackerel [*Scomberomorous cavalla*] along the US Atlantic coast; United States Department of Commerce 2002). Overall in North America, it is estimated that approximately 60% of fish caught by recreational anglers are released (e.g., United States Department of Commerce 2002; Department of Fisheries and Oceans Canada 2003).

7.1.4 What do we Know About the Fate of Released Fish in Recreational Fisheries?

Since the mid 1970's, fisheries scientists and managers have made great advances towards understanding which angling practices contribute to the injury, stress and mortality of released fish. In addition to numerous articles in the primary literature and government technical reports, proceedings from three catch-and-release symposia have also been published (Barnhart and Roelofs 1977, 1989; Lucy and Studholme 2002). By identifying and understanding the key factors associated with hooking injury and mortality of particular species (Muoneke and Childress 1994; Bartholomew and Bohnsack 2005; Cooke and Suski 2005), fisheries managers, media, competitive angling groups and conservation organisations have been able to alter angling practices to increase the probability of fish surviving catch-and-release.

The widespread involvement of people in catch-and-release fishing is predicated on the general assumption that most released fish will survive (Wydoski 1977). Since most fish that die from catch-and-release angling do so some time after release (Muoneke and Childress 1994), there is the false perception that fish which swim away after release, apparently unharmed, always survive. Although this may be the case for some species, other species experience very high, often unnoticed, rates of mortality. In a review of hooking mortality studies, Muoneke and Childress (1994) reported that the mortality rates for released fish ranged from 0 to 89% across a variety of marine and freshwater species.

Hooking mortality is usually divided into immediate (or initial) mortality and delayed mortality. Immediate (or initial) mortality is defined as capture-related death that occurs during and following capture, up to the time when the fish is released. Delayed mortality represents death at some point after the released fish swims away; this mortality is usually determined by holding released fish in cages, pens or ponds, or by acoustic telemetry. Total hooking mortality is the sum of initial and delayed mortality minus the cross-product of initial and delayed mortality (Wilde et al. 2001). There have been several major papers that have aided our understanding of hooking mortality. Muoneke and Childress' (1994) review on hooking mortality in marine and freshwater fish and suggested that total hooking mortality rates above 20% generally should be considered unacceptably high. A more contemporary review (Bartholomew and Bohnsack 2005) found that mortality rates were sufficiently high that catch-and-release angling should not be permitted in marine protected areas. The magnitude of mortalities for catch-and-release can be extensive when viewed in actual numbers. For example, in striped bass (*Morone saxatilis*) fisheries on the eastern seaboard of North America, it is believed that in excess of 12.5 million fish are landed, of which over 90% are released (Millard et al. 2003). Estimates of catch-and-release mortality are around 28% (95% confidence interval 17–44%) or approximately 3.2 million striped bass per year (Millard et al. 2003).

Another key synthesis of hooking mortality (Cooke et al. 2002a) noted that there also may be a suite of sublethal physiological, behavioural, and fitness impairments that can arise from catch-and-release angling, and these sublethal stressors are rarely considered by managers who are focused primarily on the presence, abundance and distribution of fish populations (Wydoski 1977). Although some information exists on how angling-related stress may induce mortality (Wood et al. 1983), few studies have focused on what sublethal stress means to the organism, especially in relation to long-term individual fitness (Maltby 1999; Cooke et al. 2002a). To date, most efforts have concentrated on population-level effects, but individual effects can also be important (Maltby 1999). The body of literature evaluating the impacts of catch-and-release angling is rapidly expanding, however, this research has typically focused on species or groups of fish that are economically important, readily caught by the majority of anglers, and the subject of attention by media (see Muoneke and Childress 1994). In addition, the majority of these studies have been focused on freshwater fisheries in North America (Barnhart 1989; Muoneke and Childress 1994, Cooke and Suski 2005; but see several more global examples e.g., white-spotted charr (*Salvelinus leucomaenis*) in Japan, Tsuboi et al. 2002, and cichlids in Africa, Thorstad et al. 2004).

Also potentially important is the mortality of fish that escape from the hook before being brought to the boat or fish that are removed from the hook by predators – so-called 'drop offs' (Lawson and Sampson 1996). Lawson and Sampson (1996) developed a model that suggests that drop-off mortality could be as important as hook-and-release mortality. This type of mortality is poorly understood and has not been researched in detail. Efforts in this chapter are restricted to those fish that are caught, landed and then released by recreational anglers.

7.1.5 Why is it Important to Reduce Discard Mortality in Recreational Fisheries?

Beyond obvious conservation- and ethics-based considerations, there are a number of reasons why fisheries managers must strive towards reducing the discard mortality from recreational fisheries. Many of the current management strategies employed depend on the regulated release of certain individuals with the notion that fish will be able to be captured multiple times and will attain greater sizes and have greater fitness (i.e., more opportunity to produce viable offspring) because they live longer (Wydoski 1977). Regulations are also imposed in response to overfishing and a need to increase spawner-per-recruit (i.e., spawner biomass) levels (Waters and Huntsman 1986).

Target species are often protected using input controls such as minimum and maximum size and slot limits (when a range of fish lengths is designated for either harvest or protection). Legal size regulations often are based on known relationships between reproductive maturity and size, and are typically set at a size that allows fish to reproduce at least once before removal by fishing (Martell and Walters 2004). However, undersize fish (i.e., by-catch) are caught by anglers and thus the effectiveness of length-based limits as a management tool depends on the fishes' survival after release (e.g., Shetter and Allison 1955; Mason and Hunt 1967). Similar problems exist for slot limits where both undersize and trophy size fish must be released. Success of harvest regulations depends on low discard mortality within mandated size ranges (Waters and Huntsman 1986; Muoneke and Childress 1994). Indeed, often catch-and-release research is conducted in support of this premise (St John and Moran 2001).

Similarly, creel limits (i.e., possession limits) may also be ineffective if discarded fish die, thus inflating the number of fishery removals indirectly resulting in exploitation. This also includes fisheries where it is assumed that all individuals are released and few die. Management of such fisheries requires that mortality rates be maintained at low levels. Ultimately, high levels of discard mortality could lead to a reduction in the size and abundance of fish (Wydoski 1977), resulting in lower catch rates, alterations to populations, community structures and potentially the value of the fishery. In productive waters, even moderate levels of discard mortality may not affect population structures (Wydoski 1977). However, even very low levels of discard mortality (i.e., 1 to 5%) could have devastating effects on populations of long-lived species with low rates of population increase, such as giant sea bass (*Stereolepis gigas*; see Schroeder and Love (2002) for case study).

The increasing use of aquatic protected areas as a management tool has further prompted interest in understanding and reducing discard mortality from recreational fisheries. The premise of a 'no-take' protected area is that fish are not harvested. However, some have suggested that this may not preclude activities such as catch-and-release angling if there are negligible discard mortalities. At present, there is controversy regarding the compatibility of catch-and-release angling with the premise of closed areas and this will likely continue to be a contentious topic as the creation of aquatic protected areas increases around the world (Cooke et al. 2006).

7.2 Factors Influencing the Fate of Released Fish

When a fish is hooked and released by an angler, there are many factors that can affect its fate. Ideally, the released individual will survive, recover quickly and experience no long-term sublethal impairments. Although many anglers strive for such a positive outcome, it is often more probable that there will be at least some negative impacts. Some of the factors that may affect the fate of released fish are intrinsic such as gender, age, previous exposure to stressors, maturity, condition, size and the degree of satiation. Often these intrinsic factors cannot be controlled or altered by the angler to benefit the fish and, indeed, few of these factors have been studied with sufficient rigor to provide any conclusive recommendations for any species.

The environment in which the fish is caught and released can also affect its ultimate fate. Pertinent environmental conditions include abiotic factors such as water temperature, hypoxia, depth, or habitat complexity, as well as biotic factors such as predator burden (i.e., number of predators at a site that could potentially injure or kill a released fish). Although these factors cannot be controlled by anglers, most can be readily assessed and, if deemed to be detrimental, the angler could release captured fish at alternative locations. The remaining factors that typically influence the outcome of an angling event can be controlled by the angler, including the choice of fishing gear and angling practices.

The above factors rarely act independently to cause mortalities, and will most likely manifest as a series of cumulative stressors (Wood et al. 1983; Cooke et al. 2002a). As an example, angling mortality in salmonids has been suggested to be a two-stage process, which emphasises the interrelated and cumulative nature of fishing impacts. Gjernes and Kronlund (1993) observed that injury location was affected by hook and barb type at the first stage, and mortality was affected by injury location and species at the second stage.

Below, we review the issue of discarded by-catch in the context of recreational fishing. We focus our efforts on reducing discard mortality and sublethal disturbances by discussing both angling gear and practices (including factors such as environmental conditions). In our opinion, many of the issues associated with angling gear and practices are unique and require separate treatments. For our review, we focus on the literature that was published since the review by Muoneke and Childress (1994) with reference to historical examples.

7.3 Gear

A growing interest in catch-and-release angling has led to gear developments intended to reduce the injury and mortality of released fish. These gear developments are discussed below in the context of reducing discard injury, mortality and sublethal effects.

7.3.1 Hook Types

Mortality in catch-and-release angling can arise from a number of factors including cumulative sublethal physiological disturbance, physical injury and bleeding (Muoneke and Childress 1994). Hooks play little role in physiological disturbances other than when the hook type influences the difficulty of removal, leading to increased air exposure (e.g., Cooke et al. 2001), and this factor is discussed elsewhere. Hook type, however, does play a major role in mortality arising from direct hooking injury, and almost all of the studies we examined considered mortality as an important endpoint. Indeed, the review by Muoneke and Childress (1994) focuses on hooking-related mortalities. The different types of hooks discussed in this paper are presented in Fig. 7.1.

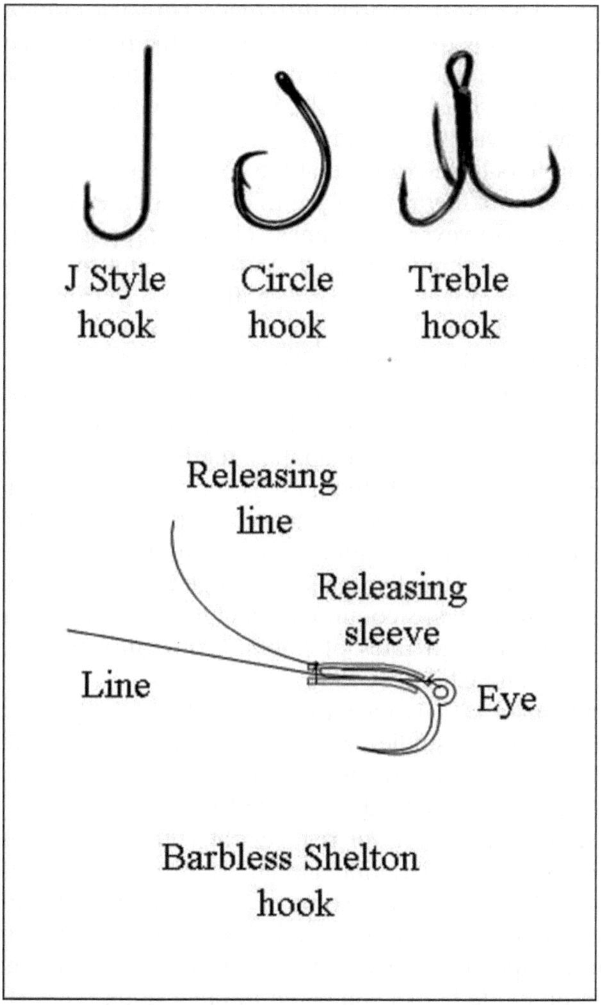

Fig. 7.1. Schematic of different hook types discussed in this chapter. The Shelton self-releasing hook (bottom) enables the angler to grip the releasing line without touching the fish (See Jenkins 2003)

7.3.1.1 Circle versus J Style Hooks

Circle hooks have become popular among recreational anglers in a number of fisheries based largely on the assumption that they reduce hook ingestion and, therefore, mortality of released fish. Owing to their geometry, circle hooks are intended to penetrate and lodge in the jaw, which typically results in fewer mortalities than when hooks are ingested. Unfortunately, the assumption that the widespread use of circle hooks reduces capture mortality has been perpetuated by anecdotal reports with few supporting scientific data. The most obvious difference between circle and conventional 'J-style' hooks is that, with a circle book, the point of the hook is oriented perpendicular to the shank, while in J-style hooks, the point is generally parallel to the shank (Fig. 7.1). In some circle hooks, the point is actually oriented towards the bend.

Some researchers have argued that the configuration of the 'circle' hook design promoted hooking as fish tried to expel bait they could not swallow (Stewart 1977). However, Johannes (1981) proposed an alternative mechanism based on simple physics. As fish attempt to consume a baited circle hook, the fish moves away, or a gentle pressure from the angler pulls the hook to the side of the mouth – thus hooking the fish superficially rather than in the gut. For circle hooks to function effectively, fishers must therefore modify their angling technique. Because circle hooks are used mostly with live bait, the premise is that an angler allows fish to ingest the bait including the hook, and then applies gentle but steady pressure as the hook and fish are reeled in. If the hook is set with the normal vigour used for conventional hooks, the hook either will not capture the fish at all, or is more likely to hook fish at locations that are injurious (such as the roof of the mouth or the eye). Usually, the species that are targeted for circle hook research are those that are commonly captured on live or dead bait and those that exhibited high rates of hooking mortality using conventional hooks (Muoneke and Childress 1994).

Cooke and Suski (2004) recently reviewed existing research on circle hooks in more than 40 studies. Here, we provide a brief summary of the main findings. Mortality, arising both from direct assessment and from projections/estimations, ranged between 0 and 34% for fish caught with circle hooks, and 0 and 46% for fish caught with J-style hooks. Although there was considerable variation among studies, overall mortality rates were consistently lower (approximately 50% lower overall) for circle hooks than J-style hooks (Cooke and Suski 2004). For example, in the United States, striped bass have consistently shown reduced mortality rates when caught on circle hooks compared to other hook types (Caruso 2000; 3% circle, 16% J), Maryland (Lukacovic 1999; 1% circle, 9% J; Lukacovic 2000; 2% circle, 9% J), and North Carolina (Hand 2001; 6% circle, 18% J).

Salmonids exhibited similar patterns with coho salmon (*Oncorhynchus kisutch*; McNair 1997; 3% circle, 24% J) and chinook salmon (*O. tshawytscha*; McNair 1997; 0% circle, 15% J; Grover et al. 2002; 31% circle, 46% J) having reduced hooking mortality rates when caught on circle hooks.

There were also instances, however, in which there were no differences in mortalities between circle hooks and J-style hooks. For example, in Canada, Cooke et al. (2003a) noted no mortality among rock bass (*Ambloplites rupestris*) caught using circle hooks or any of three other conventional hook designs (aberdeen, widegap, baitholder). Cooke *et al.* (2003c) also assessed mortality in bluegill (*Lepomis macrochirus*) and pumpkinseed (*L. gibbosus*) and found that mortality was negligible for all hook types (circle, aberdeen, widegap, baitholder). No mortality was observed for pumpkinseed, and only 1% of captured bluegill died, evenly among the circle hooks and the three other hooks types. Mortality rates were also similar for a study of largemouth bass (*Micropterus salmoides*) in the United States between fish caught on circle and conventional octopus hooks (Cooke et al. 2003b). In a study of summer flounder (*Paralichthys dentatus*), Malchoff et al. (2002) reported that mortality was similar between circle, widegap and sproat hooks.

There is no doubt that in some marine fisheries such as those for tuna, billfish and striped bass, catching efficiency remains high and injury and mortality rates are drastically reduced when circle hooks are used. However, in other freshwater species (e.g., bluegill), injury can actually be more severe from circle hooks compared with other hook types (Cooke et al. 2003c). In species such as largemouth bass, circle hooks have minimal conservation benefit, but have reduced catching efficiencies compared to conventional hook designs. Factors such as hook size, fishing style, fish feeding mode and mouth morphology all appear to affect the effectiveness of circle hooks. For these reasons, it is difficult to promote the adoption of circle hooks as a solution for all fish and fisheries. Instead, we recommend that, as is the case for most gear modifications, management agencies should focus on recommending circle hooks only where appropriate scientific data exist.

7.3.1.2 Shelton Releasing Hooks

A new hook design that shows promise for reducing or eliminating handling is the 'self-releasing' Shelton hook (Jenkins 2003; See Fig. 7.1). These hooks can be removed without handling the fish when the angler pulls on a tag line that activates a release mechanism. In a study of rainbow trout (*Oncorhynchus mykiss*), mortality rates of fish caught on barbless circle hooks that were removed were four times greater than fish caught on Shelton self-releasing hooks (Jenkins 2003).

7.3.1.3 Single versus Treble Hooks

Muoneke and Childress (1994) reported that single hooks tend to be more deeply ingested than treble hooks. However, if treble hooks are ingested, they almost certainly result in massive injury or mortality. In a meta-analysis of salmonids, Taylor and White (1992) failed to demonstrate a difference in mortality between these two hook types. Diodati and Richards (1996) also determined that treble hooks were associated with lower mortality rates than single hooks for striped bass because the latter were more likely to be swallowed, resulting in a greater occurrence of gut hooking. DuBois and Dubielzig (2004) reported that treble hooks hooked and held more brown trout (*Salmo trutta*), rainbow trout, and brook charr (*Salvelinus fontinalis*) than single hooks, but that there were no differences in the frequencies of severe injuries (i.e., in the eye or gullet) or mortalities. Similarly, Jenkins (2003) reported that treble hooks and single baited hooks lodged in the esophagus of rainbow trout at similar frequencies. Conversely, Ayvazian et al. (2002) investigated the effects of different hook designs on hooking injury and mortality of tailor (*Pomatomus saltatrix*) in Western Australia. The authors reported that treble hooks resulted in a significantly greater mortality rate than did other hook types. The authors concluded that their current management strategies, including discouraging the use of treble hooks, should be effective in ensuring the survival of a high proportion of discarded tailor.

7.3.1.4 Barbed versus Barbless Hooks

Using barbless hooks is one of the most common strategies employed to minimise discard injuries and mortalities. They are easier to remove from fish and so reduce the time required to remove hooks (Diggles and Ernst 1997; Schaeffer and Hoffman 2002; Cooke et al. 2001; Meka 2004) and tissue damage at the point of hooking (e.g., Cooke et al. 2001; Meka 2004). Cooke et al. (2001) also evaluated the effects of different handling periods (i.e., short for barbless and long for barbed) on the cardiovascular disturbance of rock bass and revealed that subtle differences in hook removal time translated to significant differences in physiological disturbance. Similar findings have been reported in a marine fishery in the Gulf of Mexico where unhooking times were shorter and injuries were reduced with barbless hooks (Schaeffer and Hoffman 2002). DuBois and Dubielzig (2004) studied stream-caught trout (rainbow, brown and brook trout) and showed that barbless single hooks were quicker to remove than other hook types (treble barbless, treble barbed and single barbed), but the difference was insufficient to reduce mortality.

One perceived concern among anglers associated with using barbless hooks is a reduced hooking efficiency. Schaeffer and Hoffman (2002) compared barbed and barbless hooks in a nearshore marine fishery in the Gulf of Mexico. Bait loss, catch-per-unit effort, and mean length of captured fish did not differ between hook types. However, anglers landed 22% more fish with barbed hooks. DuBois and Dubielzig (2004) and Meka (2004) also noted that anglers using barbed hooks hooked and retained more trout than those using barbless hooks.

Schill and Scarpella (1997) summarised the results of past studies that directly compared the hooking mortality of salmonids caught and released with barbed or barbless hooks. The authors determined that barbed hooks caused less hooking mortality in 2 of 4 comparisons with flies and in 3 of 5 comparisons with lures, however, only 1 of 11 comparisons resulted in statistically significant differences in hooking mortality. The authors concluded that the use of barbed or barbless flies or lures played no role in the mortality of trout caught and released by anglers. In fact, the authors concluded that, because natural mortality rates for wild trout in streams commonly range from 30 to 65% annually, a 0.3% mean difference in hooking mortality for the two hook types was irrelevant at the population level, even when fish were subjected to repeated capture. Others have also suggested that barbless hooks provide little benefit and are really just a 'social issue', generating substantial controversy (e.g., Taylor and White 1992; Schill and Scarpella 1997; Turek and Brett 1997). However, sublethal injuries and physiological disturbance (due to longer handling times) are more extensive with barbed hooks and, for these reasons, barbless hooks can be considered an effective conservation and management tool (Cooke and Suski 2005).

7.3.3.5 Hook Size

Among conventional hook types, the relationship between hook size, fish size and hook performance has varied widely among studies (Muoneke and Childress 1994). Taylor and White (1992) conducted a meta-analysis on factors associated with hooking mortality in salmonids and concluded that hook size did not influence mortality rate. Similarly, Savitz et al. (1995) found no effect of hook size on the mortality of coho or chinook salmon in the Laurentian Great Lakes. However, Carbines (1999) studied the relationship between mortality and hook size in blue cod (*Parapercis colias*) and observed no deaths among fish caught with 6/0 hooks, but noted significant mortality (25%) among those captured with smaller, 1/0 hooks.

Cooke et al. (2005) reported that size may be more important for circle hooks than other hook types. To function properly, the entire circle hook needs to be ingested by a fish prior to 'setting the hook'. This could pose some challenges if the optimal hook size for the targeted fish causes substantial injury in individuals that are released as by-catch. Cooke et al. (2005) caught bluegill on each of five different-sized circle hooks (1/0, 2, 6, 10, and 14). Jaw hooking rates generally increased with decreasing hook size, whereas hooking rates in the roof of the mouth decreased. Gullet hooking was restricted to the three smallest hook sizes. Beckwith and Rand (2005) found similar results for red drum (*Sciaenops ocellatus*) with fewer injuries associated with intermediate- to large-sized circle hooks. Circle hooks function most effectively when the entire hook can fit in the mouth of the fish and when the shank-to-point distance (gape) is large enough to permit jaw hooking (Beckwith and Rand 2005; Cooke et al. 2005).

7.3.3.6 Offset versus Non-offset Hooks

An important consideration with respect to hooks is the degree to which the point is offset from the shank. This is particularly important for circle hooks. Offset hooks would superficially appear to increase the potential for deep hooking and injury due to the exposed point. However, there is contradictory evidence regarding the importance of non-offset hooks for minimising injury and mortality. For example, in a study of striped bass, Hand (2001) compared offset and non-offset circle hooks and determined that offset hooks were more damaging than non-offset hooks. Bleeding and deep-hooking rates were 7 and 13%, respectively, for offset circle hooks compared to 0 and 6% for non-offset circle hooks. In contrast, Lukacovic (2001) concluded that there was no difference in the rate of deep hooking for striped bass between offset (3% all fish and 2% sublegal) and non-offset (2% all fish and 2% sublegal) hooks. Projected mortality rates (based on the degree of injury to vital tissues) for striped bass were also similar for all fish and sublegal fish between offset (1% all fish and 0.4% sublegal) and non-offset (0.6% all fish and 0.6% sublegal) circle hooks. Malchoff et al. (2002) reported that severe offset circle hooks (i.e., 15°), used in their study of summer flounder, may have affected high jaw hooking rates. Due to the inconclusive data regarding the importance of offset versus non-offset hooks, it is difficult to provide any clear management direction on these hook types at this time but, in general, severely offset hooks (i.e., $> 5^{\circ}$) tend to cause more injuries than non-offset hooks.

7.4 Bait

Another important factor is the choice of bait. Artificial lures or flies are highly regarded for superficially hooking fish, with minimal damage to the vital organs or tissues of the fish (Muoneke and Childress 1994). Organic baits, including live bait (but excluding artificial flies), are typically ingested deeper than artificial lures – resulting in more time required to remove hooks and a greater potential for mortality (Siewert and Cave 1990; Cooke et al. 2001).

Since the review by Muoneke and Childress (1994), there have been several comparisons of bait types. For example, Diggles and Ernst (1997) evaluated the effects of different lure and bait types on the hooking mortality of the yellow stripey (*Lutjanus carponotatus*) and the wire netting cod (*Epinephelus quoyanus*). Baitfishing with single hooks caused a significantly greater post-release mortality rate (5%) than did lure fishing with treble or single hooks (0.4%), and was the method most likely to cause bleeding and damage to vital organs. Similarly, Pauley and Thomas (1993) revealed that mortality rates of cutthroat trout were generally greater for fish caught on worm-baited hooks (40 to 58%) compared to those captured on lures (11 to 24%). Conversely, studies of both ling cod (*Ophiodon elongates*; Albin and Karpov 1998) and weakfish (*Cynoscion regalis*; Malchoff and Heins 1997) did not find any differences in mortality between those fish caught on natural baits or those caught on artificial lures.

Studies of flies versus lures and baits have been consistent in that flies tend to be less injurious and have a lower chance of causing mortality. For example, Schisler et al. (1996) compared the hooking mortality of fish caught on flies and lures and determined that mortalities were lowest by several fold for fly-caught fish. Meka (2004) also determined that rainbow trout caught on spinning gear tended to be injured more frequently than fish caught by fly fishing.

7.4.1 Fishing Techniques and Rigging

Although not well studied, angler experience and technique have been shown to be important predictors of catch-and-release mortality for some species. Diodati and Richards (1996) and Meka (2004) reported that mortality among fish caught by more experienced anglers was less than that observed among fish caught by less experienced anglers. Dunmall et al. (2001) found a greater incidence of deeply-hooked smallmouth bass (*Micropterus dolomieu*) among those caught by experienced anglers, which would lead one to expect greater mortality among fish released by experienced anglers.

The manner in which specific baits or lures are rigged and used also affects the mortality of hooked fish. Schisler et al. (1996) observed greater mortality among rainbow trout caught on artificial baits (slip-rigged artificial eggs) that were actively fished than among fish caught with the same bait fished passively. Similarly, Schill (1996) found that the frequency of deep hooking was greater among rainbow trout caught on a 'slack line' than a 'tight line'. The orientation of bait on hooks affected the survival of drift-caught chinook salmon (Grover and Palmer-Zwalhlen 1996), with greater mortalities observed when the bait was hooked with the head down as opposed to upwards. Persons and Hirsch (1994) evaluated hooking mortality for lake trout (*Salvelinus namaycush*) caught through the ice by jigging and by set-lining with dead baits. Seventy percent of the lake trout caught by set-lining were hooked in the gills or gut, compared with 9% of fish caught by jigging. These differences in hooking location were reflected in mortality: 32% for fish captured by set-lining and 9% for jig-caught fish.

Dedual (1996) examined the effects of four different trolling techniques on injury and mortality of rainbow trout. The author reported cumulative mortalities of 15% for fish caught on downriggers, 14% for those caught using wire line, 8% for those captured on lead line, and 2% for fish caught by harling (fly fishing gear trolled near the surface). The differences in mortality were related to the depth of capture and fishing gear: fish caught on downriggers generally were played with lighter lines than were fish caught on lead and wire lines.

7.4.2 Gear Summary

Although we have a reviewed a number of specific gear types and styles, there is a growing body of literature across a variety of species that indicates hooking location is perhaps the single greatest gear-related factor in determining the outcome of an angling event for a fish. For example, anatomical hooking location has been identified as the primary factor determining the mortality of striped marlin (*Tetrapturus audax*; Domeier et al. 2003), yellow stripey (*Lutjanus carponotatus*; Diggles and Ernst 1997), wire netting cod (*Epinephelus quoyanus*; Diggles and Ernst 1997), largemouth bass (Pelzman 1978), rainbow trout (Schisler et al. 1996) and snook (*Centropomus unidecimalis*; Taylor et al. 2001). In fact, because hooking mortality varies with anatomical hooking location, some researchers have developed models to estimate the mortality of spring adult chinook salmon in Oregon (Lindsay et al. 2004). The authors modelled hooking mortality rates for each of five anatomical locations (jaw, 2%; tongue, 18%; eye, 0.0%; gills, 82%; and esophagus-stomach, 67%) using recaptures of tagged fish and from the frequency of these anatomical locations in the sport fishery determined by creel surveys (jaw, 82%; tongue, 5%; eye, 0.4%; gills, 5%; and esophagus-stomach, 8%). This work also estimated total hooking mortality rates of 12% for wild chinook salmon caught-and-released in the sport fishery and 3% for the entire run of wild chinook salmon based on a mean encounter rate of 26%. The question remains as to how different gear types influence hooking location.

Of all current gear developments, only circle hooks have consistently had a demonstrable positive effect on anatomical hooking location (Cooke and Suski 2004; McEachron et al. 1985; Woll et al. 2001). The recent interest in circle hooks has been beneficial for stimulating interest and research on the role of hook designs in reducing hooking related injury and mortality. The challenge is to develop additional hook designs or configurations that reduce or eliminate hooking of vital tissues or deep regions. There is no doubt that other gear-related factors can also be important – such as barbless hooks (in reducing handling and air exposure time), but these do little to alter the location where fish are hooked. To date there have been few novel gear developments that have revolutionised the recreational fishing industry with respect to reducing discard injuries or mortality. This contrasts strongly with the commercial sector where considerable effort has been devoted towards developing gear that reduces by-catch and discard mortality. We encourage tackle manufacturers to continue to develop new hook designs that have the potential to provide conservation benefits to caught-and-released fish.

7.5 Practices

Fishing practices refer to events that are largely under the control of the angler and do not include gear-related decisions. For example, on a seasonal basis, anglers must make decisions regarding if and when they will fish, knowing that water temperature or life-history stages of the targeted species are potentially important factors. In theory, angling practices should be easy to change since they depend on an individual making a change in their behaviour. However, change by anyone, including anglers, takes time and is never as straightforward as one may hope, even when scientific data are compelling.

7.5.1 Fighting Time

There is a general consensus among the current body of catch-and-release research that the duration of an actual angling event experienced by a fish correlates positively with the magnitude of physiological disturbance and the time required for recovery (Kieffer 2000). Angling is essentially a combination of aerobic and anaerobic exercise for fish that results in a series of physiological changes, including a depletion of energy stores and an accumulation of lactate, as well as acid/base changes and osmoregulatory disturbances (Wood 1991).

Evidence supporting the concept that the duration of angling influences the degree of sublethal disturbances can be found for several fish species, and the general physiological processes that result in this response should be consistent for most fishes. Gustaveson et al. (1991) determined that the length of angling duration for largemouth bass (varying between 1 and 5 minutes) was correlated with the degree of physiological disturbance measured by variables such as blood cortisol and plasma lactate. Similar haematological disturbances (increases in plasma lactate and decreases in blood pH) were observed to be correlated with the duration of angling in Atlantic salmon (*Salmo salar*; Thorstad et al. 2003). In a study of smallmouth bass, Kieffer et al. (1995) determined that white muscle disturbance, including increases in metabolites and decreases in energy stores, were more severe in fish angled for 2 minutes than those angled for only 20 seconds. Similar patterns were observed in a marine fish, red drum, where plasma glucose, cortisol, lactate and osmolality all increasing according to the duration of angling (varying between 10 seconds and 6 minutes; Gallman et al. 1999). In addition, striped bass angled for long durations in Maryland also had more severe physiological disturbances (in terms of plasma pH, O_2, and CO_2) compared to briefly-angled individuals (Thompson et al. 2002).

Beyond the magnitude of disturbance, the time needed for recovery can also be prolonged with longer angling durations. For example, Schreer et al. (2001) reported that smallmouth bass exposed to brief simulated angling in a swim tunnel recovered more rapidly than those fish exercised until exhaustion. The heart rate and cardiac output returned to resting values twice as rapidly for briefly-angled smallmouth bass compared to exhaustively-angled individuals. Extended angling duration can also result in death through mechanisms outlined in Black (1958) and Wood et al. (1983). Indeed, Thompson et al. (2002) noted that the mortality of striped bass increased 3-fold when angling duration increased from 1 to 3 minutes at 26°C. Interestingly, at 8°C no mortality was observed when fish were angled for similar durations, highlighting the important role of water temperature and the concept that stressors rarely act alone.

The duration of the angling event primarily depends on the type of tackle used and size of fish caught, but can also be affected by water temperature and habitat (especially depth). Larger individuals within a species may require longer periods of time to land – as observed for Atlantic salmon (Thorstad et al. 2003). In this study, the duration of the angling events ranged from 1 to 49 minutes with fish undertaking between 0 and 10 runs (mean of 3.7 runs). Plasma lactate increased and plasma pH decreased with increased angling duration. A recent study by Meka and McCormick (2005) revealed that plasma cortisol and lactate were greater in large fish that took longer than 2 minutes to land compared to smaller fish that were landed in shorter periods (Thorstad et al. 2003). In addition, Meka (2004) determined that experienced anglers took longer to land fish than novices because they tended to capture larger individuals. Thus, factors such as fish size and angler experience can affect the duration of angling and subsequent physiological responses (Meka and McCormick 2005). In some cases, fish landed rapidly (< 20 seconds) have even been used as 'unangled controls' in physiological studies (Kieffer et al. 1995). Collectively, the trends in the literature point towards increased physiological disturbance and risk of mortality as fish are fought for longer durations. These effects appear to be pronounced when combined with multiple stressors such as high water temperatures. Based on this evidence, we conclude that anglers should attempt to land fish as rapidly as possible to minimise the duration of exercise and the concomitant physiological disturbances.

7.5.2 Landing

The processes of landing a fish and removing the hook present several opportunities for fish to experience injury and sublethal physiological disturbances. Landing the fish is usually accomplished by hand or with the aid of a device (e.g., a landing net, a gaff or a Boca Grip for holding fish by the lower jaw). All of these techniques have the potential to injure fish. Landing fish by hand can result in disruption or removal of the external mucous covering, which may increase the risk of pathogenic infections, especially those associated with fungi. However, some fish such as the centrarchids can be landed safely by gripping the fish by the lower jaw. Commercially-available gripping devices such as the Boca Grip may also be effective for safely restraining large (or toothy) fish.

Although landing nets are widely used, they can be detrimental to fish. A recent study (Barthel et al. 2003) involving freshwater fish determined that the use of landing nets can result in physical injury and increased risk of mortality compared with to that observed in fish landed by hand. In addition, the degree of injury (including dermal disturbance and fin fraying) varies with the type of mesh in the landing net, with knotless nylon and rubber being the least injurious and knotted, large/coarse mesh being the most damaging (Barthel et al. 2003). We are unaware of any studies that explicitly evaluated the effects of gaffing on released fish, presumably because this practice is generally viewed as incompatible with live release.

Decisions regarding how to land fish will be influenced by the species, environment, fishing gear used, etc., but should also include consideration of what will be best for the fish. The key is to restrain the fish sufficiently to enable hook removal and then release it safely without excessive injury.

7.5.3 Air Exposure and General Handling

Among all species of recreational fishes examined thus far, exposure to air is harmful. In recreational fisheries, air exposure occurs after capture when anglers remove hooks, weigh and measure fish, and/or hold fish for photographs and causes hypoxia to the fish. During hypoxia, gill lamellae collapse leading to adhesion of the gill filaments (Boutilier 1990) which cause several major physiological changes. For example, in rainbow trout, blood oxygen tension and the amount of oxygen bound to haemoglobin were lowered by over 80% during brief air exposure, causing severe anoxia (Ferguson and Tufts 1992). Furthermore, those fish exposed to air typically experienced greater acid/base disturbance than those fish that were exercised but not exposed to air (Ferguson and Tufts 1992). Several researchers have also monitored cardiovascular variables for fish exposed to air. Cooke et al. (2001) subjected rock bass to either 30 seconds or 3 minutes of air exposure. When fish were exposed to air for longer periods, all cardiac variables measured (cardiac output, stroke volume, heart rate) took significantly longer to return to base levels. Similar studies on smallmouth bass (Cooke et al. 2002b) determined that the duration of air exposure was correlated with the time required for cardiovascular variables to recover. Extended exposure to air eventually results in permanent tissue damage beyond some threshold. Mortality rates can also be increased by exposing fish to air. Short-term mortality (12 hours) was negligible for control rainbow trout, and low for trout that were exercised to exhaustion but not exposed to air (12%; Ferguson and Tufts 1992). When trout were exposed to air for either 30 or 60 seconds following exhaustive exercise, mortality increased to 38 and 72%, respectively. In a recent study, the swimming performance of brook trout was not impaired following short duration air exposure (e.g., less than 60 sec; Schreer et al. 2005). However, exposure to air for 2 minutes led to swimming performance being reduced by 75%.

Based on these studies, it appears that air exposure, especially in fish that have experienced physiological disturbances associated with angling, can be extremely harmful. Although different fish species will vary in their sensitivity to air exposure, we recommend that whenever possible, anglers attempt to eliminate air exposure by handling fish that are to be released in the water. When fish must be exposed to air, we urge that anglers do everything possible to minimise the duration of air exposure.

The manner in which fish are held, particularly for photographs or other displays, has implications for post-release survival and well-being. Although undocumented in the literature, there are numerous anecdotal accounts of large largemouth bass and other fish, when held by their lower jaws without additional support, sustaining debilitating injuries including broken jaws.

7.5.4 Hook Removal

As discussed above, fish hooked deep in the esophagus or stomach have an increased chance of mortality. This increased mortality has been attributed to the nature and severity of hooking wounds and to the difficulty of, and increased handling time attributed to, removing ingested hooks. Consequently, there has been some discussion as to whether it is better to remove, or leave in place, ingested hooks. Diggles and Ernst (1997) left hooks in one specimen each of two Australian reef fishes, captured on bait, and hooked in the gut or esophagus. In both instances, the fish survived and subsequently regurgitated the hook during the observation period. Removing hooks was found to result in increased mortality among brown trout (Hulbert and Engstrom-Heg 1980) and red drum (Jordan and Woodward 1994). In the latter study, there was little difference in mortality due to hook removal, among fish hooked in the esophagus (41% if hook was removed and 50% if hook was left in); however, among fish hooked in the gills, mortality was greater if hooks were removed (79%), than if they were left in place (54%).

A number of studies have presented evidence that leaving hooks in place increases the survival of deeply-hooked fish. Schill (1996) determined that cutting the line on deep-hooked rainbow trout, rather than removing the hook, reduced post-release mortality by 36% in a hatchery setting and 58% among wild-caught fish. Among surviving fish in which the line was cut, hooks were shed by 74% of the hatchery fish and 60% of wild-caught rainbow trout during the two-month study period. Similarly, Schisler and Bergersen (1996) reported 55% mortality among rainbow trout when the hook was removed by hand and only 21% when the hook was not removed. Among surviving fish in which the hook was not removed, 25% of fish shed their hooks during the 3-week observation period. Taylor et al. (2001) removed hooks from 12 deeply-hooked common snook and left hooks in place in another 12 fish. Mortality was 25% among fish from which hooks were removed and 0% among those in which the line was cut. This difference was not statistically significant, however, which Taylor et al. (2001) attributed to low statistical power.

The above studies have a common limitation. They were not specifically designed to examine the mortality associated with hook removal and, consequently, suffer from small sample sizes and little power. Wilde and Sawynok (unpublished manuscript) examined capture and recapture records (n = 248,010) for 27 species of Australian fishes collected as part of a large cooperative angler-tagging program. Anglers left hooks in 1% of released fish and the overall recapture rate, across species, was 9%. Wilde and Sawynok (unpublished manuscript) used relative risk, the probability of an event (recapture) in one group (fish with hooks not removed) divided by the probability of an event in a second group (fish with the hook removed), to assess the potential effects of leaving hooks in released fish. Relative risk did not differ significantly from 1.0 for any species; thus there was no evidence that hook removal affected recapture probability. Pooling results across all species and habitats yielded an overall relative risk of 1.18 (with a 95% confidence interval of 1.02 to 1.36), which suggests that survival was 18% greater, on average, among fish in which hooks were not removed. Wilde and Sawynok (unpublished manuscript) concluded there was no clear benefit to removing hooks from deeply-hooked fish and recommended that anglers use their best judgment in when deciding whether to remove hooks.

If hooks are removed, using de-hooking tools may help to reduce mortality. Survival of fish from which anglers remove hooks also can be increased by educating anglers in best practices. Meka (2004) noted that training was required to promote proper hook removal techniques to minimise injury and that even barbless hooks can injure fish if not removed properly.

7.5.5 Short-Term Retention (Fish Baskets and Keep Nets)

Catch-and-release angling sometimes involves the retention of fish for a period of time (usually hours) prior to release as anglers assess whether they will harvest individuals, or in competitive events when fish are retained for later enumeration at a weigh-in. Professional anglers often hold fish in aerated live-wells, whereas recreational anglers commonly use more affordable, readily available and convenient methods, including fish baskets and keep nets. Research has investigated the effects of keeping fish in keep nets on the growth, survival (Raat et al. 1997), stress response and recovery (Pottinger 1997 1998) of various cyprinid species. Additional research has focused on changes in water quality in keep nets during retention (Pottinger 1997). Collectively, these studies suggest that retention is stressful to fish, but if provided with adequate water quality, mortality and sublethal disturbances are minimised. Cooke and Hogle (2000) compared 6 retention methods on smallmouth bass for 3–5 hour periods: metal stringer through the lip, metal stringer through the gill arch, cord through the lip, cord through gill arch, wire fish basket and nylon keep net. Control fish exhibited very little mortality (3%) and had negligible physical injury in all sampling periods. Most (95%) fish retained experienced some form of injury or mortality. In general, injury and mortality increased with high water temperatures. Survival and injury varied among retention methods, but gill damage or fungal lesions associated with abrasion, and the cumulative stress of angling and retention appeared to be the precursor to most deaths. Details on live-wells (vessels for holding fish in water aboard a boat) are provided in the case study on black bass below.

7.5.6 Fishing Locations and Environment

The habitat where fish reside, and the environmental conditions faced by fish at the time of angling, can also affect the outcome of angling events. Here, we briefly discuss the role of these factors (i.e., water temperature, oxygen, water hardness, depth, salinity, other habitat features and predation).

7.5.6.1 Water Temperature

In species for which data exist across a gradient of water temperatures, angling at extreme water temperatures (especially high) is correlated with increased physiological disturbances and the probability of mortality. This is not surprising considering that beyond some thermal optima, fish performance is constrained (e.g., Farrell et al. 1996; Schreer et al. 2001; Farrell 2002). Since fish are poikilothermic, changes in ambient water temperatures are realised throughout the animal, and can have pronounced impacts on cellular function (Prosser 1991), protein structure (Somero and Hoffman 1996), enzyme activity, diffusion rates and metabolism (Fry 1971).

There are many examples in temperate recreational fisheries where temperature has been identified as an important determinant of the degree of sublethal disturbance and mortality (See Muoneke and Childress 1994). For example, mortality among Atlantic salmon was reduced when fish were caught in water temperatures between approximately 8 and 18°C, but as water temperatures increase above 18°C, the risk of angling-induced mortality increased exponentially (Thorstad et al. 2003). Similar patterns were observed for largemouth bass captured in fishing tournaments, for which there was a strong positive correlation between water temperature and mortality (Wilde 1998). Underlying the mortality of Atlantic salmon at high temperatures are limitations in maximal cardiovascular performance as fish approach their maximal metabolic rate (Anderson et al. 1998) and extreme biochemical alterations (Wilkie et al. 1996). Wilkie et al. (1997)

determined that whereas warmer water may facilitate post-exercise recovery of white muscle metabolism and acid-base status in Atlantic salmon, extremely high temperatures increased their vulnerability to mortality. Greater oxygen debt may also be correlated with higher water temperatures (McKenzie et al. 1996). In tropical marine fish, most studies have been conducted at moderate temperatures and thermal relationships are not as obvious (e.g., there was no effect of minor changes in water temperature on hooking mortality of snook, Taylor et al. 2001).

Catch-and-release angling at extremely cold water temperatures has also been suggested as potentially challenging to fish. However, Persons and Hirsch (1994) concluded that the lack of mortality for lip-hooked lake trout captured under ice suggested that catching and handling fish in cold (i.e., sub-zero) temperatures had little effect on mortality.

Individual species exhibit different thermal tolerances (Beitinger et al. 2000) and this must be considered for each species, population and location. However, there is a period of the year where water temperatures are at their highest, and it is during this period that catch-and-release angling has the potential to be particularly harmful. Under these scenarios, if anglers do continue to fish, both the duration of the fight and handling time should be minimised. Because water temperature influences most physiological processes in fish (Fry 1971), extreme water temperatures lead to fish being particularly susceptible to mortality. Ideally, fishing should be restricted during such periods of extreme water temperature.

7.5.6.2 Oxygen

Temperature is also negatively correlated with oxygen availability. At present, we are unaware of any studies that evaluate the role of low dissolved oxygen in the natural environment on caught-and-released fish. However, there are several studies that have revealed the importance of providing fish with adequate water quality during live well retention to minimise the lethal effects of hypoxia (e.g., Hartley and Moring 1995; Furimsky et al. 2003).

7.5.6.3 Water Hardness

A recent study examined the influence of environmental water hardness (40 mg/L versus 100 mg/L $CaCO_3$) on the physiology and survival of exhaustively-exercised Atlantic salmon (Kieffer et al. 2002). In softer water, exhaustive exercise caused a significantly greater elevation in post-exercise blood lactate concentrations and a larger acid-base disturbance compared with fish caught in hard water. Post-exercise survival of Atlantic salmon in softer water was directly related to environmental water hardness, and those that succumbed failed to exhibit any post-exercise correction of their extra-cellular acid-base disturbance. In contrast, all fish captured in hard water survived.

7.5.6.4 Depth

When brought to the surface rapidly, the gasses in swimbladders of physoclistous fish rapidly expand to the point that the fish are unable to achieve neutral buoyancy, maintain equilibrium, and may even have their stomachs protruding from their mouths or anus (because of the expanded swimbladder pushing out the viscera; Burns and Restrepo 2002). The fish may also experience embolisms and blood-gas disturbances (Morrissey et al. 2005). Different species respond to capture at depth differently and each also has their own threshold regarding which depths are problematic. For example, depth of capture was the major source of mortality in a study of pink snapper in Australia (St John and Moran 2001). Mortality after 3 days increased with depth, but not linearly. Mortality was drastically lower at the shallow sites (4% at 15 m and 7% at 30 m) than at the deeper sites (71% at 45 and 84% at 65 m).

One obvious, but draconian, option for anglers to avoid these problems is to not fish in deep waters. However, an alternative solution can involve anglers venting the swimbladder with a needle to release the gas and enable the fish to swim back to depth (Keniry et al. 1996; Collins et al. 1999; Kerr 2001, Burns and Restrepo 2002). However, St John and Moran (2001) found that such venting failed to reduce mortality. This research topic requires more work before definitive answers can be provided.

7.5.6.5 Salinity

There are few studies, and none that have been done in a quantitative manner, that have evaluated how salinity affects either fish mortality or sublethal impairments. For example, Gallman et al. (1999) evaluated responses to exercise across salinity values of 17 to 33‰ but did not include salinity as a factor in analyses. However, research with striped bass (Diodati and Richards 1996) and red drum (Jordan and Woodward 1994) suggests that for marine species, survival is related to salinity, presumably because fish that were caught, handled and maintained in isotonic environments were under less stress than those similarly handled in less dilute environments.

7.5.6.6 Habitat Features and Predation

The habitats where fish are caught and released may also affect their ability to survive a catch-and-release angling event. For example, Schill (1996) concluded that stream locations where bait anglers catch fish and the general habitat characteristics of a stream could influence bait-related hooking mortality based on empirical data on rainbow trout. Similar data do not exist for other systems or fisheries.

The habitat where fish are released can influence exposure to predators. For example, mortality rates of bonefish in the Bahamas exceeded 40% due to post-release predation by sharks (Cooke and Philipp 2004). However, mortality was related to the density of sharks. Survival was greater in areas with few sharks and in complex, shallow mangrove habitats. Whilst shark predation was greatest in areas with significant densities of predators, these also tended to be areas near deep water and with little cover. Edwards (1998) observed limited predation on caught-and-released tarpon (*Megalops cyprinoides*) but suggested that the predation was associated with the use of light tackle that resulted in severe exhaustion, which thus made fish more susceptible to predation. In a study of cichlids in the Zambezi River, Africa, Thorstad et al. (2004) reported that catch-and-release angling may increase the risk of predation before recovery: they located a telemetry transmitter from a tagged threespot tilapia (*Tilapia andersoni*) under a tree used as a roost by an African fish eagle (*Haliaeetus vocifer*).

7.5.7 Seasonality, Sensitivity and Biologically-Intrinsic Factors

There are a number of factors other than those discussed above, that can affect the fate of discarded fish. For example, in addition to temperature, different seasons may also affect the sensitivity of fish due to their reproductive status. Lowerre-Barbieri et al. (2003) determined that common snook subjected to catch-and-release angling did not immediately leave a spawning aggregation and there was no obvious negative consequences of angling during this period. Brobbel et al. (1995) compared the physiological response to angling in Atlantic salmon at two different stages of migration (kelts and bright salmon). This demonstrated large differences in the degree of physiological disturbance that derived from angling in these two migratory stages, as well as differences in angling-induced mortality.

Mortality (Meals and Miranda 1994; Thorstad et al. 2003) and physiological disturbance (Kieffer 2000) can also vary with the size of individuals of the same species – larger individuals generally experience more substantial physiological disturbance. In addition, the gender of individual fish may also play an important role, but there are few tests of that supposition. In fact, there has been insufficient research on all of these topics and work needs to continue on these (i.e., sex, life-stage, reproductive status, etc.) and other factors that are not typically considered.

7.5.8 Expediting Recovery

Recent research, primarily focused on salmonids, indicates that, after capture and exposure to air, slow speed swimming can expedite recovery (Milligan et al. 2000). This knowledge is being applied to reduce by-catch mortality of fishes captured in the commercial troll-fishery for salmonids and to facilitate the recovery of tournament-caught largemouth bass (Cory Suski, unpublished data). Live-wells used in freshwater fishing tournaments were once regarded as stressful but, if provided with adequate water quality and if fish are kept at low densities, some fish can actually recover tissue energy stores and reduce cardiac output (Cooke et al. 2002b; Suski et al. 2004).

7.6 Case Studies

In addition to the above review of the many factors associated with by-catch in recreational fisheries, we felt that it would be useful to present three brief relevant case studies. The species covered in these case studies are very different (Atlantic salmon, marine pelagics and black bass) and provide an opportunity to explore specific issues further (e.g., predation, hook technology, stress, fitness impacts) using these well-studied species.

7.6.1 Atlantic Salmon Case Study

Atlantic salmon (Fig. 7.2A) are a highly-valued recreational species in North America and Europe that has experienced population declines that have been partially attributed to recreational fishing mortality. Recognising the importance of Atlantic salmon and its sensitivity to fishing-induced mortality, much effort has been devoted to catch-and-release research for this species. This species is an appropriate model for a case-study due to the high levels of mortality that are believed to occur after angling. Indeed, there are many jurisdictions where strict management regulations have been applied that require the release of some or all of these fish (e.g., O'Connell et al. 1992). Here we present a brief case-study on the catch-and-release of Atlantic salmon focusing on the effects of two issues on post-release survival: migratory disruptions and thermal effects. Although there are a number of other issues such as handling, air exposure, use of barbless hooks and type of bait, migratory disruptions and thermal effects are particularly important for Atlantic salmon. More detailed evaluation of all catch-and-release issues facing Atlantic salmon may be found in DFO (1998) and Tufts et al. (2000).

Like many other species, Atlantic salmon tend to be targeted during their migration. As such, research efforts have focused on the consequences of angling these fish during migration, particularly en route to spawning grounds. For example, Whorisky et al. (2000) studied the effects of catch-and-release fishing on Atlantic salmon in Russia's Ponoi River. This highly-developed salmon sport fishery has estimated angler exploitation rates of between 10 and 19%, resulting in concern for the sustainability of this activity. The authors determined that released salmon had high rates of survival, and anglers recaptured about 11% of released fish per year. The authors also held 62 angled fish for 24 hours in a cage to evaluate rates of delayed mortality. Only one of the 62 fish died, and it was heavily scarred with gillnet marks. Approximately 10% of released fish were angled and released twice, and about 0.5% were angled and released three times. No significant biases were detected in the post-angling movement patterns of these fish. The multiple captures and lack of differences in movement patterns suggest that fish behaviour was altered little by the angling experience.

Conversely, in Finland, when subjected to catch-and-release, Atlantic salmon migrating upriver actually moved downstream after release causing delays in their net upstream migration (Makinen et al. 2000). Thorstad et al. (2003) conducted a study on catch-and-release of adult Atlantic salmon in a Norwegian river. At intermediate water temperatures (10–14.5°C), a high proportion of the radio-tagged salmon (97%) survived hook-and-release and stayed in known spawning areas during the spawning period. However, behaviour after release was altered by the angling event. The authors attributed increased playing time, increased number of runs during the angling event, hooking in the throat, bleeding at the hook wound, increased handling time, air exposure and water temperature to contributing to a cumulative negative effect. Dempson et al. (2002) evaluated the effects of catch-and-release angling on survival of Atlantic salmon at Conne River, Newfoundland. The authors determined that, overall, 8% of salmon caught-and-released died, and mortality rates increased to 12% at water temperatures greater than, or equal to, 17.9°C. Interestingly, there were no significant differences between salmon that survived or those that died due to the time associated with angling, exposure to air, tagging, transfer to holding cages, nor total handling time.

Wilkie et al. (1997) determined that while warmer water may facilitate post-exercise recovery of white muscle metabolism and acid-base status in Atlantic salmon, extremely high temperatures increased their vulnerability to mortality. More recently, Anderson et al. (1998) used heart rate telemetry to evaluate the response of Atlantic salmon in Newfoundland. Heart rate, after angling, was found to increase but not vary with temperature, but the magnitude of the increase was similar among temperatures. Time to recovery was assessed as the return to the observed resting heart rate for each individual fish and was found to be similar for both the 8°C and 16.5°C angled groups (approximately 16 hours). However, approximately 80% of the fish died at the higher water temperature.

Collectively, data for Atlantic salmon indicated that they tend to be sensitive to angling during their spawning migration. In part, this is due to the stress associated with migration and reproduction. However, the more pervasive factor appears to be water temperature which can be lethal when combined with exercise and the stress associated with recreational angling. In eastern Canada, Atlantic salmon rivers are temporarily closed to recreational angling during excessive water temperatures (Department of Fisheries and Oceans 1998).

7.6.2 Marine Pelagics

Marine pelagic recreational fish (Fig. 7.2B) include some of the most iconic, yet imperiled, marine ichthyofauna, owing in part to their large size and value, but also to their low reproductive output. Examples of important recreational marine pelagic species include marlin, sailfish, tuna and sharks. Despite their diffuse distribution throughout the world, these fish have become the frequent target of certain recreational fisheries. Here, we explore several catch-and-release issues that are particularly relevant to marine pelagics. Specifically, we discuss developments in hook technology (i.e., circle hooks) and issues associated with stress and predation.

For years, hooking mortality rates in marine pelagics were assumed to be high due to frequent deep hooking (Muoneke and Childress 1994). Recent developments in gear technology have resulted in efforts to assess the role of circle hooks in potentially reducing injury and mortality. For example, Prince et al. (2002) determined that Pacific sailfish caught on J style hooks were 21 times more likely to experience bleeding than those hooked with circle hooks. Atlantic bluefin tuna (*Thunnus thynnus*) also had reduced mortality rates when circle hooks (4%) were used instead of conventional J hooks (28%; Skomal et al. 2002). A similar study on white marlin (*Tetrapturus albidus*) revealed that circle hooks resulted in no mortality, but J hooks resulted in 35% mortality (Horodysky and Graves 2005). Domeier et al. (2003) used satellite archival tags to assess the effects of catch-and-release angling on striped marlin (*Tetrapturus audax*) using live bait. The authors compared circle hooks and J hooks and determined that circle hooks were equally effective in hooking and landing striped marlin and far less likely to cause bleeding or deep hooking. Also, non-offset and 5° offset circle hooks had very similar performances (unlike the findings of

Prince et al. 2002). Depth and temperature records allowed the authors to assess the fate of individual marlin following release. All mortality (26%) occurred within 5 days of release with injury being the best predictor of mortality; all of the fish that were bleeding from the gill cavity died, and 63% of deeply-hooked fish died. Data generated by these studies on circle hooks are currently being used to develop angling guidelines and legislation.

When targeting powerful and toothy marine pelagics, it is common for fish to escape capture by biting off the line. Borucinska et al. (2001) found that a retained fish hook in a single blue shark (*Prionace glauca*) led to peritonitis and pericarditis. This was the first documentation of a pathogenic affect of a retained fish hook. In a more exhaustive survey, Borucinska et al. (2002) found retained fishing hooks from previous capture events in 6 of 211 blue sharks off Long Island New York. The hooks were embedded within the esophagus or perforated the gastric wall and lacerated the liver. Collectively, tissue damage led to lesions including esophagitis, gastritis, hepatitis and proliferative peritonitis. Because circle hooks tend to be hooked more superficially (e.g., in jaw tissue), they may help to reduce the chance of internal damage.

The capture of large marine pelagics is stressful to the fish as the duration of the fight can be more than an hour. Wild kahawai (*Arripis trutta*) exhibited immediate increases in muscle and plasma lactate after angling while cortisol peaked about 1 to 2 hours later (Davidson et al. 1997). Lowe and Wells (1996) studied both primary and secondary stress responses to line capture in blue mao mao (*Scorpis violaceus*). They also noted increases in lactate and cortisol that correlated to the increase in time after angling and intensity of the exercise intensity. Marine pelagic fish including bluefin tuna, yellowfin tuna (*Thunnus albacares*), blue shark and white marlin are angled for long durations (up to 1 hour) and usually experience pronounced acedemia and high plasma lactate that increase with the duration of angling (Skomal and Chase 2002).

When marine pelagics are in poor condition, they tend to become targets of predators. Jolley and Irby (1979) noted that an Atlantic sailfish (*Istiophorus albicans*) released after angling, and which had an eye injury from the hook, was attacked by a shark within 6 hours after release. Horodysky and Graves (2005) determined that the mortality of white marlin in the western north Atlantic occurred between 10 minute and 64 hours after release. Pepperell and Davis (1999) evaluated the post-release behaviour of black marlin (*Makaira indica*) off the Great Barrier Reef in Australia using acoustic telemetry and observed that 5 of 6 tagged fish survived, but one was attacked and killed by a shark. Graves et al. (2002) found that at least 8 of 9 blue marlin tagged off Bermuda survived the monitoring period (assessed with satellite tags), but provided no information on whether or not the dead animal was eaten.

Fig. 7.2. Photographs illustrating key fishes discussed in the three case-studies including (a) a trophy Atlantic salmon being released by a fly angler in Norway (photo credit, Eva Thorstad), (b) a white marlin jumping during the angling event in the Atlantic Ocean (photo credit, Greg Skomal), and (c) largemouth bass tournament in Ontario, Canada

7.6.3 Black Bass

Black bass (*Micropterus* spp.), including the largemouth and smallmouth bass, represent some of the most popular recreational fish in North America. Black bass are also the frequent focus of competitive angling events, most of which are catch-and-release. Since the inception of catch-and-release angling in 1954, regulations mandating the release of black bass and numerous other fish species have increased (Barnhart 1989). Even voluntary catch-and-release angling has increased substantially (Quinn 1996). Due to their popularity and the frequency with which they are released, black bass are among the most well-studied species with respect to post-release survival. Here, we present a brief case-study on black bass, emphasising several unique characteristics of their fishery including competitive angling events and the consequences of catch-and-release.

The first unique characteristic of the black bass is that they are extremely popular targets of many competitive angling events in North America (Fig. 7.2C). A survey done by the American Fisheries Society's Competitive Fishing Committee over a decade ago estimated the number of inland and marine events to be 31,000 annually, of which 73% targeted black bass (Schramm et al. 1991a). Most of these events are catch-and-release but require that fish be held in live-wells for extended periods of time until the fish are brought to the weigh-in prior to being released (Holbrook 1975). Recently, Wilde (1998) conducted a meta-analysis of tournament-associated mortality of black bass and found that it was important to measure both initial and delayed mortality to determine total mortality accurately. Delayed (hours to weeks) mortality rates were highly variable within and among black bass species, 0–77% for largemouth bass and 0–47% for smallmouth bass.

The increasing popularity of competitive angling and concerns for its effects on released fish (Schramm et al. 1991b) have spurred recent research into improving competitive angling events to reduce post-release mortality fish and other sublethal effects. Suski et al. (2003) revealed that black bass were stressed (e.g., blood and white muscle disturbances) when sampled at tournament weigh-ins, suggesting an opportunity to improve such effects. Cooke et al. (2002b) and Suski et al. (2004) determined those components of a competitive angling event which are most stressful to the fish. When angled, fish exhibited alterations in blood and muscle biochemistry, but after confinement in a live-well, the fish had actually recovered. However, when these fish were exposed to weigh-in procedures, stress indicators again became elevated and in tournaments, fish are generally released immediately after the weigh-in. The authors showed that maintaining good live-well conditions are essential to enhance the survival of released fish.

As mentioned, the effects of retaining black bass in live-wells (Plumb et al. 1988; Hartley and Moring 1993; Steeger et al. 1994), as well as other tournament procedures may alter survival (Weathers and Newman 1997) or invoke sublethal stress (Cooke et al. 2002b; Suski et al. 2004). Water quality parameters like temperature and dissolved oxygen in live-wells have been consistently deemed as important influences on mortality (Carmichael et al. 1984). Meals and Miranda (1994) studied pre-release mortality at major fishing tournaments on Sardis Reservoir, Mississippi, and found that mortality increased with water temperature and the mean number of fish per boat. In contrast, Schramm et al. (1985) concluded that largemouth bass mortality during tournaments depended more on poor live-well management practices (e.g., not aerating frequently) by some anglers than on the density of fish in live-wells. Cooke et al. (2002b) found that the cardiac and locomotory activity of smallmouth bass was elevated when fish were first put into live-wells after angling. When only one fish was in the live-well, cardiac activity slowly decreased. However, when fish were held at greater densities (i.e., 2, 4 or 6 fish), cardiac variables remained high. Clearly, not only does high fish biomass and density lead to greater demands on available oxygen (as evidenced by increased cardiac output and heart rate), but it may also affect the activity levels of fish as they interact with each other. Interestingly, Furimsky et al. (2004) determined that smallmouth and largemouth bass had different hypoxia sensitivity providing an explanation for the greater tournament mortality rates observed for smallmouth bass.

Water conditioners and antibacterial treatments have been added to live-wells in an attempt to reduce the mortality and stress of fish during confinement. Research has provided contradictory conclusions as to the effects of water conditioners on black bass. Plumb et al. (1988) reported that adding a commercially-available water conditioner to live-wells enhanced survival, but Cooke et al. (2002b) determined that commercial live-well conditioner and salt delayed the recovery of smallmouth bass compared to fish that were held in unmodified water. Most research has suggested that antibiotics do not improve survival (Plumb et al. 1975; Seidensticker 1975; Schramm et al. 1987), although Welborn and Barkley (1974) and Archer and Loyacano (1975) reported improved survival rates when antibiotics were used. Hartley and Moring (1993) recommended continuous aeration in live-wells and this is consistent with the findings from other physiological analyses (e.g., Cooke et al. 2002b; Suski et al. 2004).

Since the reproductive period of a species is essential for generating offspring for subsequent populations, it is logical to do everything possible to minimise sublethal stress during this phase. Black bass are perhaps the best studied species with respect to the sublethal effects of catch-and-release angling. Black bass provide sole male parental care and evidence suggests that when nesting males are angled from the nest, even for a short period, the unprotected offspring are quickly consumed by predators (Philipp et al.

1997). Further, even if fish are released after angling, Cooke et al. (2001) determined that when nesting males return to the nest, they exhibit impairments in locomotory activity for over 24 hours. Suski et al. (2003) also found that angling reduced the level of care provided to offspring by the attending male. Ostrand et al. (2004) determined that largemouth bass exposed to a simulated fishing tournament immediately prior to the spawning period produced fewer and smaller offspring than control fish. In the aquaculture-based literature, there is overwhelming data suggesting that salmonids exposed to acute and chronic stressors exhibit endocrine alterations that depress fitness (Campbell et al. 1992). Similar hormonal changes have been shown to occur in largemouth bass and walleye following bouts of angling (Suski et al. 2003), but the extent to which these hormonal changes can affect fitness have yet to be explored. At other times of the year, it is less clear if such effects on fitness occur. Pope and Wilde (2004) used growth as an indicator of fitness and found no differences among angled fish and controls. Conversely, Siepker (2004) found that simulated tournaments led to reduced food intake by black bass. Bioenergetic simulations suggest that this would result in long-term reductions in growth which would be contrary to most fisheries management objectives.

In some states in the north of the United States and several Canadian provinces, seasonal closures are used to restrict angling and/or the harvest of black bass during their reproductive period (Quinn 1993). However, in some jurisdictions, catch-and-release angling for nesting bass is permitted. Compliance with such regulations has been observed to be minimal in many areas (Schneider et al. 1991; Kubacki 1992; Philipp et al. 1997), probably because anglers often assume that as long as the fish are released, they will return to the nest and raise a successful brood.

7.7 Summary and Synthesis

Recreational fisheries share four important characteristics with commercial fisheries: (i) fish are harvested and removed from the population; (ii) non-targeted species and sizes of fishes are captured and subsequently discarded, of which varying proportions of fish survive; (iii) the gears and fishing practices used by fishers have a substantial influence on the nature and magnitude of such survival; and (iv) the interests of all stakeholders are best served by a high survival rate among released fishes, although this must be considered within the context of the economic and social costs of various release practices and/or regulations. Simply put, both commercial and recreational fisheries seek to maximise economic, social and biological goals. There are, however, fundamental differences between recreational and commercial fisheries. In particular, the most important goal of commercial fisheries is to maximise economic returns, whereas in recreational fisheries, the most important goal generally is to maximise social, or psychological, returns. This simple difference has profound implications for how these fisheries are managed and how they are scientifically investigated.

Given the value of commercial fishes and the economic gains derived from their harvest, it is relatively easy for management agencies to determine fees and regulations governing the harvest of commercially-important species. In the case of recreational fisheries, however, managers are reluctant to disaffect anglers, because there is no latent source of participants. Therefore, recreational fishery managers, particularly in fresh waters, historically have been hesitant to restrict angling gears and behaviours. Instead, they have relied on the slow, voluntary adoption of improved gears and techniques among anglers.

In the extreme, this approach can fail as is best illustrated by competitive fishing events for black bass in the United States. Wilde (1998) reviewed studies of the mortality of fishes caught and released in black bass tournaments over the 30-year period between 1965–1995. Few estimates of mortality were available prior to 1975, but he found no change in the initial (or total) mortality of released fishes from the 1980s and the 1990s. However, in a subsequent study, Wilde et al. (2002) reported that in tournaments conducted by the Bass Anglers Sportsman Society (B.A.S.S.), initial mortality had shown a dramatic decrease. In the case of B.A.S.S. tournaments, where a direct economic value is derived from the recreational fishery and where there is a real desire to minimise effects of by-catch, adoption of various handling improvements was rapid and effective. In contrast, in the case of the general recreational fishery, in which fishery managers have suggested the possibility for improvement, none was observed. Recent efforts have attempted to develop a general understanding of catch-and-release that can be broadly applied to most species (Cooke and Suski 2005; Box 3).

The second obvious difference between commercial and recreational fisheries is that, in the latter, there is an incentive for, and indeed often a high incidence of, voluntary release of harvestable fish. This difference may not alter how we regulate fisheries, or how we minimise effects of a fishery, but it does affect the need for leadership by management agencies. In many recreational fisheries, anglers have often assumed the leadership role, by default, and directly affect fisheries by their individual decisions to keep or release fish. Management agencies have often adopted a role of setting fishery regulations that have been shown to have little direct effect on fishery characteristics (e.g., Wilde 1997).

Box 3: Proposed general strategies for minimising the effects of catch-and-release angling

In a recent paper, Cooke and Suski (2005) explored the need for developing species-specific guidelines for catch-and-release angling. They reasoned that there were some generalities that could be derived from existing studies that could be broadly applied to most species. However, they also cautioned that the diversity in the function and form of fish, and the techniques used by anglers for different species, requires some level of specialised guidelines. In the coming years, we suggest that there will be a greater need for species-specific guidelines to reduce mortality and sublethal disturbances further. Until then, Cooke and Suski (2005) provided five generalisations based on research conducted to date on catch-and-release that should be applicable to virtually any catch-and-release fishery. These generalisations should reduce the application of inappropriate data from one species to another and include: (i) the duration of the angling event increases the physiological disturbance; (ii) air exposure is harmful to fish and should be minimised; (iii) excessive water temperatures magnify the level of disturbance and angling should be avoided at those temperatures; (iv) barbless hooks and artificial lures or flies can greatly reduce handling time, hooking injuries, and the likelihood of mortality; and (v) angling immediately prior to, or during the fish's reproductive period can affect fitness and should be avoided.

Beyond these five generalisations, there are few others that can be broadly applied to catch-and-release fisheries. Data in support of each of these generalisations is presented in this review and in Cooke and Suski (2005).

Despite some fundamental differences (e.g., in why fish were released), Cooke and Cowx (2006) concluded that by-catch/discard mortality issues were similar in both the recreational and commercial fisheries sectors. Our synthesis reveals that there are a number of changes that can be made with respect to gear and practices that could benefit recreational fishes and these same opportunities also exist in the commercial sector. Indeed, many of the stresses affecting fish that are associated with recreational and commercial fisheries are identical, such as handling and air exposure (Alverson 1998; Davis 2002; Cooke and Suski 2005). Because both sectors have the common goal of returning more unwanted fish alive after capture and handling (Hall et al. 2000; Cooke and Suski 2005), Cooke and Cowx (2004) suggest that it is intuitively apparent that progress could be gained from common research programs. Indeed, efforts to solve by-catch problems in the recreational fisheries may be best served by pursuing catch-and-release research in a systematic fashion using the framework developed for commercial by-catch reduction by Kennelly and Broadhurst (2002; See Box 4 for a modified framework specific to recreational fisheries). The framework that we have modified for catch-and-release recognises the important role of the angler and the angling industry to ensure that the research is relevant to the recreational angling community.

Box 4: Framework for solving by-catch problems in the recreational fishery. Modified from Kennelly and Broadhurst (2002)

Identify and quantify the problem

- catch, harvest, by-catch, discard, effort and angler behaviour studies done in the field using creel surveys, angler diary programs, log books, etc.
- empirical mortality studies done via observation and experimentation

Identify species/fisheries of concern

- link fishing mortality with other population/community parameters to identify critical issues of concern
- determine required rates of reduction to ensure sustainability using modeling approaches

Develop modifications (in gear or practices) to reduce injury, mortality, and sublethal disturbances

- ideas based on scientists' input, the literature, etc.
- angler (or related stakeholders, e.g., tournament organisers, guides) ideas and experiences, knowledge of gear
- gear technology in recreational fisheries tends to be driven by consumer demand and potential profit rather than by government or non-industry scientists so industry, social/environmental conscience, marketing, sales and new product development are critical
- N.B. can bypass point 1 and 2 if issues are led by industry or anglers

Test modifications (in gear or practices) to reduce injury, mortality and sublethal disturbances

- scientists conducting field experiments
- industry-based experiments (field testing, including outdoor media opportunities)
- angler experiences involved in experiments to ensure their practical application

Implementation of appropriate modifications

- scientists and managers disseminating information (delivering presentations, writing scientific papers, outreach, internet postings)
- angler communication (sharing experiences, peer pressure, voluntary adoption)
- outdoor media, government outreach efforts, regulations, brochures, product marketing (note that sometimes gear or practices are broadly or inappropriately implemented and which can be driven by misinformation)

Catch-and-release recreational angling has become very popular as a conservation strategy among anglers and as a tool for fishery management in a diverse array of fisheries. Implicit in catch-and-release angling strategies, is however, the assumption that released fish experience low mortality and minimal sublethal effects. Despite the importance of these premises, research on this topic has mostly focused on a relatively small number of popular North American sportfish species, with negligible efforts directed towards understanding catch-and-release angling effects on other species, especially in developing countries. Clearly, there is a need to conduct additional research on several key topics in this field as outlined in Box 5. The sustainability of recreational fisheries in the future will largely depend on effective catch-and-release angling and it is our hope that catch-and-release research will focus not only on minimising sublethal and fitness-related disturbances, but also on facilitating or enhancing recovery (See Box 5). Only when constituents are provided with access to reliable information on how to properly execute catch-and-release angling while minimising lethal and sublethal effects, can we hope to manage sustainable recreational fisheries in the long-term.

The studies reviewed in this chapter demonstrate that a diversity of factors influence survival and the subsequent well-being and performance of fish caught and released by recreational anglers. These factors are well known, having been rather conclusively documented in a general form by Muoneke and Childress (1994) and in subsequent efforts. It is apparent, however, that except among highly-specialised or invested anglers, adoption of these improvements by the general angling public has been poor. In the short term, this affects fishing quality and, in the long term, it may affect whether or not we fish. It is time for all scientists, managers, and participants in these fisheries, to come together and face an unpopular challenge – we have the knowledge and technology necessary to reduce the effects of discarding, but it is yet to be determined whether we will use them to alter the way recreational fisheries are conducted.

Box 5: Research Agenda for Recreational Fisheries Release Issues

Although there is now a substantial background of research associated with catch-and-release angling and the reduction of release mortality and sublethal effects, most of this research tends to focus on a few popular species. In fact, Cooke and Suski (2005) suggested that there are only five species of fish for which we have a reasonable understanding of catch-and-release angling effects, all of which are freshwater (largemouth bass, *Micropterus salmoides*; walleye, *Stizostedion vitreum*) or anadromous (rainbow trout, *Oncorhynchus mykiss*; striped bass, *Morone saxatilis*; Atlantic salmon, *Salmo salar*). Indeed, these species are some of the most popular and heavily managed fish in the world. However, we contend that there is still much research to be done as the data from these species are obviously not representative of the vast diversity of morphologies, life-histories, physiologies, habitats, etc. associated with fish that are subjected to recreational capture and release. Below we briefly outline the key research questions that we believe need to be addressed if we are to understand and minimize the effects of recreational angling on released fish.

Construct baseline information on blood and muscle biochemistry and determine how these parameters are affected by angling for a variety of marine and freshwater species

There are few studies (see Thorstad et al. 2003; Suski et al. 2003) that evaluate the baseline and post-angling blood and muscle biochemistry profiles of recreationally important fish. Factors worthy of initial investigation include the effects of gear type, duration of angling event and water temperature.

Conduct controlled experiments to document the disturbance and recovery trends of blood and muscle biochemistry, hormones and the cardio-respiratory system

Controlled laboratory experiments can be used to manipulate factors such as the duration of air exposure, degree of exhaustion and water temperature to determine how these factors may contribute to sublethal disturbances or mortality, and how they alter recovery duration. Such experiments would most likely involve cannulation to collect serial plasma samples or cardiovascular monitoring devices to record cardio-respiratory activity. Such research is essential if we are to establish time-course recovery profiles for angled fish.

Evaluate the fate and behaviour of released fish at multiple temporal and spatial scales

When a fish is captured and released, its behaviour is almost certainly affected. What does this mean to short- and long-term survival? Some species are susceptible to post-release predation and mortality, whereas others survive with negligible negative effects. Researchers must apply techniques that enable them to evaluate the mechanisms associated with different outcomes (i.e., what are the physiological and behavioural correlates of those fish species that tend to die after release).

Box 5. (cont.)

Assess the effects of different strategies for facilitating the recovery of angled fish

There has been recent interest in trying to develop strategies that actually facilitate the recovery of commercial by-catch and caught-and-released fish. It would be useful to know if short-term retention in live-well devices could provide captured fish adequate time to recover such that they would be able to evade predators upon release.

Evaluate the performance of novel hook types and fishing techniques in the context of their potential to reduce mortality or injury and other sublethal effects

Recent advances in terminal tackle show promise for reducing injury and mortality of released fish. New tackle developed by anglers, industry and scientists will need to be evaluated for their potential conservation benefits.

Consider catch-and-release angling from an animal welfare perspective

Much research in aquaculture has recently focused on an assessment of welfare correlates. There is a need for similar research activities in the recreational fishing sector. In reality, the concepts associated with considering the welfare of angled fish are identical to those associated with ensuring that fish are released in the best possible condition (See Cooke and Sneddon – in press).

Assess the sublethal effects of angling-related behaviour on growth and other fitness-related variables

Growth and other fitness-related indices can be affected by catch-and-release angling either directly through reduced food intake or indirectly through sublethal acute or chronic stress (See Cooke et al. 2002a). There is an important need for research that evaluates how different angling-related stressors affect factors such as the quality and quantity of gametes, reproductive behaviour, viability of offspring, etc. (See Cooke et al. 2002a for comprehensive list of possible fitness alterations).

Acknowledgements

SJC thanks Cory Suski, Bruce Tufts, Scott Hinch, Tony Farrell, David Philipp, David Wahl, Mike Siepker, Kenneth Ostrand, Jason Schreer, Karen Dunmall, Tony Goldberg, Andy Danylchuk, and Brandon Barthel for helping to develop the ideas presented in this paper. Andrea Weckworth assisted with editorial activities. We also thank several anonymous reviewers for their contributions to the paper. Cory Suski, Eva Thorstad, and Greg Skomal kindly provided photographs. The Illinois Natural History Survey, the University of British Columbia, the Izaak Walton Killam Foundation, Carleton University, the Canadian Foundation for Innovation, and the Natural Sciences and Engineering Research Council of Canada, and the Ontario Research Fund provided support and funding to SJC.

References

Aas Ø, Thailing CE, Ditton RB (2002) Controversy over catch-and-release recreational fishing in Europe. In: Pitcher TJ, Hollingworth C (eds) Recreational Fisheries: Ecological, Economic, and Social Evaluation. Blackwell Science, Oxford, United Kingdom, pp 95–106

Albin D, Karpov KA (1998) Mortality of lingcod, *Ophiodon elongatus*, related to capture by hook and line. Marine Fisheries Review 60:29–34

Alverson DL (1998) Discarding Practices and Unobserved Fishing Mortality in Marine Fisheries: An Update. Washington SeaGrant. June 1998, 76 pp

Alverson DL, Freeberg MK, Murawski SA, Pope JG (1994) A global assessment of fisheries bycatch and discards. FAO Fish Tech Pap no. 339, Rome, 233 pp

Anderson WG, Booth R, Beddow TA, McKinley RS, Finstad B, Økland F, Scruton D (1998) Remote monitoring of heart rate as a measure of recovery in angled Atlantic salmon, *Salmo salar* (L.) Hydrobiologia 371/372:233–240

Archer DL, Loyacano HA Jr (1975) Initial and delayed mortalities of largemouth bass captured in the 1973 National Keowee B.A.S.S. tournament. Proceedings of the 28th Annual Conference of the Southeast Association of Game and Fish Commissioners 28:90–96

Arlinghaus R, Cooke SJ (2005) Global impact of recreational fisheries. Science 307:1561–1562

Arlinghaus R, Mehner T, Cowx IG (2002) Reconciling traditional inland fisheries management and sustainability in industrialized countries, with emphasis on Europe. Fish Fish 3:261–316

Ayvazian SG, Wise BS, Young GC (2002) Short-term hooking mortality of tailor (*Pomatomus saltatrix*) in Western Australia and the impact on yield per recruit. Fish Res 58(2):241–248

Bache SJ (2003) Bycatch mitigation tools: Selecting fisheries, setting limits, and modifying gear. Ocean Coast Manag 46(1–2):103–125

Balon EK (2000) Defending fishes against recreational fishing: An old problem solved in the new millennium. Environ Biol Fishes 57:1–8

Barnhart RA (1989) Symposium review: Catch-and-release fishing, a decade of experience. N Amer J Fish Manag 9:74–80

Barnhart RA, Roelofs TD (1977) Catch-and-release fishing as a management tool. Arcata, California. Humboldt State University, California Cooperative Fisheries Research Unit

Barnhart RA, Roelofs TD (1989) Catch-and-release fishing, a decade of experience. Arcata, California. Humboldt State University, California Cooperative Fisheries Research Unit

Barthel BL, Cooke SJ, Suski CD, Philipp DP (2003) Effects of landing net mesh type on injury and mortality in a freshwater recreational fishery. Fish Res 63:275–282

Bartholomew A, Bohnsack JA (2005) A review of catch-and-release angling mortality with implications for no-take reserves. Rev Fish Biol Fish 15(1–2): 129–154

Beckwith GH, Rand PS (2005) Large circle hooks and short leaders with fixed weights reduce incidence of deep hooking in angled adult red drum. Fish Res 71(1):115–120

Beitinger TL, Bennett WA, McCauley RW (2000) Temperature tolerances of North American freshwater fishes exposed to dynamic changes in temperature. Environ Biol Fishes 58:237–275

Black EC (1958) Hyperactivity as a lethal factor in fish. J Fish Res Bd Can 15:573–586

Borucinska J, Martin J, Skomal G (2001) Peritonitis and pericarditis associated with gastric perforation by a retained fishing hook in a blue shark. J Aquat Anim Health 13:347–354

Borucinska J, Kohler N, Natanson L, Skomal G (2002) Pathology associated with retained fishing hooks in blue sharks, *Prionace glauca* (L.), with implications for their conservation. J Fish Dis 25:515–521

Boutilier RG (1990) Control and co-ordination of gas exchange in bimodal breathers. In: Boutilier RG (ed) Advances in Comparative and Environmental Physiology, 6. Vertebrate Gas Exchange: From Environment to Cell. Springer-Verlag, New York, USA, pp 280–346

Brobbel MA, Wilkie MP, Davidson K, Kieffer JD, Bielak AT, Tufts BL (1995) Physiological effects of catch and release angling in Atlantic salmon (*Salmo salar*) at different stages of freshwater migration. Can J Fish Aquat Sci 53:2036–2043

Burns KM, Restrepo V (2002) Survival of reef fish after rapid depressurization: Field and laboratory studies. Am Fish Soc Symp 30:148–151

Campbell PM, Pottinger TG, Sumpter JP (1992) Stress reduces the quality of gametes produced by rainbow trout. Biol Reprod 47:1140–1150

Carbines GD (1999) Large hooks reduce catch-and-release mortality of blue cod *Parapercis colias* in the Marlborough Sounds of New Zealand. N Amer J Fish Manag 19:992–998

Carmichael GJ, Tomasso JR, Simco BA, Davis KB (1984) Confinement and water quality induced stress in largemouth bass. Trans Am Fish Soc 113:767–777

Caruso PG (2000) A comparison of catch and release mortality and wounding for striped bass (*Morone saxatilis*), captured with two baited hook types. Sportfisheries Research Project (F-57-R), Completion Report for Job 12. Massachusetts Division of Marine Fisheries. 13 pp

Chuenpagdee R, Morgan LE, Maxwell SM, Norse EA, Pauly D (2003) Shifting gears: Assessing collateral impacts of fishing methods in US waters. Front Ecol Environ 1(10):517–524

Coleman FC, Figueira WF, Ueland JS, Crowder LB (2004) The impact of United States recreational fisheries on marine fish populations. Science 305: 1958–1960

Collins MR, McGovern JC, Sedberry GR, Meister HS, Pardieck R (1999) Swim bladder deflation in black sea bass and vermilion snapper: Potential for increasing postrelease survival. N Amer J Fish Manag 19:828–832

Cooke SJ, Cowx IG (2004) The role of recreational fisheries in global fish crises. BioScience 54:857–859

Cooke SJ, Cowx IG (2006) Contrasting recreational and commercial fishing: Searching for common issues to promote unified conservation of fisheries resources and aquatic environments. Biol Conserv 128:93–108

Cooke SJ, Hogle WJ (2000) Effects of retention gear on the injury and short-term mortality of adult smallmouth bass. N Amer J Fish Manag 20:1033–1039

Cooke SJ, Philipp DP (2004) Behavior and mortality of caught-and-released bonefish (*Albula* spp.) in Bahamian waters with implications for a sustainable recreational fishery. Biol Conserv 118:599–607

Cooke SJ, Sneddon LU (in press) Animal welfare perspectives on catch-and-release recreational angling. Appl Anim Behav Sci

Cooke SJ, Suski CD (2004) Are circle hooks effective tools for conserving freshwater and marine recreational catch-and-release fisheries? Aquat Conserv Mar Freshwater Ecosys 14:299–326

Cooke SJ, Suski CD (2005) Do we need species-specific guidelines for catch-and-release recreational angling to conserve diverse fishery resources? Biodivers Conserv 14:1195–1209

Cooke SJ, Schreer JF, Dunmall KM, Philipp DP (2002a) Strategies for quantifying sublethal effects of marine catch-and-release angling – insights from novel freshwater applications. Am Fish Soc Symp 30:121–134

Cooke SJ, Schreer JF, Wahl DH, Philipp DP (2002b) Physiological effects of catch-and-release angling practices on largemouth bass and smallmouth bass. Am Fish Soc Symp 31:489–512

Cooke SJ, Barthel BL, Suski CD (2003a) Effects of hook type on injury and capture efficiency of rock bass, *Ambloplites rupestris*, angled in southeastern Ontario. Fish Manag Ecol 10:269–271

Cooke SJ, Suski CD, Siepker MJ, Ostrand KG (2003b) Injury rates, hooking efficiency and mortality potential of largemouth bass (*Micropterus salmoides*) captured on circle hooks and octopus hooks. Fish Res 61:135–144

Cooke SJ, Suski CD, Barthel BL, Ostrand KG, Tufts BL, Philipp DP (2003c) Injury and mortality induced by four hook types on bluegill and pumpkinseed. N Amer J Fish Manag 23:883–893

Cooke SJ, Philipp DP, Dunmall KM, Schreer JF (2001) The influence of terminal tackle on injury, handling time, and cardiac disturbance of rock bass. N Amer J Fish Manag 21:333–342

Cooke SJ, Barthel BL, Suski CD, Siepker MJ, Philipp DP (2005) Influence of circle hook size on hooking efficiency, injury, and size selectivity of bluegill with comments on circle hook conservation benefits in recreational fisheries. N Amer J Fish Manag 25:211–219

Cooke SJ, Danylchuk AD, Danylchuk SA, Suski CD, Goldberg TL (2006) Is catch-and-release recreational fishing compatible with no-take marine protected areas? Ocean Coast Manag 49:342–354

Cowx IG (2002) Recreational fisheries. In: Hart PBJ, Reynolds JD (eds) Handbook of Fish Biology and Fisheries, Volume II. Blackwell Science, Oxford, United Kingdom, pp 367–390

Davidson GW, Thorarensen HT, Lokman M, Davie PS (1997) Stress of Capture and Captivity in Kahawai *Arripis trutta* (Bloch and Schneider) (Perciformes: Arripidae). Comp Biochem Physiol 118A:1405–1410

Davis MW (2002) Key principles for understanding fish bycatch discard mortality. Can J Fish Aquat Sci 59(11):1834–1843
Dayton PK, Thrush SF, Agardy MT, Hofman RJ (1995) Environmental effects of marine fishing. Aquat Conserv Mar Freshwater Ecosys 5:1–28
Dedual M (1996) Observed mortality of rainbow trout caught by different angling techniques in Lake Taupo, New Zealand. N Amer J Fish Manag 16(2): 357–363
de Leeuw AD (1996) Contemplating the interest of fish: The angler's challenge. Environ Ethics 18:373–390
Dempson JB, Furey G, Bloom M (2002) Effects of catch and release angling on Atlantic salmon, *Salmo salar* L., of the Conne River, Newfoundland. Fish Manag Ecol 9(3):139–147
DFO (Department of Fisheries and Oceans) (1998) Effects of hook-and-release angling practises – Atlantic salmon. Department of Fisheries and Oceans Stock Status Report D0-03, Ottawa, Canada, 7pp
DFO (Department of Fisheries and Oceans) (2003) Summary of 2000 recreational fishing statistics in Canada. Ottawa, Canada. (20 January 2004; Availble online at http://www.dfo-mpo.gc.ca/communic/statistics/RECFISH/new2002/sum2000_e.HTM)
Diodati PJ, Richards LA (1996) Mortality of striped bass hooked and released in salt water. Trans Am Fish Soc 125:300–307
Diggles BK, Ernst I (1997) Hooking mortality of two species of shallow-water reef fish caught by recreational angling methods. Mar Freshw Res 48(6): 479–483
Domeier ML, Dewar H, Nasby-Lucas N (2003) Mortality rate of striped marlin (*Tetrapturus audax*) caught with recreational tackle. Mar Freshw Res 54(4):435–445
DuBois RB, Dubielzig RR (2004) Effect of hook type on mortality, trauma, and capture efficiency of wild stream trout caught by angling with spinners. N Amer J Fish Manag 24:609–616
Dunmall KM, Cooke SJ, Schreer FJ, McKinley RS (2001) The effect of scented lures on hooking injury and mortality of smallmouth bass caught by novice and experienced anglers. N Amer J Fish Manag 21:242–248
Edwards RE (1998) Survival and movement patterns of released tarpon (*Megalops atlanticus*). Gulf Mex Sci 16:1–7
FAO (Food and Agriculture Organization of the United Nations) (1997) Inland Fisheries. FAO Fisheries Department Technical Guidelines for Responsible Fisheries no. 6 (Technical guidelines for the sustainable management of inland fisheries). FAO, Rome
Farrell AP (2002) Cardiorespiratory performance in salmonids during exercise at high temperature: Insights into cardiovascular design limitations in fishes. Comp Biochem Physiol 132A:797–810
Farrell AP, Gamperl AK, Hicks JMT, Shiels HA, Jain KE (1996) Maximum cardiac performance of rainbow trout, *Oncorhynchus mykiss*, at temperatures approaching their upper lethal limit. J Exp Biol 199:663–672

Ferguson RA, Tufts BL (1992) Physiological effects of brief air exposure in exhaustively exercised rainbow trout (*Oncorhynchus mykiss*): Implications for 'catch and release' fisheries. Can J Fish Aquat Sci 49:1157–1162

Fry FEJ (1971) The effect of environmental factors on the physiology of fish. In: Hoar WS, Randall DJ (eds) Fish Physiology, Volume VI: Environmental relations and behavior. Academic Press, New York, USA, pp 1–98

Furimsky M, Cooke SJ, Suski CD, Wang X, Tufts BL (2003) Respiratory and circulatory effects of hypoxia in largemouth bass and smallmouth bass: An application to 'live-release' angling tournaments. Trans Am Fish Soc 132:1065–1075

Gallman EA, Isely JJ, Tomasso JR, Smith TIJ (1999) Short-term physiological responses of wild and hatchery-produced red drum during angling. N Amer J Fish Manag 19:833–836

Gjernes T, Kronlund AR, Mulligan TJ (1993) Mortality of chinook and coho salmon in their first year of ocean life following catch and release by anglers. N Amer J Fish Manag 13:524–539

Graves JE, Luckhurst BE, Prince ED (2002) An evaluation of pop-up satellite tags for estimating postrelease survival of blue marlin (*Makaira nigricans*) from a recreational fishery. Fish Bull 100:134–142

Greenstreet SPR, Rogers SI (2000) Effects of fishing on non-target fish species. In: Kaiser MJ, de Groot SJ (eds) Effects of fishing on non-target species and habitats. Blackwell Science, Oxford, United Kingdom, pp 217–234

Grover A, Zwahlen ML (1996) Hooking mortality of sublegal chinook salmon caught by drift mooching with various terminal gear. Report to Pacific Fishery Management Council

Grover AM, Palmer-Zwahlen ML, Mohr MS (2002) Hook-and-release mortality of Chinook salmon from drift mooching with circle hooks: Management implications for California's ocean sport fishery. Am Fish Soc Symp 30:39–56

Gustaveson AW, Wydowski RS, Wedemeyer GA (1991) Physiological response of largemouth bass to angling stress. Trans Amer Fish Soc 120:629–636

Hall MA (1996) On bycatches. Rev Fish Biol Fish 6(3):319–352

Hall MA, Alverson DL, Metuzals KI (2000) By-catch: Problems and solutions. Mar Pollut Bull 41(1–6):204–219

Hand RG (2001) Evaluation of circle hooks as a means of reducing catch and release mortality of Roanoke River stripes bass. Federal Aid in Fish Restoration Project F-22. North Carolina Wildlife Resources Commission, Division of Inland Fisheries, Raleigh, North Carolina

Hartley RA, Moring JR (1993) Observations of black bass (Centrarchidae) confined during angling tournaments: A cautionary note concerning dissolved oxygen. Aquacult Fish Manage 24:575–579

Hartley RA, Moring JR (1995) Differences in mortality between largemouth and smallmouth bass caught in tournaments. N Amer J Fish Manag 15:666–670

Hilborn R, Branch TA, Ernst B, Magnussson A, Minte-Vera CV, Scheuerell MD, Valero JL (2003) State of the world's fisheries. Annu Rev Environ Resour 28:359–399

Holbrook JA (1975) Bass fishing tournaments. In: Clepper H (ed) Black Bass Biology and Management. Sport Fishing Institute, Washington DC, USA, pp 408–415

Horodysky AZ, Graves JE (2005) Application of pop-up satellite archival tag technology to estimate postrelease survival of white marlin (*Tetrapturus albidus*) caught on circle and straight-shank ('J') hooks in the western North Atlantic recreational fishery. Fish Bull 103:84–96

Hulbert PJ, Engstrom-Heg R (1980) Hooking mortality of worm-caught hatchery brown trout. New York Fish and Game Journal 27:1–10

Jenkins TM (2003) Evaluating recent innovations in bait fishing tackle and technique for catch and release of rainbow trout. N Amer J Fish Manag 23:1098–1107

Johannes RE (1981) Words of the lagoon: Fishing and marine lore in the Palau district of Micronesia. University of California Press, Los Angeles, California, USA

Jolley JE Jr, Irby EW Jr (1979) Survival of tagged and released Atlantic sailfish (*Istiophorus platypterus*: Istiophoridae). Bull Mar Sci 29:155–169

Jordan SR, Woodward AG (1994) Survival of hook-caught red drum. Proc of the Annual Conf Southeastern Assoc Fish Wildl Agencies 46:337–344

Keniry MJ, Brofka WA, Horns WH, Marsden JE (1996) Effects of decompression and puncturing the gas bladder on survival of tagged yellow perch. N Amer J Fish Manag 16:201–206

Kennelly SJ, Broadhurst MK (2002) By-catch Begone: Changes in the philosophy of fishing technology. Fish Fish 3:340–355

Kerr SJ (2001) A Review of 'Fizzing' – A Technique for Swim Bladder Deflation. Ontario Ministry of Natural Resources, Peterborough, Ontario, USA, 13 pp

Kieffer, JD (2000) Limits to exhaustive exercise in fish. Comp Biochem Physiol 126A:161–179

Kieffer JD, Kubacki MR, Phelan FJS, Philipp DP, Tufts BL (1995) Effects of catch-and-release angling on nesting male smallmouth bass. Trans Am Fish Soc 124:70–76

Kieffer JD, Rossiter AM, Kieffer CA, Davidson K, Tufts BL (2002) Physiology and survival of Atlantic salmon following exhaustive exercise in hard and softer water: Implications for the catch-and-release sport fishery. N Amer J Fish Manag 22:132–144

Kubacki MR (1992) The effectiveness of a closed season for protecting nesting largemouth and smallmouth bass in southern Ontario. MSc Thesis. University of Illinois, Urbana, Illinois, USA

Lawson PW, Sampson DB (1996) Gear-related mortality in selective fisheries for ocean salmon. N Amer J Fish Manag 16:512–520

Lewison RL, Crowder LB, Read AJ, Freeman SA (2004) Understanding impacts of fisheries bycatch on marine megafauna. Trends Ecol Evol 19(11):598–604

Lindsay RB, Schroeder RK, Kenaston KR, Toman RN, Buckman MA (2004) Hooking mortality by anatomical location and its use in estimating mortality of spring Chinook salmon caught and released in a river sport fishery. N Amer J Fish Manag 24:367–378

Lowe TE, Wells RMG (1996) Primary and secondary stress responses to line capture in the blue mao mao. J Fish Biol 49:287–300

Lowerre-Barbieri SK, Vose FE, Whittington JA (2003) Catch-and-release fishing on a spawning aggregation of common snook: Does it affect reproductive output? Trans Am Fish Soc 132:940–952

Lucy JA, Studholme AL (eds) (2002) Catch and release in marine recreational fisheries. Am Fish Soc Symp 30. Bethesda, Maryland, USA, 275 pp

Lukacovic R (1999) Hooking mortality of deep and shallow-hooked striped bass under different environmental conditions in Chesapeake Bay. In: Weinrich DR, Piavis PG, Pyle BH, Jarzynski AA, Walstrum JC, Sadzinski RA, Webb EJ, Rickabaugh HW, Zlokovitz E, Mower JP, Lukacovic R, Whiteford KA (eds) Stock assessment of selected resident and migratory recreational finfish species within Marylands Chesapeake Bay. Federal Aid Project F-54-R. Annual Report, Department of the Interior, Fish and Wildlife Service

Lukacovic R (2000) Hooking mortality of deep and shallow hooked striped bass under different environmental conditions in Chesapeake Bay. In: Weinrich DR, Piavis PG, Pyle BH, Jarzynski AA, Walstrum JC, Sadzinski RA, Webb EJ, Rickabaugh HW, Zlokovitz E, Mower JP, Lukacovic R, Whiteford KA (eds) Stock assessment of selected resident and migratory recreational finfish species within Marylands Chesapeake Bay. Federal Aid Project F-54-R. Annual Report, Department of the Interior, Fish and Wildlife Service

Lukacovic R (2001) An evaluation of deep hooking rates and relative hooking efficiency of several styles of circular configured hooks. In: Weinrich DR, Piavis PG, Pyle BH, Jarzynski AA, Walstrum JC, Sadzinski RA, Webb EJ, Rickabaugh HW, Zlokovitz E, Mower JP, Lukacovic R, Whiteford KA (eds) Stock assessment of selected resident and migratory recreational finfish species within Marylands Chesapeake Bay. Federal Aid Project F-54-R. Annual Report, Department of the Interior, Fish and Wildlife Service

Malchoff MH, Heins SW (1997) Short-term hooking mortality of weakfish caught on single-barb hooks. N Amer J Fish Manag 17:477–481

Malchoff MH, Gearhart J, Lucy J, Sullivan PJ (2002) The influence of hook type, hook wound location, and other variables associated with post catch-and-release mortality in the U.S. summer flounder recreational fishery. Am Fish Soc Symp 30:101–105

Makinen TS, Niemela E, Moen K, Lindstrom R (2000) Behaviour of gill-net and rod-captured Atlantic salmon (*Salmo salar* L.) during upstream migration and following radio tagging. Fish Res 45:117–127

McCaughran DA (1992) Standardized nomenclature and methods of defining bycatch levels and implications. In: Schoning RW, Jacobson RW, Alverson DL, Gentle TG, Auyong J (eds) Proceedings of the National Industry Bycatch Workshop, February 4 – 6 1992, Newport, Oregon. Natural Resources Consultants, Inc., Seattle, Washington, USA, pp 200–201

McNair D (1997) Effect of gear and methods on marine sports salmon hooking mortality rates. NEAP Final Research Report 68.2. Pacific Fisheries Management Council

McPhee DP, Leadbitter D, Skilleter GA (2002) Swallowing the bait: Is recreational fishing ecologically sustainable? Pacific Conserv Biol 8:40–51

Maltby L (1999) Studying stress: The importance of organism-level responses. Ecol Appl 9:431–440

Mason JW, Hunt RL (1967) Mortality rates of deeply hooked rainbow trout. Prog Fish-Cult 29:87–91

McEachron LW, Green AW, Matlock GC, Saul GE (1985) A comparison of trotline catches on two hook types in the Laguna Madre. Management Data Series no. 86. Texas Parks and Wildlife Department

McKenzie DJ, Serrini G, Piraccini G, Broni P, Bolis CL (1996) Effects of diet on responses to exhaustive exercise in Nile tilapia (*Oreochromis nilotica*) acclimated to three different temperatures. Comp Biochem Physiol Part A 114: 46–50

Meals KO, Miranda LE (1994) Size-related mortality of tournament-caught largemouth bass. N Amer J Fish Manag 14:460–463

Meka JM (2004) The influence of hook type, angler experience, and fish size on injury rates and the duration of capture in an Alaskan catch-and-release rainbow trout fishery. N Amer J Fish Manag 24:1309–1321

Meka JM, McCormick SD (2005) Physiological response of wild rainbow trout to angling: Impact of angling duration, fish size, body condition, and temperature. Fish Res 72:311–322

Millard MJ, Welsh SA, Fletcher JW, Mohler J, Kahnle A, Hattala K (2003) Mortality associated with catch and release of striped bass in the Hudson River. Fish Manag Ecol 10:295–300

Milligan CL, Hooke B, Johnson C (2000) Sustained swimming at low velocity following a bout of exhaustive exercise enhances metabolic recovery in rainbow trout. J Exp Biol 203:921–926

Morrissey MB, Suski CD, Esseltine KR, Tufts BL (2005) Incidence and physiological consequences of impacts of decompression in smallmouth bass after live-release angling tournaments. Trans Amer Fish Soc 134:1038–1047

Muoneke MI, Childress WM (1994) Hooking mortality: A review for recreational fisheries. Rev Fish Sci 2:123–156

Nevill Jon (2005) The impacts of spearfishing: Notes on the effects of recreational diving on shallow marine reefs in southern Australia. OnlyOnePlanet Australia, Hampton Melbourne. 32pp. Available online at http://www.ids.org.au/~cnevill/marineSpearfishing.doc

O'Connell MFO, Dempson JB, Reddin DG (1992) Evaluation of the impacts of major management changes in the Atlantic salmon (*Salmo salar* L.) fisheries of Newfoundland and Labrador, Canada, 1984–1988. ICES J Mar Sci 49: 69–87

Ostrand KG, Cooke SJ, Wahl DH (2004) Effects of stress on largemouth bass reproduction. N Amer J Fish Manag 24:1038–1045

Pauley GB, Thomas GL (1993) Mortality of anadromous coastal cutthroat trout caught with artificial lures and natural bait. N Amer J Fish Manag 13:337–345

Pelzman, RJ (1978) Hooking mortality of juvenile largemouth bass, *Micropterus salmoides*. California Fish and Game 64:185–188

Pepperell JG, Davis TLO (1999) Postrelease behavior of black marlin, Makairaindica, caught off the Great Barrier Reef with sportfishing gear. Mar Biol 135:369–380

Persons SE, Hirsch SA (1994). Hooking mortality of lake trout angled through ice by jigging and set-lining. N Amer J Fish Manag 14:664–668

Philipp DP, Toline CA, Kubacki MF, Philipp DBF, Phelan FJS (1997) The impact of catch-and-release angling on the reproductive success of smallmouth bass and largemouth bass. N Amer J Fish Manag 17:557–567

Pitcher TJ, Hollingworth C (2002) Recreational fisheries: Ecological, economic, and social evaluation. Blackwell Science, Oxford, United Kingdom, 288 pp

Plumb JA, Gaines JL, Gennings M (1975) Experimental use of antibiotics in preventing delayed mortality in a bass tournament on Lake Seminole, Georgia. Proceedings of the 28th Annual Conference of the Southeast Association of Game and Fish Commissioners 28:87–90

Plumb JA, Grizzle JM, Rodgers WA (1988) Survival of caught and released largemouth bass after containment in live wells. N Amer J Fish Manag 8: 325–328

Policansky D (2002) Catch-and-release recreational fishing: A historical perspective. In: Pitcher TJ, Hollingworth CE (eds) Recreational Fisheries: Ecological, Economic and Social Evaluation. Blackwell Science, Oxford, United Kingdom, pp 74–94

Pope KL, Wilde GR (2004) Effect of catch-and-release angling on growth of largemouth bass, *Micropterus salmoides*. Fish Manag Ecol 11:39–44

Post JR, Sullivan M, Cox S, Lester NP, Walters CJ, Parkinson EA, Paul AJ, Jackson L, Shuter BJ (2002) Canada's recreational fisheries: The invisible collapse? Fisheries 27(1):6–17

Pottinger TG (1997) Changes in water quality within anglers' keepnets during the confinement of fish. Fish Manag Ecol 4:341–354

Pottinger TG (1998) Changes in blood cortisol, glucose and lactate in carp retained in anglers' keepnets. J Fish Biol 53:728–742

Prince ED, Ortiz M, Venizelos A (2002) A comparison of circle hook and 'J' hook performance in recreational catch-and-release fisheries for billfish. Am Fish Soc Symp 30:66–79

Prosser CL (1991) Environmental and metabolic animal physiology. Wiley-Liss Inc., USA

Quinn S (1993) Bass seasons, conservation measure or needless regulation? In-Fisherman 18(6):30–38

Quinn S (1996) Trends in regulatory and voluntary catch-and-release fishing. Am Fish Soc Symp 16:152–162

Raat AJP, Klein-Breteler JGP, Jansen SAW (1997) Effects on growth and survival of retention of rod-caught cyprinids in large keepnets. Fish Manag Ecol 4:355–368

Schaeffer JS, Hoffman EM (2002) Performance of barbed and barbless hooks in a marine recreational fishery. N Amer J Fish Manag 22:229–235

Schill DJ (1996) Hooking mortality of bait-caught rainbow trout in an Idaho trout stream and a hatchery: Implications for special-regulation management. N Amer J Fish Manag 16:348–356

Schill DJ, Scarpella RL (1997) Barbed hook restrictions in catch-and-release trout fisheries: A social issue. N Amer J Fish Manag 17:873–881

Schisler GJ, Bergersen EP Schisler GL (1996) Postrelease hooking mortality of rainbow trout caught on scented artificial baits. N Amer J Fish Manag 16:570–578

Schneider JC, Waybrant JR, O'Neal RP (1991) Results of early season, catch-and-release bass fishing at six lakes. Michigan Department of Natural Resources, Fisheries Technical Report 91-6, Ann Arbor, USA

Schramm HL Jr, Haydt PJ, Bruno NA (1985) Survival of tournament-caught largemouth bass (*Micropterus salmoides floridanus*) in two Florida (USA) lakes. N Amer J Fish Manag 5:606–611

Schramm HL Jr, Haydt PJ, Portier KM (1987) Evaluation of prerelease, postrelease, and total mortality of largemouth bass caught during tournaments in two Florida lakes. N Amer J Fish Manag 7:394–402

Schramm HL Jr, Armstrong ML, Funicelli NA, Green DM, Lee DP, Manns RE Jr, Taubert BD, Waters SJ (1991a) The status of competitive sport fishing in North America. Fisheries 16(3):4–12

Schramm HL Jr, Armstrong ML, Fedler AJ, Funicelli NA, Green DM, Hahn JL, Lee DP, Manns RE Jr, Quinn SP, Waters SJ (1991b) Sociological, economic, and biological aspects of competitive fishing. Fisheries 16(3):13–21

Schreer JF, Cooke SJ, McKinley RS (2001) Cardiac response to variable forced exercise at different temperatures – an angling simulation for smallmouth bass. Trans Amer Fish Soc 130:783–795

Schreer JF, Resch D, Gately M, Cooke SJ (2005) Swimming Performance of brook trout following simulated catch-and-release angling: Looking for air exposure thresholds. N Amer J Fish Manag 25:1513–1517

Schroeder DM, Love MS (2002) Recreational fishing and marine fish populations in California. CalCOFI Rep 43:182-190

Seidensticker EP (1975) Mortality of largemouth bass for two tournaments utilizing a 'don't kill your catch' program. Proceedings of the 28th Annual Conference of the Southeast Association of Game and Fish Commissioners 28:83–86

Shetter DS, Allison LN (1955) Comparison of mortality between fly-hooked and worm-hooked trout in Michigan streams. Michigan Department of Conservation, Institute for Fisheries Research, Miscellaneous Publication no. 9, Ann Arbor, USA

Siepker M (2004) Effects of recreational angling on black bass. MSc Thesis. Department of Natural Resources and Environmental Sciences, University of Illinois, Urbana, Illinois, USA

Siewert HF, Cave JB (1990) Survival of released bluegill, *Lepomis macrochirus*, caught on artificial flies, worms, and spinner lures. J Freshwater Ecol 5: 407–411

Skomal GB, Chase BC (2002) The physiological effects of angling on post-release survivorship in tunas, sharks, and marlin. Am Fish Soc Symp 30:135–138

Skomal GB, Chase BC, Prince ED (2002) A comparison of circle hook and straight hook performance in recreational fisheries for juvenile Atlantic bluefin tuna. Am Fish Soc Symp 30:57–65

Somero GN, Hofmann GE (1996) Temperature thresholds for protein adaptation: When does temperature change start to 'hurt'? Soc Exp Biol Semin Ser 16: 1–24

Steeger TM, Grizzle JM, Weathers K, Newman M (1994) Bacterial diseases and mortality of angler-caught largemouth bass released after tournaments on Walter F George Reservoir, Alabama-Georgia. N Amer J Fish Manag 14: 435–441

St John J, Moran M (2001) Release mortality of line-caught pink snapper – a case study in depth. Western Australia Department of Fisheries Occasional Publications 2001(5):179

Stewart H (1977) Indian Fishing – Early methods on the Northwest coast. University of Washington Press, Seattle, Washington, USA

Suski CD, Svec JH, Ludden JB, Phelan FJS, Philipp DP (2003) The effect of catch-and-release angling on the parental care behavior of male smallmouth bass. Trans Amer Fish Soc 132:210–218

Suski CD, Killen SS, Cooke SJ, Kieffer JD, Philipp DP, Tufts BL (2004) Physiological significance of the weigh-in during live-release angling tournaments for largemouth bass. Trans Amer Fish Soc 133:1291–1303

Taylor MJ, White KR (1992) A meta-analysis of hooking mortality of nonanadromous trout. N Amer J Fish Manag 12:760–767

Taylor RG, Whittington JA, Haymans DE (2001) Catch-and-release mortality rates of common snook in Florida. N Am J Fish Manage 21:70–75

Thompson JA, Hughes SG, May EB, Harrell RM (2002) Effects of catch and release on physiological responses and acute mortality of striped bass. Am Fish Soc Symp 30:139–143

Thorstad EB, Naesje TF, Fisker P, Finstad B (2003) Effects of hook and release on Atlantic salmon in the River Alta, northern Norway. Fish Res 60:293–307

Thorstad EB, Hay CJ, Naesje TF, Chanda B, Okland F (2004) Effects of catch-and-release angling on large cichlids in the subtropical Zambezi River. Fish Res 69:141–144

Tsuboi JI, Morita K, Matsuishi T (2002) Effects of catch-and-release angling on growth, survival and catchability of white-spotted charr *Salvelinus leucomaenis* in wild streams. Bull Jap Soc Sci Fish/Nissuishi 68:180–185

Tufts BL, Davidson K, Bielak AT (2000) Biological implications of 'catch-and-release' angling of Atlantic salmon. In: Whoriskey FG, Whelan KE (eds) Managing Wild Atlantic Salmon. Atlantic Salmon Federation, St Andrews, New Brunswick, USA, pp 195–225

Turek SM, Brett MT (1997) Comment: Trout mortality from baited barbed and barbless hooks (and reply). N Amer J Fish Manag 17:807

U.S. Department of Commerce (2002) Fisheries of the United States 2001. National Marine Fisheries Service, Fisheries Statistics and Economics Division, Silver Springs, Maryland, USA

Waters JR, Huntsman GR (1986) Incorporating mortality from catch and release into yield-per-recruit analyses of minimum size limits. N Amer J Fish Manag 6:63–471

Weathers KC, Newman MJ (1997) Effects of organizational procedures on mortality of largemouth bass during summer tournaments. N Amer J Fish Manag 17:131–135

Welborn TL Jr, Barkley JH (1974) Study on the survival of tournament released bass on Ross R Barnett Reservoir, April 1973. Proceedings of the 27th Annual

Conference of the Southeast Association of Game and Fish Commissioners 27:512–519

Whoriskey FG, Prusov S, Crabbe S (2000) Evaluation of the effects of catch-and-release angling on the Atlantic salmon (*Salmo salar*) of the Ponoi River, Kola Peninsula, Russian Federation. Ecol Freshw Fish 9:118–125

Wilde GR (1997) Largemouth bass fishery responses to length limits. Fisheries 22(6):14–23

Wilde GR (1998) Tournament-associated mortality in black bass. Fisheries 23(10):12–22

Wilde GR, Sawynok W (unpublished manuscript) Effect of hook removal on recapture rates in 27 species of angler-caught fishes in Australia.

Wilde GR, Pope KL, Strauss RE (2001) The correct calculation of delayed and total mortality in catch-and-release and tournament studies: You only live once. Abstract, American Fisheries Society 2001 Annual Meeting, Pheonix, Arizona, USA

Wilde GR, Shavlik CE, Pope KL (2002) Initial mortality of black bass in B.A.S.S. fishing tournaments. N Amer J Fish Manag 22:950–954

Wilkie MP, Davidson K, Brobbel M, Kieffer J, Booth RK, Bielak AT, Tufts BL (1996) Physiology and survival of wild Atlantic salmon following angling in warm summer waters. Trans Am Fish Soc 125:572–580

Wilkie MP, Brobbel MA, Davidson K, Forsyth L, Tufts BL (1997) Influences of temperature upon the post-exercise physiology of Atlantic salmon (*Salmo salar*). Can J Fish Aquat Sci 54:503–511

Woll AK, Boje J, Holst R, Gundersen AC (2001) Catch rates and hook and bait selectivity in longline fishery for Greenland halibut (*Reinhardius hippoglossoides*, Walbaum) at east Greenland. Fish Res 51:237–246

Wood CM (1991) Acid-base and ion balance, metabolism, and their interactions, after exhaustive exercise in fish. J Exp Biol 160:285–308

Wood CM, Turner JD, Graham MS (1983) Why do fish die after severe exercise? J Fish Biol 22:189–201

Wydoski RS (1977) Relation of hooking mortality and sublethal hooking stress to quality fishery management. In: Barnhart RA, Roelofs TD (eds) Catch-and-release fishing as a management tool. Humbolt State University, Arcata, California, USA, pp 43–87

8 Working with Fishers to Reduce By-catches

MARTIN A. HALL, HIDEKI NAKANO, SHELLEY CLARKE, SIMON THOMAS, JANICE MOLLOY, S. HOYT PECKHAM, JOHATH LAUDINO-SANTILLÁN, WALLACE J. NICHOLS, ERIC GILMAN, JIM COOK, SEAN MARTIN, J.P. CROXALL, K. RIVERA, C.A. MORENO, STEPHEN J. HALL

8.1 Introduction

(by Martin A. Hall)

Fifty years ago, when the oceans' stocks of fish were thought to be inexhaustible, there were no so-called 'by-catches'. Marine scientists studying fisheries were mostly limited to the monitoring of landings, and they developed the methods used in fisheries science from this perspective. Discards and by-catches were not part of the equation. By-catches in the context of this chapter mean dead discards; and because discarding happens at sea, land-based monitors could not see this component of the fishing process. What were the consequences of this very incomplete picture?

For species that were the targets of fisheries, when there were discards of undersized individuals, or high grading, etc., there was an additional unaccounted harvest of the population. The figures used to determine how populations were doing were therefore incorrect, and underestimated impacts.

For non-target species, there were several issues, but one of the biggest problems seemed to be when by-catches involved low productivity species mixed in with the higher productivity target species. For example, a tuna that begins to reproduce at 1.5 years of age, and may produce 100,000,000 eggs per year cannot be compared with a dolphin that begins to reproduce at age 10, and can only produce one calf every second year. When dolphins, sea turtles or seabirds, are taken in fisheries targeting tunas, anchovies, squids, etc., the level of fishing that could be sustainable for the target species is far greater than what the by-catch species can sustain. The dilemma is therefore to reduce fishing to the level that is adequate for the by-catch species, or to try to break the coupling of the target and by-catch species via selective fishing.

S. J. Kennelly (ed.), By-catch Reduction in the World's Fisheries, 235–288.

At sea, fishers were facing this by-catch issue from their own perspective. Seabirds taking bait from hooks were reducing fishing opportunities; the fact that once in a while some would get caught on a hook added to the aggravation because the bird had to be removed. By-catches of fish species increased the work on deck in order to discard unwanted individuals. Some of the by-catch species are also popular among fishers (although those species that take bait or target species are seldom popular), and this brings an additional incentive to avoid incidental mortality of such species.

The first attempts to improve selectivity in fishing gears were simple changes in mesh size, with the objective of releasing smaller individuals from the net and retaining only those individuals of desired sizes. But during the 1950s and 1960s, more scientists and technicians started going out to sea frequently, and other impacts became known. Two very different issues related to by-catch took prominence. One of them was the potential utilisation of the by-catch from shrimp trawls. Here by-catch was considered an issue of wastage, a problem of protein harvested but not consumed and many studies, particularly in tropical areas, discussed the potential utilisation of that by-catch. The other issue was the realisation by the public of the large incidental mortalities of dolphins in the tuna purse-seine fishery in the eastern Pacific. When the public knew that some very charismatic species were being killed in large numbers by this fishery, reaction came swiftly. What followed were attempts by the industry's participants to 'sweep the issue under a huge carpet', denying the existence of the problem, or trying to argue that the mortalities were sustainable, under a naive belief that the mortality of a 'few hundred thousand dolphins,' even if it were deemed sustainable, could be accepted by the public. Years of lobbying and developing political connections by industry, amounted to little in the face of this new movement that scared and confused the industry. The conflict was quite bitter, and by-catch became a dominant issue in the management of the tuna fishery. Other by-catch cases soon followed, involving charismatic components of the ecosystem (sea turtles and seabirds), as well as other cases, involving not-so-charismatic species (such as sharks, juvenile fish, etc.).

To reduce by-catches, we always have two options: 'fish less or fish better'. The option of fishing less, that at the extreme, leads to banning some fishing gear or practices entirely, is frequently preferred by some sectors, but very rarely by the fishing community. Given the social and economic situations of many countries, it is unlikely that they would accept the economic impacts, and especially the social costs, caused by increased unemployment. So for them, the preferred option to reduce by-catch is usually to find ways to 'fish better'. To achieve this goal, we need to find ways to encourage the fishers' cooperation and participation in the process. This is a necessary step because: (i) fishers know more about fishing than anybody else; (ii) fishers produce practical solutions, as the case-studies in this chapter will show, whilst academics produce diagnoses, but seldom practical solutions; and (iii) because modifying the behaviour of fishers at sea is frequently part of the solutions, they must be engaged in the process, rather than forced into it.

In this chapter we provide a variety of case-studies that illustrate the evolution of fishers, environmental advocates, fisheries managers and others, in dealing with by-catch issues. What we have learnt from these pioneer experiences should prove useful in facing future by-catch problems. These case studies offer a variety of views in different fisheries, regions and conditions that should help inform anyone trying to implement a program to reduce by-catches in fisheries. It is by no means a complete picture, and efforts such as those of TAMAR in Brazil, Karumbe in Uruguay, Parrish and Melvin in the Pacific Northwest and Alaska, and those of Kennelly and Broadhurst in Australia, should be examined for how to successfully integrate fishers, scientists and managers in dealing with by-catch issues. In this chapter, we have elected to concentrate on fisheries and by-catch issues concerning seabirds, turtles and dolphins – i.e., the charismatic by-catch issues. Other chapters in this book concentrate more on the non-charismatic by-catches associated with trawling, dredging and hooking.

Our focus here is not on the legal, engineering or scientific aspects of by-catch issues, but on the development of constructive and responsive interfaces between fishers, technicians, scientists and managers to succeed in dealing with by-catch problems. We have not tried to homogenise the contents of these case-studies: the voices of the storytellers have been respected and personality and cultural differences have been retained. Most people working with fishers on by-catch issues are good communicators, and there is little point in second-guessing the style and language of their choice.

8.2 Case Study 1 – Learning to Work with Fishers after Twenty Years in the Eastern Pacific Fisheries: The Tuna-dolphin Case

(by Martin A. Hall)

Almost 20 years ago, a young Latin American boat owner, Mr. Carlos Arbelaez, with a fleet of several purse-seiners in his stable, walked into my office. He had seen once again the gory videos of dolphins rolling down a purse-seine net. It was the same shot that had been shown over and over on different TV channels and programs. He realised the impact the video would have on the public and, in spite of the doubts many people in the industry had about the authenticity of the video (a strange boat that had been inactive for years, with a clueless and callous captain and crew, etc.), we felt that it would be a waste of time to question the images. The behaviour of the crew shown in the video was very far from 'typical' behaviour on the boats, as shown in years of scientific observer records, but it was not impossible that a crew like that existed. Dolphin mortality was happening, and the figures were quite high (high being defined not in a population-sense but in a public-perception sense, where numbers have a psychological value). He asked me what we could do to reduce mortality, and I knew he meant business.

Over the previous months, we had been studying our observer data, and we had identified a number of factors that were affecting the level of dolphin mortality during fishing operations. There were environmental and mechanical factors, the availability of gear and its condition, etc. But the skill, experience, and motivation of captains and crews played a major role, as shown by differences in the performance of similar vessels operating in more or less the same areas with similar gear. The willingness of owners to provide their vessels with the right gear and equipment was also important. Approximately 20% of the vessels caused close to 80% of the mortality of dolphins.

However, it was not just technology that was leading to high dolphin mortality. The effect of individual differences among fishers was also significant. In fact, the performance of Carlos' fleet was the worst of the eastern Pacific. Captains that were new to the fishery on dolphins were trying to grasp the new techniques and equipment, and that learning was costly. When I showed Carlos the statistics for his fleet, and compared those to data for the other fleets, he was shocked, and right then and there he decided to do something about it. Several of his boats were at sea, and the captains that rotated with those at sea were in Basque country, but he put his money where his mouth was. He called everyone in, and told me I had 3 days to show them how to lower dolphin mortalities. Very few boat owners would have made that decision; it was a combination of the belief that something needed to be done, with some trust that we may be able to produce a change, and the economic courage to put up a considerable sum of money to back those beliefs. In less than 3 years, that fleet had the lowest dolphin mortality rates of all those operating in the eastern Pacific and in 12 years, the incidental mortality of dolphins for the whole international fleet had been cut to 1% of the original level. After that first effort, we have been organising workshops for tuna fishers for almost 20 years. This is the story of how we learned to work with the fishers, and how they learned to work with us.

Role playing

Can you really put yourself in somebody else's shoes? We had to show people who spent most of their time at sea the way their activity was being portrayed, and therefore perceived, by the public. We had to explain to them that, even though the dolphins were not in danger of extinction, the public response was strong enough to create a need for the industry to respond to mitigate the problem.

We had to decide how to use this opportunity to communicate with them in a very effective way.

The first thing we did was to show them all the videos seen on TV newscasts and documentaries, newspaper clippings, magazine articles, flyers and pamphlets. Even though they complained bitterly at the way they were being portrayed, they understood.

Round 1 finished with the acknowledgement that they had to face the problem, and an awareness of the possible consequences of not doing so. Nobody could promise them a solution if things changed, but it was quite obvious that only a major change could give them a fighting chance to keep their jobs and their industry in operation.

Fishers need to understand the problem they are facing, and believe in the proposed solutions. In this case, it was believing in their own ability to change the impact of the fishery. We showed them that some boats were doing very well in reducing dolphin mortality, and those vessels had no significant differences from the others in equipment or in their productivity.

Round 2 began with putting together the necessary building blocks, by firstly providing the fishers with an introduction to the species involved – in this case, the dolphins and the tunas. A lot of judgment is required to decide what they need to know – what could be helpful for them to understand these aspects of the ecology and behaviour of tunas and dolphins that are important and perhaps even to anticipate the circumstances that lead to incidental captures. They don't need to become biologists, and the person in charge of the presentation is not there to show off how much he/she knows. No jargon, no Latin names, no complicated sentences. Clear, useful information and concepts, briefly and well explained. Why state the obvious? Because many people seem to have a major difficulty communicating directly. For many, scientific training results in an increasing inability to convey concepts without a heavy load of jargon.

The next component of the discussion with the fishers was an understanding of what we know about the factors that cause or increase by-catches. These ranged from environmental factors (e.g., strong currents), to gear and operational factors (e.g., execution of release manoeuvres, availability of rescue equipment, etc.), and the skill and motivation of captains and crews. Parallel to the identification of each problem, we developed the responses that had originated from the fishers themselves over the years. This review of factors causing by-catch was an excellent opportunity to bring to the table their individual experiences and perceptions in sometimes heated discussions. This was an excellent learning time for everyone.

We also discussed the performance of the fleet and, in private with each captain, their individual performances. In the highly-competitive environment of this tuna fleet, looking bad in front of their peers is something fishers would all like to avoid. At the same time, their understanding that a few captains were responsible for the image of the entire fleet and the majority of the problem was very useful to build a management model based on recognising those differences. The captains were always strong supporters of management schemes that separated 'good' from 'bad' fishers.

For over 10 years now, the fleet has operated with an overall dolphin mortality limit, but that limit is divided by the number of participating vessels, and each vessel receives an individual dolphin mortality limit for a year. If a vessel's limit is exceeded, it has to stop setting on dolphins for the rest of the year. Fishers always liked this scheme, because they didn't want to be the victims of others' lack of skill or motivation. Individual responsibility in management is an excellent concept when it is feasible; it is fair and equitable, and with time it results in a selective process for better captains and crews. Most of the captains who were involved in the higher-mortality trips are now gone from the fishery. When the boat owners realised that the better captains were not only those that filled the boats quickly, but that did so without compromising the fate of the vessel with carelessness about dolphins, the changes happened.

To reduce conflicts, we also clarified the role of observers, and finished by presenting to the fishers the problems we are still trying to solve, and asked for their impressions and suggestions, plus criticisms about the way we are proposing to work. And we listen. Sometimes there are simpler ways to achieve the same ends; sometimes the proposed solutions have unintended consequences. Once the workshops started, many fisheries authorities decided to follow our model, and today these workshops organised by the IATTC or by national dolphin program staff take place several times a year, in different countries and ports.

At the end of the workshops, private meetings are held with the fishing captains present to review their records of performance. Very frequently, the reasons for poor performances become evident from these records. Gear availability and use, problems with release manoeuvres, and risk-taking tendencies, are all described one-on-one. You don't want to embarrass proud and very independent people, as these fishers are, but you need to show them why their performances are below par. Sometimes, they may share their 'score' with others, but it remains their choice to do so. Their competitive instincts are heightened by the interactions with their peers, which often consist of using a sense of humour as a pointed stick, to jab at those bringing problems to the others. This is a male society; there are no women captains in the eastern Pacific. In many cases the ports are far away from the captains' homes, and all their contacts are limited to a small world composed of captains, navigators, deck bosses, boat owners and the staffs of the national fisheries agencies and the IATTC. The social networks in which these fishers work are quite limited in membership, but they are crucial in the formation of opinions.

At the workshops, we emphasise the issues for which we have no answers yet, (e.g., by-catches of other species) and we ask them to start thinking about those problems. We usually show them gear changes and innovations from other fisheries that are of potential interest in our fishery, which may later be tested and introduced, and seek their views. The communication among fishers from different regions is quite weak, and we try to remedy that by serving as a channel for those ideas. As an example of this, we have started showing the fishers the sorting grids developed in the Norwegian mackerel and saithe fisheries to release smaller fishes alive, and those used in Canada to release smaller salmon. This is always accompanied with questions about their perception of the usefulness of those ideas in the tuna fishery.

Acknowledgement of the good performers is as important as identification of those responsible for most of the problems. Each year, the captains with the best performance in reducing dolphin mortality are recognised. We make sure to highlight the examples of leadership, responsibility and consistency among the captains.

8.3 Case Study 2 – Sea Turtles, Longlines, and the Artisanal Fisheries of the Eastern Pacific

(by Martin A. Hall)

The critical condition of several of the populations of leatherback turtles in the Pacific Ocean led to an increasing level of concern in the late 1900s and early 2000s. In spite of years of nest-protection programs, and the implementation of Turtle Excluder Device (TED) programs, the populations continued to decline. By-catch in fisheries was considered to be one of the reasons, if not the main reason, for the decline. Information was scarce, and clearly insufficient to assess the level of mortality caused by coastal gillnets, industrial and artisanal longliners, etc., in a rigorous way. In any case, longline fisheries were in the sights of many who thought that the only way to save the turtles was a moratorium on all fisheries that contributed to the decline in their populations. We started drawing the attention of governments and industry leaders to this crisis, and the International Fishers Forum II, held in Hawaii in 2002, was a great opportunity to show them the problems and possible solutions, and to identify a global effort that was developing to save both the turtles and the fisheries involved.

The more visionary and better informed sector of the industry was persuaded by a technical advisor, Ingro. Guillermo Morán who attended that Forum, that it was in their best interests to face the problem, and work at finding a solution that could ensure the survival of their industry. But the good intentions of the industry needed an echo in the government, and the Under Secretary of Fisheries of Ecuador at the time, Mrs. Lucia De Genna, had the vision to see the problem, and more importantly, the courage to go forward. Why courage? Because every time that a fishery is opened to scrutiny for any reason, its fishing practices and their impacts become exposed, and some of them may cause negative reactions from the public, managers, etc. Very seldom has openness been rewarded. Governments and cooperatives of fishers, were keenly aware of the potential impact for their economies and employment levels. Hundreds of thousands of workers depend on longlining for their livelihood, and they are already in marginal economic and social situations, with very few options available to them. Pressure was clearly evident to find a solution that would allow the survival of the industry, and keep the fishers employed. This was one of the main ingredients that led to action.

When the IATTC received a request from the Under Secretary of Fisheries Resources of Ecuador, strongly supported by the Association of Exporters and the National Federation of Fishers Cooperatives of Ecuador, it became necessary to search for solutions, and for a strategy to implement them. Researchers from NOAA had been testing a wider type of hook, a circle hook, that reduced sea turtle mortality in two ways: (i) by reducing hooking rates, and (ii) by changing the way the turtles are hooked, increasing the survival of the turtles that did get hooked. The hooks also did not reduce the catch rates of the target species, and in some cases even increased catch rates.

It seemed that changing the type of hook was a reasonable thing to do, so the next problem faced was the development of an implementation strategy. The necessary steps were:

1. Show that the circle hooks are an effective way of reducing sea turtle mortality.
2. Show the fishers that they can continue making a living with the new technology, i.e., that the catch rates with circle hooks would be at least equivalent to current levels with conventional J hooks.

3. Make sure that the adoption of the circle hooks was economically viable.

Since you can't expect a fisher to agree to change the basic fishing instrument based on experiments performed in other fisheries and regions, it became obvious that they needed to test the hooks in their own fishing conditions; in their boats, with their baits, in their fishing grounds, etc.

The decision was to facilitate these tests by providing the circle hooks free of charge, and inviting the fishers to compare the new hooks with the old ones in comparative trials. We obtained the very willing cooperation of the NOAA authorities and researchers, of the Western Pacific Regional Fishery Management Council (WPRFMC), and of the World Wildlife Fund (WWF), to develop a program to begin these tests. We offered the fishers the opportunity to exchange some of their hooks for the new ones free of charge. After testing them for a trip or two, they had the option to undo the exchange, return the circle hooks and recover their J hooks. With these same partners (NOAA, WPRFMC, IATTC, and WWF), with the addition of the Overseas Fishery Cooperation Foundation (OFCF) from Japan, with contributions from The Ocean Conservancy, and Defenders of Wildlife-Mexico, and with the participation and support from government fisheries agencies in all countries, national conservation organisations, and national industry and fish workers organisations, the program has rapidly expanded to all countries operating in the eastern Pacific.

The key issues in this process were:

- fishers' participation was voluntary, after they were explained the situation and the reasons to join the program;
- the hook exchange was partial, so we reduced the risk involved;
- the exchange was free, but the operation was not a charity;
- fishing effort was not increased;
- the results were monitored through an observer program, and made available to all fishers;
- instruments and technical training to reduce mortality of hooked turtles were provided to the fishers.

The program was introduced using workshops, modelled from our experience with the tuna fleet outlined earlier in this chapter. We explained what was needed, why, and the way we were proposing to go about it.

After the experiments were begun, we followed up with frequent contacts with the fishers to assess the performance of the hooks, and the difficulties they caused. We learned about the difficulties for baiting and storage posed by the use of hooks of different sizes and shapes on the same line, and we helped find options to reduce these difficulties. We also worked with them in finding the right hook with respect to size, design and materials. For circle hooks, there are different materials and designs available. When hooks rusted quickly, or when there was some breakage of hooks, they were replaced by other types, brands or materials. Following their evaluations, we explored the various options available, and settled on one that could help the turtles without harming the fishers' catches.

Solutions were not imposed, but were developed with their active participation. Frequent contacts are needed to keep the flow of information going both ways: we received their feedback and suggestions for adaptations of the program, and we provided them with the results for the whole group of vessels involved in the work.

During the communication process with the fishers, we gathered a host of ideas about possible ways to reduce sea turtle mortality. For example, seeing the tendency of the turtles to approach the float, and become entangled near them, they suggested replacing the lines connecting the float and the line by cable, or stiffer materials, changing the colour of the floats, using fewer floats, etc. We are currently in the process of setting up these experiments.

In the case of artisanal fish workers, their organisations are an important point of contact, and we have had the support, and the presence in workshops of the leaders of FENACOPEC (Ecuador's National Federation of Fishers' Cooperatives) and later in Peru of the sister organisation, the Frente Integrado Unico de Pescadores Artesanales del Peru (FIUPAP). A message presented by the government's fisheries authorities, industry, exporters, environmentalists, scientists, and their own elected leaders, has much more power to influence people than the isolated effort from any one of these sectors.

At the same time that we recognise the major role of the fish workers' organisations, we have to remember that in many cases a large proportion of fishers do not belong to any organisation. This means that our efforts should not be channelled exclusively through them, but must also include the participation of independent groups and individuals.

Another important difference with respect to tuna captains was that the roles of the family unit and of the community were very important. While the men are concerned with the day-to-day needs and problems of their operations, the women in these families are the ones interacting with fish buyers, governments and other sectors, and they understand well the impact that different market problems could have. They are also operating on a longer time-horizon that the men, more concerned with the continuity of the day-to-day operations, and they will be the reminders, in the future, of what needs to be done. In these fishing communities, social interactions are important, and the size of the social networks is much larger than in the tuna fleet. First of all, large families frequently inhabit the same village, and they frequently function as a unit for the purposes of communication, formation of opinion, etc. Children begin to go to sea when they are 10 or 11 years old, during school holidays, and they can also be vehicles for change. Programs targeting schools in fishing villages could have much more rapid effects than are often seen in programs of environmental education directed to the public at large, which are prolonged, difficult to evaluate, and slow in bringing change.

The leadership of these fishing communities is different than the leadership of the fish workers' organisations. The former leaders, who are frequently women, have a significant power in the group, and their endorsement of the work is very valuable. They also have a clear perspective that the problem cannot be solved by only a few of them. Unless everyone contributes his or her share of the solution, the problem won't go away; a few careless fishers may cause the defeat of the efforts of the rest of the community – and we can offer them the example of the tuna-dolphin case outlined earlier to illustrate that.

Of course, the sea turtle by-catch issue is only one of the issues faced by the fish workers' sector on this region, and we should not lose sight of the other social and economic factors that affect these communities. A strong and active fish workers' sector is in critical need for sustainable fisheries management, and we should use every opportunity to contribute to the achievement of this larger goal. To work with fishers we need to understand and respect their organisations, and to reach out to those not belonging to them. As the fish workers are the first victims of poor fisheries management, we should empower them to become more like the custodians of the resources they harvest.

The success of the above approach resulted in an expansion of the program to cover practically all countries from the Pacific coast of America from Mexico to Peru, and the welcome addition of the support and collaboration of many other organisations from all sectors. In each country, government agencies, local environmental organisations, and industry sectors, are participating in the activities. A network of scientists and managers has also been created, linked through the common support of the NOAA and IATTC scientific staffs, and of the WWF national and regional offices involved (Peru, Colombia, Central America and Mexico) that coordinate the implementation of the program with the respective fisheries agencies. The process is built on two basic, simple premises: (i) nobody wants to kill sea turtles, nor drive them to extinction, and (ii) nobody wants to put fishers out of work. With a common ground, and building trust among the participants, we are hoping to put together a different model to face conservation problems; a model based on cooperation, in which the resources and motivations of all the sectors are brought together.

8.4 Case Study 3 – The Tori Pole in the Japanese Longline Fishery

(by Hideki Nakano and Shelley Clarke)

8.4.1 Introduction

Seabird interactions with fishing gear resulting in inadvertent by-catch and mortality occur in several Japanese longline fisheries. One of these is the Japanese southern bluefin tuna fishery operating in sub-Antarctic waters, mainly in the Indian Ocean. Fishing vessels are approximately 400 tonnes in capacity and 50 m in length, with crews of 20 to 25 usually comprised of Japanese officers and non-Japanese deck crew and seamen. Seabird by-catch, consisting mainly of 20 species of albatross, is a major issue in these fishing grounds. Given concerns raised by several conservation organisations and Japanese authorities regarding incidental catches of seabirds in longline fisheries by various nations, Japan is committed to objectively and scientifically analysing the impact of its longline fisheries under a basic policy of encouraging fishers to develop creative solutions to by-catch issues. A method for reducing seabird by-catch by employing bird scaring lines, called 'tori' (Japanese for 'bird') poles, was originally implemented by Japanese fishers and has been required by the Convention for the Conservation of Southern Bluefin Tuna (CCSBT) for all longline vessels since 1991. It is believed that this device reduces the level of seabird by-catch by approximately one third.

Seabird by-catch is also an issue in the North Pacific Ocean. Of the three species of albatrosses occurring in this area, Laysan and Black-footed albatrosses comprise the majority of the by-catch. Vessels operating here can be categorised into coastal, offshore and distant water fleets. Coastal fishing vessels are less than 10 tonnes, have crews of 1 to 3, and are at sea for less than 1 week. Offshore fishing vessels are between 10 and 120 MT, have crews of less than 10 and are at sea for periods ranging from 1 week to 1 month. Their fishing grounds are located west of the international date line. Distant water longline fishing vessels are larger than 120 tonnes, have crews of 15–20, are at sea for periods of two to three months, and may range farther from Japan than the offshore vessels. Nearly all of the crews in the coastal and offshore fleets are Japanese but in the distant water fleet, most crews are non-Japanese with the exception of a few officers. When officers and crew are of different nationalities, not only do problems of communication and education regarding mitigation measures for sea birds arise, but also in such situations there is often a different perspective on fishing operations. In particular, previous traditions of passing knowledge and skill from more experienced crew members to newcomers are broken as foreign crew members are not seen as apprentices. Instead, foreign crews are considered a necessity to continue operations when economic conditions preclude the attraction of Japanese workers.

8.4.2 The Tori Pole Solution

Although it is not known who first invented the tori pole, it has been documented that a Japanese fishing master working in the southern bluefin tuna fishing grounds was deploying the device as early as 1988. The tori-pole system involves a solid line towed from a pole installed at the stern of the fishing vessel, equipped with a curtain of streamers and bird-avoidance tapes, aimed at deterring seabirds from taking baited hooks. Since albatrosses have poor in-flight manoeuvrability, their feeding behaviour is disrupted when obstacles are set above the area where baited hooks are cast onto the water surface. The tori pole was initially designed to prevent seabirds from stealing fish from baited longline hooks and therefore increase the catch of target species, as well as minimise seabird interference with line retrieval. In addition to these objectives, some fishers may have welcomed the tori pole because they believe that seabirds are an incarnation of the gods and that seabirds indicate good fishing grounds, therefore avoiding the killing of seabirds will bring good luck. For these reasons, the tori pole conformed perfectly to fishers' own interests and thus spread on its on accord throughout the fishery. It was subsequently adopted as a regulatory requirement under the CCSBT as a means of protecting and conserving seabirds, but it is important to recognise that for Japanese fishers, it was not originally intended specifically for that purpose.

8.4.3 Remaining Problems with Seabird By-catch

The implementation of the tori pole in the southern bluefin tuna fishing grounds has been highly successful because it reduced seabird by-catch by one third. Nevertheless, by-catch in this fishery still results in the mortality of seabirds and thus further by-catch reduction is desirable. It has been documented in field trials that other by-catch reduction methods, such as making bait less viable using a harmless blue dye, can be even more effective than the tori pole. However, the introduction of blue-dyed bait faces some obstacles in acceptance and implementation. Firstly, although the cost of the blue dye is low, the crew cannot dye the bait themselves on deck due to the rough weather conditions of sub-Antarctic waters and it is thus necessary to order pre-dyed bait from suppliers, primarily located in China and Vietnam. At present there is insufficient demand for blue-dyed bait to make its cost competitive with standard baits. Fishers are accustomed to changing bait suppliers frequently in order to achieve cost savings, and therefore the additional effort required in acquiring blue-dyed bait is seen as both an additional expense and an inconvenience. The key issue in promoting the use of blue-dyed bait will be to change bait market dynamics so that demand for blue-dyed bait increases, resulting in greater availability and lower prices.

Other potential mitigation measures include weighting of branch lines, setting lines underwater, avoiding disposal of offal from the vessels during line setting, using automatic bait-casting machines and properly thawed bait, setting lines at night, using water-jet devices, and setting from the side of the vessels. These techniques have undergone various types of testing and implementation, and have been shown to have different degrees of effectiveness and acceptability to fishers.

Japanese longline vessels in the North Pacific do not employ the tori pole widely despite its proven effectiveness and acceptance in the southern ocean. There are several reasons for this. The most apparent is that there is a relatively lower abundance of threatened seabirds such as albatrosses in the North Pacific compared to the South Pacific and as yet there is no mandate for tori pole usage, nor any other seabird by-catch mitigation measures in the North Pacific. Furthermore, fishers have resisted calls for voluntary implementation saying that the design of the tori pole would need to be scaled down for use on the smaller vessels employed in the coastal and offshore longline fleets, which conduct most of the Japanese fishing operations in the North Pacific. In addition to the re-scaling of the device's design, simultaneous operation of the tori pole, and setting and retrieval of longline gear, poses a significant challenge due to the smaller crew size in these fleets. Impediments to widespread adoption of blue-dyed bait also exist in the North Pacific. Onboard dyeing may be possible in some cases but the purchase of pre-dyed bait from suppliers is likely to be preferable given operational constraints such as deck space and crew size. Since Japanese longliners in the North Pacific use domestic suppliers and may prefer to maintain long-standing supplier contracts, different market incentives may be required to influence the availability of blue-dyed bait for North Pacific fleets.

8.4.4 Characteristics of the Japanese Situation

One of the strengths of the Japanese political system is its ability to act quickly to achieve resolution of problems that are raised. However, the range of possible actions that can be taken by government in response to by-catch issues is limited due to its historical relationship with the fishing industry. As a traditionally coast-oriented nation, many decades ago, Japan evolved a system of fishing rights management based on mutual agreements between communities. In later years, as the central government grew stronger, its role was limited to adjusting these agreements as necessary rather than regulating with a firm hand. Japan's heavy reliance on coastal resources, in combination with rapid population growth, also created a need for distant water fishing activities to meet food requirements much earlier than in other countries. As a result, fishing communities have maintained a strong sense of independence and self-governance and the government usually considers it best that new policies be initiated by the fishing community itself.

The maintenance of this historical system during the development of modern Japan has also resulted in a strong hierarchical structuring of the fishing community and its various interest groups. The fishing sector is characterised by a number of industry organisations which serve as channels of information to fishers. While such organisations may facilitate dissemination of information, the large number of layers between government or scientific staff and the fishers themselves can prevent direct communication. In one way, this may result in fishers failing to appreciate international conservation concerns due to no direct experience with such issues, compounded by cultural or language differences, and a lack of attention by the Japanese media to conservation topics. On the other hand, the situation may hinder the recognition of fishers' own innovations by government and the rewarding of such innovations with incentives.

In recent years, the Japanese fishing industry feels it has suffered from a number of negative influences. As certain fisheries have closed (for example the drift net fisheries in the early 1990s) some fishers have converted to other gear types, but many now find their new fisheries are under pressure from a combination of over-capacity and limited resources. In many cases, foreign lobby groups are seen as contributing to fishers' hardships and thus the fishing industry may be reluctant to share information freely. As described above, many vessels are now crewed by a combination of Japanese officers and non-Japanese workers in order to reduce operating costs. Nevertheless, some of these vessels are managing only to cover basic costs and are not otherwise profitable, hence such vessels are unwilling to make any significant investment in by-catch mitigation gear or training.

8.4.5 Outlook and Conclusion

It is likely that several factors, working in concert, will be necessary to resolve seabird by-catch issues in Japanese longline fisheries. Current initiatives by government and scientists to provide educational materials to fishing industry organisations in the form of laminated panels, booklets, posters and educational videos should be continued. While fishing industry representatives are responsive to these initiatives, other efforts to directly contact fishers through educational and feedback sessions in fishing ports should also be pursued.

Despite an expected increase in awareness of conservation issues, the response of the fishing sector is likely to continue to be based on economic factors. In this sense, the ongoing, gradual reform of the fishing industry through both vessel de-commissioning and the inevitable discontinuance of unprofitable operations will result in a fleet that should be able to absorb the costs of by-catch mitigation. However, further economic incentives may be necessary to subsidise by-catch mitigation, at least in the initial stages of implementation. This could take the form of government-sponsored research into tori pole re-sizing for the North Pacific or stimulating the market for blue-dyed bait.

In parallel, it is essential to continue by-catch research activities. Although most technical aspects of reducing by-catch are already well understood, further work to facilitate implementation of existing techniques by specific fleets may be required, for example improving methods for side setting of lines or new dyeing methods for bait. In addition, it is necessary to study the by-catch situation in various fleets and areas in order to identify which operations are most likely to benefit from mitigation measures.

Ultimately, successful solutions will not be achieved by top-down decision-making in Japanese fishing fleets. Mitigation measures which are effective and easy to implement, and which will diffuse through the fishery by means of the fishers themselves, provide the best hope for achieving by-catch mitigation targets while maintaining economically viable longline fisheries.

This case study has illustrated:

- The tori pole mitigation method was implemented independently by Japanese fishers in response to their own desire to reduce seabird by-catch;
- Barriers to implementation of the tori pole in other fisheries stem from important structural and economic differences in operations;
- The Japanese Government is working to distribute educational materials and sponsoring mitigation research, but does not have a history of strong intervention in fishing operations; and
- Fishers may respond most favourably to low- or no-cost measures proposed by the industry itself, particularly when incentives are provided by Government.

8.5 Case Study 4 – Southern Seabird Solutions: Conservation Through Cooperation

(by Simon Thomas and Janice Molloy)

To spread news, find a gossip; to spread new behaviour, find a role-model.

The Southern Seabird Solutions Trust developed from a workshop in Nelson, New Zealand in July 2002 that incorporated fishers and fishing company representatives, government departments, environmental NGOs and seabird researchers.

The timing was certainly right in terms of engaging the interest of fishing companies because the killing of 312 white-chinned petrels by a kingclip auto-liner seven months earlier had gained the attention of the New Zealand public and politicians. The issues of seabird by-catch and mitigation had been known about and worked on in industry circles for a long time. But this incident, and the political interest it generated, suddenly made progress more urgent. In addition to this, forward-thinking industry participants at the Nelson workshop realised that seabird kills in other parts of the world – by other fisheries – might affect them if seabird breeding populations on New Zealand's offshore islands fell as a result. Tough measures would be introduced for vessels fishing in our waters to safeguard these seabirds if this happened. As most of these seabirds actually spend much of their lives in other parts of the world, working with Southern Africa, South America and Australia was seen as critical.

Having something at stake helped engage companies, fishers and the wider industry. They could agree something needed to be done, and that it needed to happen out on the water – where both the problems and solutions lay.

It was decided that there was a need to accelerate the transfer of 'seabird-smart' attitudes and behaviours amongst the skippers and crews in a fishing fleet. Doing this required the trust and co-operation of all parties – government agencies, environmental NGOs and, above all others, fishers and fishing companies. And, surprisingly, this trustful attitude came quite readily.

All involved in the Southern Seabird Solutions group saw the issue as solvable and as something they needed to work together on. And while there were not infrequent tensions between these various parties in other areas of fisheries management, we all 'left our swords at the door' when we came together for Southern Seabird Solutions meetings.

Engendering this trustful and cooperative approach between our partners has been the cornerstone of the group's success.

Most of Southern Seabird Solutions projects involve fishers, simply because fishers are most receptive to new ideas from their peers. For instance, the group has carried out two skipper-exchange projects to date, one between Chile and New Zealand and the other between New Zealand and Reunion Island. Both have had excellent outcomes. In the Chilean exchange, the skipper returned home motivated to continue spreading good practice in his own fleets. He has attended workshops around Chile talking about his experiences in New Zealand and describing the measures he observed being used. In the Reunion Island exchange, the whole fleet has begun using a new weighted longline that sinks the line quickly out of the diving range of seabirds.

Both fishers and companies need to feel good about themselves, and about their adoption of 'seabird-smart' fishing techniques. Like everyone, they like to be seen as 'good citizens'. So with the help of environmental NGOs and government agencies, the Southern Seabirds group has helped industry mitigation efforts and successes to be celebrated publicly through the general news media, as well as through seafood trade publications.

The Southern Seabirds group focused much of its efforts during its first three years on this communication of successes; and most particularly communicating these successes within industry circles. Monthly stories were carried in the *Seafood New Zealand* magazine that celebrated role-model skippers, new 'seabird-smart' fishing technologies and information on the birds themselves. The stories had a huge effect in building and maintaining support from across the fishing community and wider industry.

A classic example of the effect of these was the reception that a new seabird mitigation advisory officer (a fisher himself) got when visiting a fleet he had never had contact with previously. He found the skippers all read *Seafood New Zealand* magazine, and knew who he was, and about the concept of 'seabird-smart' fishing. They welcomed him aboard their vessels and were eager to have him help them improve the ways they fished.

We have found this role of seabird mitigation advisory officer crucial in spreading attitudinal and behavioural change across a fleet, particularly amongst inshore fleets that may have many small vessels. And as the previous example illustrates, we found targeted communications materials that support their work helped accelerate the progress an advisory officer makes with a fleet. We found that fishers can also more easily stand in the shoes of another fisher and know how to communicate the message in a way that is meaningful. So the group has always aimed at fisher-to-fisher communication.

We recently ran workshops at different ports in northern New Zealand, aimed at inshore longliners, and hosted by their local fish-receiving shed. We brought several role-model fishers from other fleets to talk at these, as well as an environmental representative with knowledge of seabirds and their conservation. Part of our purpose was to thank fishers for their efforts to date and to encourage them to continue using seabird-smart fishing practices. The workshops were all held in the local bars, which ensured good attendance. Participants were given T-shirts with the group's logo and catch phrase 'Conservation through Cooperation'; the venues and the T-shirts helped create an atmosphere of receptiveness.

Other fisher-to-fisher work included the production of a 'seabird-smart' fishing video. This was hosted and narrated by a fisher, and largely featured skippers and vessel managers from different fleets talking about the issue. The video proved hugely popular, with copies being distributed free amongst New Zealand fleets. The video has since been translated into Spanish and a special introduction added from the Chilean fisher involved in the skipper-exchange programme.

A critical factor in the success of Southern Seabird Solutions has been developing a goal that everyone can agree on and work towards. In addition, gathering this initially loose coalition of interest groups under a name and governance structure has resulted in a cohesion and identity that members are proud to be part of. The main limiting factor to date has been securing enough resources to undertake the many additional projects we have lined up.

In summary, the key elements of success of Southern Seabird Solutions have been good timing in terms of the public profile of the issue, developing a common goal, patience, a no-surprises approach, behaviours that engender trust, use of fishers as role models and messengers, and public acknowledgement of the efforts of fishers through the media. We have had a high level of engagement from longline fleets, but have yet to achieve this with the New Zealand large-scale trawl fleets or with recreational fishers, both of whom catch seabirds. These fisheries will be among our next priorities.

8.6 Case Study 5 – Networks, Knowledge and Communication: An Integrated Approach to Empowering Fishers to Reduce Turtle By-catch

(by Hoyt Peckham, Johath Laudino-Santillán, and Wallace J. Nichols)

In August 2002, Anselmo Ruiz-Camacho, a halibut fisher from Baja California Sur, Mexico, asked, 'How can loggerhead turtles possibly be endangered? I caught thirty in my nets this morning.' We were astonished. We had come to Puerto Lopez Mateos, a small fishing village on the Pacific coast of Baja California Sur to study sea turtles, but not dead ones. 'Thirty?' we asked, hoping we had misunderstood his heavily accented Spanish. 'Thirty,' he confirmed. 'All but two dead.'

We spent the next few days offshore with Anselmo hand-catching turtles, and during that time we did our best to help him answer the question for himself. We explained how loggerheads in the North Pacific only nest in Japan and, drawing maps in the sand, we explained that they swim across the Pacific as juveniles to feed their way to maturity in the rich waters of Baja California Sur (BCS; Fig. 8.1). He protested, saying he and his friends frequently catch loggerheads in summer, that one guy caught seventy in a single day, so how, really, could they be *endangered*? Together we looked over graphs of nesting trends from Japan. Fewer than 1500 loggerheads had nested in the North Pacific the winter before, and nesting had declined 50 to 80% over the past decade (Kamezaki et al. 2003).

Anselmo's question was painfully ironic but not unusual; we've heard the same question from dozens of other fishers along his coastline. Despite local perceptions, there are now few loggerheads in the Pacific, and they are declining rapidly (Kamezaki et al. 2003). Those few left appear to be numerous to Anselmo and his fellow fishers because they regularly aggregate at unusually high densities off the Baja California peninsula. We interviewed these fishers at length and conducted surveys along a 50 km shoreline to gauge the extent and identify the cause of local turtle by-catch. Local gillnetters catch an average of four turtles per week during their four month halibut season. Most turtles are caught dead, and fishers throw the

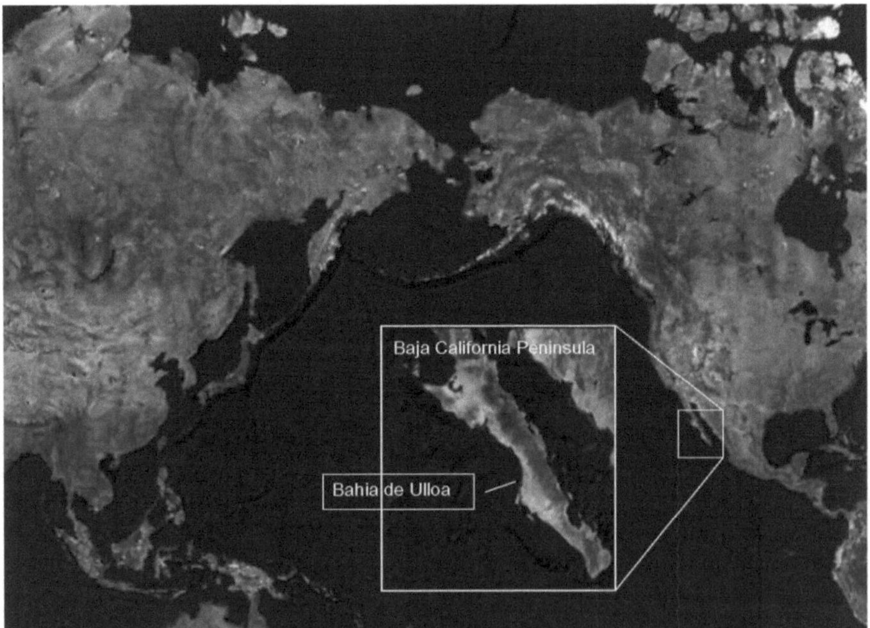

Fig. 8.1. Pacific Ocean. Inset: Bahia de Ulloa

majority of carcasses overboard after disentangling them. Somewhere between thirty to seventy *pangas* (6 to 9 m outboard-powered skiffs) fish bottom-set gillnets and longlines out of Puerto Lopez Mateos, Anselmo's homeport.

Extrapolating from these data we came to understand that by-catch along Anselmo's coast is one of the most significant known sources of loggerhead mortality in the entire North Pacific (Peckham et al. 2004). We realised that the future of loggerheads in the North Pacific lies heavily in the hands of Anselmo and other Baja California Sur fishers. Our objectives thus became clear: (i) to empower the people of this coast to answer Anselmo's question for themselves; and (ii) to partner with them to develop practical by-catch solutions.

We began with what we knew was working – personal conversations and shared experiences. Anselmo quickly grasped the reality of the by-catch problem, and he acted on it. He and his wife and four sons adopted one of the turtles we captured together and helped us fit her with a satellite transmitter and release her. They named her Esperanza ('Hope' in English) and avidly tracked her movements via regular updates we faxed and emailed them (Fig. 8.2). Anselmo left the fishery the following summer in

Fig. 8.2. Family with the loggerhead turtles they helped to catch, fit with a satellite transmitter, release and track

part to avoid catching sea turtles, and he became a spokesman for reducing turtle by-catch. We partnered with other fishers in other towns along the BCS coast and explored, through group discussions, the full costs of by-catch such as time and resources lost to disentangling turtles and repairing damaged nets. We found that once fishers appreciate the Pacific-wide impact and true costs of their local by-catch, they usually strive to reduce and eliminate that by-catch. The challenge, then, was to scale-up this success.

8.6.1 Conservation Mosaic

Based on this modest success, we began implementing a conservation mosaic strategy (Nichols 2003). The mosaic consists of three approaches to achieving conservation, each informed by an established literature and differing degrees of proven effectiveness. The novelty of the mosaic lies in strategic integration of these approaches: (i) building a conservation ***network*** of fishers, students, teachers, activists, researchers, managers and other coastal people; (ii) drawing on these partnerships to derive new ***knowledge*** to develop locally practical solutions; and (iii) ***communication*** of this knowledge in resonant and appropriate ways to avoid by-catch and foster a sustainable ethic (Fig. 8.3).

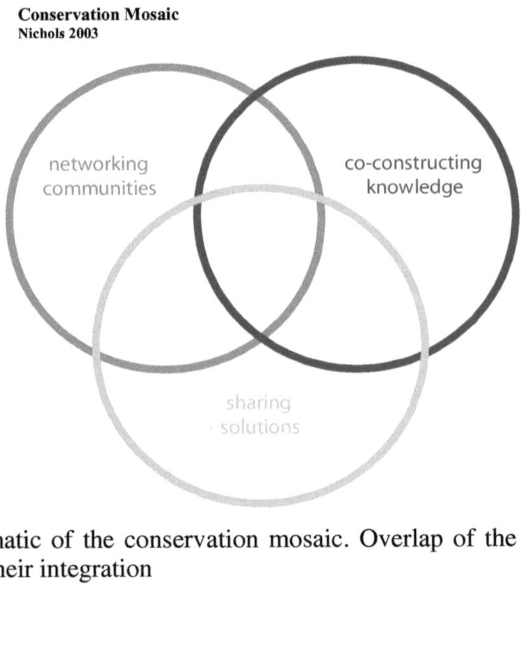

Fig. 8.3. Schematic of the conservation mosaic. Overlap of the three spheres of action reflects their integration

8.6.1.1 Community Conservation Networks

Clearly, our team doesn't have the time or resources to reach every last fisher along the vast, isolated Baja California Sur coast. But the Grupo Tortuguero, an emerging community conservation network, does (Pesenti et al. 2005). Networks are decentralised, non-hierarchical, diverse and resilient (Barabasi 2002). As such, they are ideal for addressing widespread problems and creating the social change needed to address by-catch issues in isolated fishing villages.

We build local conservation capacity by partnering with fishers like Anselmo directly, by engaging local women's and youth groups and by offering internships for local students. These conservation leaders are empowered and connected through workshops, regional meetings and international conferences. By interacting with colleagues from other towns, regions and countries, these leaders' perspectives are broadened so that they appreciate the global impact of local by-catch and learn ways to avoid it. This conservation network serves as a new social fabric that fosters and facilitates a culture of marine conservation. Among other awareness-raising initiatives to date, we have brought Spanish-speaking Japanese biologists to Baja California fishing communities. When Japanese experts share their firsthand experience of declines in nesting turtles, local leaders increasingly appreciate the importance of protecting juvenile loggerheads in their waters (Fig. 8.4). These leaders then become the local spokespeople for reducing by-catch, sharing the problem and working towards solutions with their families, friends and neighbours.

Fig. 8.4. Mizuno Kojiro (centre), a Spanish speaking Japanese biologist, shares the decline in nesting turtles he and his colleagues are witnessing in Japan through school outreach and fishers' workshops

8.6.1.2 Co-constructing Knowledge

Ecological research on turtles has been used to reduce by-catch in numerous fisheries through modification of both fishing gear (e.g., use of turtle excluder devices in shrimp trawling; Crowder et al. 1994) and practices (e.g., deeper setting of longlines; Polovina et al. 2003). Developing such solutions requires detailed knowledge of the fisheries involved and the ecology of affected species (Hall et al. 2000). Involving fishers in conservation planning can result in better solutions that account for fishers' needs and incorporate their vast local knowledge while protecting imperiled populations. Moreover, fishers' investment in the conservation process can increase subsequent adoption of conservation solutions (Nichols 2003; Santora 2003). This last point is especially important along isolated coasts such as the Baja California peninsula where enforcement is scarce and adoption of conservation solutions is largely up to fishers.

Drawing on the relationships described above, we formed a task force of local fishers, managers, community members and conservation biologists to: (i) elucidate turtle diving and feeding behaviour; (ii) collect data on stranding rates and mortalities; and (iii) experiment in modifying gillnet design and deployment. Local fishers are thus learning firsthand both the conservation process and the status of loggerhead turtles while helping to generate new knowledge such as data on turtle diets, diving and movement that are credible both locally and in scientific circles (Fig. 8.5).

Fig. 8.5. Alejandro Camacho and Victor de la Toba carry a loggerhead they have fitted with a satellite transmitter to their boat for release. Despite having accidentally caught thousands of loggerheads over his thirty year career, this was the first turtle Alejandro released alive

The task force is combining local ecological knowledge with these data to develop practical solutions. For instance, tracking indicates that turtles are utilising fine-scale foraging hotspots. Fishers are enthusiastic that they might be able to reduce by-catch by avoiding these local hotspots. In this way fishers' personal participation in deriving new ecological data and combining them with their local knowledge directly empowers them to conserve sea turtles.

8.6.1.3 Communication and Outreach

The emerging field of community-based social marketing guides our communication and outreach initiatives (MacKenzie-Mohr and Smith 1999; Jacobson 1999). The social marketing approach consists of a four-step process: (i) local attitudes and behaviours are assessed; (ii) a range of media and events are evaluated for their effectiveness; (iii) outreach campaigns are designed to inform and engage all fishers and their families; and (iv) the effectiveness of campaign components are measured in terms of changes in local attitudes and behaviours (Delgado 2005).

According to these precepts, our team designs and continually refines a suite of outreach initiatives to convey our core message of empowerment: specifically that BCS fishers and families hold the fate of the Pacific loggerhead in their hands. Informative workshops for fishers and curriculum enrichment for schoolchildren convey the facts about by-catch behind the message. To supplement these experiences across whole communities, Grupo Tortuguero offers a range of locally resonant media including comic books, children's books, neighbourhood murals, informative brochures and local radio programming. Public events such as regional festivals, holiday parades, sports competitions and puppet shows are offered to celebrate sea turtles as natural treasures to be cherished and protected. Moreover, the network is working closely with ecotour operators to explore the feasibility of offshore turtle tours. Because loggerhead and olive ridley turtles aggregate in certain areas at extraordinarily high abundance, offshore trips could offer unprecedented experiences with foraging turtles for ecotourists and alternatives to gillnetting for boatmen. In all of these ways, fishers and their families are informed, engaged and empowered to protect sea turtles and the ecosystems they inhabit.

8.6.2 Success to Date

As a result of their personal participation in this research and their recognition of the Pacific-wide impacts of their local by-catch, the fishers of Puerto López Mateos, BCS declared the loggerhead high-use area off their coast a 'Fishermen's Turtle Reserve' in February 2006, thus self-limiting turtle by-catch in this region. In this way fishers' personal participation in deriving new ecological data from their local knowledge directly empowered them to effect conservation change. Currently, the fishers of Puerto López Mateos are seeking federal legislation to officially protect their reserve.

The novelty and strength of this approach has yielded a conservation constituency among fishers and their families characterised by local pride, empowerment and stewardship. Three years into this 5 year initiative, preliminary results indicate decreased turtle by-catch and poaching, changes in local attitude and an emerging 'sea ethic'. Enforcement agents from PROFEPA, SAGARPA and local councils are pursuing turtle violations that in the past were ignored. Increasing numbers of fishers are self-enforcing turtle protection amongst themselves and between and within their cooperatives. Fishers, students and their families are celebrating sea turtles through festivals, artwork and music. All of this translates into turtles saved and steps toward the recovery of turtle populations. Finally, there are indications that this emerging 'sea ethic,' borne by people's increasing interest in turtle conservation, is leading them to manage fisheries such as lobster and abalone more sustainably, an unexpected but welcome result.

8.6.3 Summary: Global Impacts of Small-scale Fishing

Small-scale fisheries such as the one described herein are ubiquitous to the coastal waters of developing nations. Because small-scale fishers are often unlicensed, their boats are usually unregistered and their catch and by-catch are rarely quantified. This means that the impact of these fisheries has gone virtually unnoticed. But as this case study shows, by-catch in these fisheries may jeopardise both fishers' livelihoods and endangered species as much as, and perhaps more than, any other fishing sector.

Because regulation and enforcement of such fisheries is often lacking and/or ineffective, conservation can therefore depend almost entirely on small-scale fishers' direct participation. Our collective challenge then is to empower small-scale fishers around the world to conserve shared marine resources. We suggest that our conservation model could be employed in other regions to build grassroots constituencies among fishers and their families characterised by local pride, empowerment and stewardship to conserve marine species and their ecosystems.

8.7 Case Study 6 – Working with Hawaii-based Longline Fishers to Abate Fisheries By-catch

(by Eric Gilman, Jim Cook and Sean Martin)

8.7.1 Introduction

Hawaii-based pelagic longline fisheries are faced with strong incentives to reduce by-catches of sensitive species, including sea turtles and albatrosses. Here we highlight several approaches, some effective, others not, to engage Hawaiian longliners in getting directly involved in trying to abate fisheries by-catch.

In 2004, there were 125 active Hawaii-based longline tuna and swordfish vessels, which made 1,338 trips, setting about 32 million hooks. Table 8.1 summarises target species catch-per-unit-of-effort for the combined Hawaii-based longline tuna and swordfish fisheries from 1999 to 2004. In 2004, the Hawaiian longline fisheries landed approximately 8,200 tonnes and generated ex-vessel revenues estimated at $US 42.6 million with tuna (*Thunnus spp.*) the dominant components of landings.

Because of concerns over turtle interactions, the Hawaii-based longline swordfish fishery was closed for over two years and is now subject to strict management measures. Measures include prescribed use of large circle hooks and fish bait, restricted annual effort, caps on turtle captures, 100% onboard observer coverage, required possession and use of specialised turtle de-hooking equipment and mandatory attendance of annual protected

Table 8.1 Hawaii pelagic longline tuna and swordfish fisheries catch-per-unit-of-effort (CPUE), number of fish per 1,000 hooks, 1999 – 2004 (U.S. National Marine Fisheries Service Pacific Islands Regional Office unpublished data, March 2005).

Year	Tuna CPUE	Sharks CPUE	Billfish CPUE	Other CPUE [a]
1999	9.21	4.59	3.9	4.8
2000	8.18	3.91	2.88	4.8
2001	8.64	2.1	1.61	4.21
2002	7.48	1.87	0.98	4.27
2003	6.33	2.32	1.77	4.58
2004	6.42	2.34	1.24	5.49

[a] mahimahi, moonfish, oilfish, pomfret, wahoo

species workshops by vessel operators and owners. If seasonal limits on turtle interactions are reached, the fishery is closed for the year, and if a threshold is exceeded, federal resource management agencies consult to determine if additional restrictions on the fishery are warranted. Furthermore, the Hawaiian longline swordfish and tuna fleets are each authorised to take annually, through injury or mortality, only one endangered short-tailed albatross. If more than one short-tailed albatross is observed to interact with gear of the Hawaiian longline tuna or swordfish fleet in a single year, resource management agencies consult to determine if the fleet should be required to employ additional seabird avoidance measures.

Tens of Laysan and black-footed albatrosses are now annually captured by the fleet, down from thousands that were caught before the fleet was required to employ seabird avoidance methods and restrictions on swordfish fishing effort. The fleet has not had any observed captures of short-tailed albatrosses. Since June 2001, management authorities have required the Hawaiian longline tuna and swordfish fisheries to use a number of measures intended to reduce seabird by-catch, including weighted branch lines, thawed and dyed bait, offal discards, and night setting in certain geographical areas for certain components of the fleet. Interactions between the fleet and false killer whales is another issue that has received recent attention. While there have been claims that this is causing population-level effects, in reality, there is little understanding of the status and trends of false killer whale populations nor of the consequence of interactions with longline gear.

8.7.2 Litigation

Over the past five years, there have been numerous lawsuits filed against the United States fishery management authority by environmental organisations and the Hawaii Longline Association over the by-catch of sea turtles, seabirds and whales by Hawaii-based longline fisheries. There have been a number of positive results from the litigation, but overall we believe that this has not been a wise long-term approach or efficient use of money, time, or energy to address fisheries by-catch.

There was little attention paid to reduce by-catch of sea turtles in the Hawaiian longline fisheries since the fisheries inception until the litigation began in 2000 (which aimed to close the fisheries) brought about substantial improvements involving changes in fishing gear, fishing practices and methods to handle and release caught turtles. Turtle by-catch levels are now much lower than in the past, and turtles are being released with less injury and a greater chance of survival.

Another positive result of the litigation was increased cohesiveness of Hawaii Longline Association members. The numerous ethnic groups comprising the fishery came together to counter efforts to eliminate their source of livelihood and denigrate the reputation of the Hawaiian longline fisheries. The industry is now in a much better position to represent their interests.

However, even after substantial improvements were adopted by fishery management authorities and the longline industry, the litigation continued, as some environmental groups continued to pursue their goal of permanently closing the fishery. The result was that the fishers became bitter, were much less receptive to collaborating with outside groups, and lost the drive to pursue voluntary initiatives to innovate new by-catch solutions, which might also be exportable to longline fleets internationally. Other environmental groups, that had a goal of reducing fisheries by-catch and reducing this source of turtle mortality by working with fishers, had a much harder time gaining industry's trust to work with them as a result of the actions of the groups that were working to close the fishery. In fact, the efforts to close the Hawaiian fleet may have actually increased turtle and bird mortality: During a four-year closure of the Hawaii longline swordfish fishery due to concerns over by-catch of sea turtles, swordfish supply to the United States marketplace traditionally met by the Hawaiian fleet was replaced by imports from foreign longline fleets, including fleets from Mexico, Panama, Costa Rica, and South Africa, which have substantially higher ratios of sea turtle captures to unit weight of swordfish catch and less stringent or no measures to manage seabird by-catch. Groups that wanted to pursue collaborative work with the Hawaiian longline fleet to make the Hawaiian fleet a model fishery, and to export identified solutions internationally, were frustrated by the misplaced efforts to close the Hawaiian fisheries.

The Hawaii Longline Association spent over $US 1.6 million and innumerable staff hours over the past five years as a result of involvement in this litigation. If this money, plus the funds spent by the United States Government and environmental groups on the litigation, had instead been used to conduct research to find effective and commercially viable solutions in the Hawaiian fleet and abroad, this might have saved many more turtles' lives.

As we will describe next, collaborative, industry-led research has been effective at reducing seabird by-catch in Hawaiian longline fisheries and substantially more progress has been made to find effective and practical solutions to seabird by-catch than turtle by-catch in Hawaiian pelagic longline gear, without litigation as a motive, and at a cost an order of magnitude lower than that spent on law suits.

8.7.3 Collaborative Research and Commercial Demonstrations to Reduce By-catch

Between 1999 and 2003, the Hawaii Longline Association collaborated with fishery management authorities and an environmental organisation to conduct three experiments and commercial demonstrations of various strategies (blue-dyed bait, towed buoy, offal discards, streamer line, underwater setting chute, and side setting) to reduce seabird by-catch in longline gear (Fig. 8.6). The United States Western Pacific Regional Fishery Management Council was the driving force behind the initial experiment, and researcher Brian McNamara was an excellent choice, as he quickly gained the trust of Hawaii longline fishers who worked with him to make the initial trials of various seabird avoidance methods a success. Two subsequent cooperative experiments were initiated by Eric Gilman, a scientist initially employed by an environmental organisation called the National Audubon Society and later a new organisation called the Blue Ocean Institute, who took the initiative to approach industry and fishery managers to work together to plan, fund and implement the project. Hawaii Longline Association representatives Sean Martin, Jim Cook, and Scott Barrows; Western Pacific Regional Fishery Management Council director Kitty Simonds; and United States National Marine Fisheries Service scientist Dr. Chris Boggs, joined the team to plan and implement the cooperative research project. Nigel Brothers, a consultant recently retired from the Tasmania Parks and Wildlife Service, Australia, joined the team and was key to securing the eventual success of these two latter experiments. Nigel's extensive experience working on longline vessels around the world, understanding of albatross behaviour, approach to working with fishers, stubbornness and perseverance to find effective and viable solutions to seabird by-catch, greatly contributed to the success of these experiments.

From these experiments we determined that several seabird by-catch avoidance methods are capable of nearly eliminating bird captures in longline fisheries when effectively employed. Our industry-led experiments focused on identifying the most effective seabird by-catch abatement methods that are also economically viable and practical. Fishery management authorities recently amended regulations on measures for the Hawaii longline fleet to reduce seabird by-catch based on results from this most recent research.

Fig. 8.6. Industry lead research on an underwater setting chute (left panel) and side setting (right panel), two promising techniques to reduce seabird by-catch, in the Hawaiian pelagic longline fisheries

Longline fishers are some of the most qualified people to develop and improve seabird by-catch mitigation techniques. They have a large repository of knowledge and information related to by-catch, which can be tapped to contribute to finding effective and practical solutions. This has been demonstrated by the successful research initiatives in Hawaii and elsewhere. Mitigation methods that effectively avoid seabirds, do not reduce fishing efficiency, or better yet, increase fishing efficiency and provide operational benefits, have the highest chance of being accepted by industry. The longline association became an active participant to address seabird by-catch problems by instituting and participating in research and commercial demonstrations and supporting adoption of regulations based on the best available science before restrictions, embargoes and possible closures were imposed on the fleet. This bottom-up approach fostered a sense of industry ownership for effective seabird mitigation methods, and resulted in high compliance with the resulting rules mandating the use of seabird avoidance methods. By being directly involved in the development and testing of seabird avoidance methods, Hawaii longline fishers developed a sense of ownership for these tools and now support their required use.

8.7.4 Economic Viability, Practicality and Enforceability Considerations in Research Designs

The experiments on techniques to reduce seabird by-catch in the Hawaii-based longline fisheries provide an example of how research can be designed to collect information on economic viability, practicality and enforceability. Analysing differences in the effects of alternative seabird avoidance methods on bait retention, hook setting rates and catch-per-unit-of-effort of targeted fish; operational benefits and costs; time and money to adopt and employ; and enforceability is of great interest to industry, fishery management authorities and other stakeholders.

Given the political context and management frameworks of the majority of the world's longline fisheries, there is a need to focus on the commercial viability of by-catch reduction methods in order to catalyse changes in fishing methods and gear and regulatory measures that will abate longline by-catch. To resolve global fisheries by-catch problems, there is a need to identify and institute the broad use of methods that not only have the capacity to minimise by-catch of sensitive species, but which are also practical and convenient and provide crew with incentives to employ them consistently and effectively. That is, it is critical to account for economic and social values of longline fisheries to achieve changes that abate by-catch.

For instance, because the loss of bait to seabirds and concomitant reductions in the catch of fish can be significant, the use of seabird avoidance measures is expected to lead to cost savings for longline fisheries. However, most longline fleets do not employ effective seabird avoidance methods despite the availability of effective methods that also increase fishing efficiency (Brothers et al. 1999a; Gilman 2001; FAO 2003). Reasons for this may be: (i) low industry awareness of the availability, effectiveness and practicality of these methods; (ii) few national fishery management authorities manage interactions between seabirds and longline vessels or require employment of effective seabird avoidance methods (Brothers et al. 1999a; BirdLife International 2003; FAO 2003; Gilman and Freifeld 2003); and (iii) lack of a sufficiently strong economic incentive for industry to change long-standing fishing practices. Recognising that this context also applies to many global commercial marine fisheries, maximising industry's sense of ownership for using effective by-catch avoidance measures and providing industry with incentives for voluntary compliance are needed. Commercial fishing industries respond best to economic incentives and disincentives (Gilman et al. 2002). By-catch mitigation methods that increase fishing efficiency and have operational benefits have the best chance of being accepted by industry. Eco-labeling and certification programs can also provide industry with strong market-based and social incentives to meet criteria to be certified as a sustainable fishery, including the employment of effective by-catch reduction methods, but requires adequate marketing of the label to make it economically viable for industry to participate (Gilman et al. 2002). Additionally, if regulations requiring the use of by-catch avoidance methods are effectively enforced and carry sufficient economic consequences for noncompliance, broad industry compliance can be achieved.

8.7.5 Outreach, Capacity-building and Disseminating the Lessons Learnt

The Hawaii longline association, in partnership with fishery management authorities and environmental conservation groups, has produced a number of educational materials on methods to abate fisheries by-catch. These include a poster (Fig. 8.7) and pamphlet on side setting to reduce seabird by-catch, a poster on best practices to handle and release incidentally caught seabirds in longline gear and methods to reduce seabird capture, and a booklet on methods to reduce sea turtle by-catch in pelagic longline gear and practices to handle and release captured turtles. The Hawaii Longline Association is also able to disseminate lessons learnt from experiments and commercial demonstrations and learn from by-catch research in other fisheries through participation in, and providing financial support for, conferences such as the International Fishers Forum series.

The Hawaii Longline Association is working with management authorities and the Blue Ocean Institute to implement a dockside technical assistance program for longline vessels to convert deck designs from the conventional setting position from the stern, to the side of the vessel to reduce seabird by-catch. Deck conversion requires considering the deck position for setting, selection of main line shooter hinges and hydraulics, line pullers, motor and mounting plate design for starboard setting, and the design, construction and installation of a bird curtain. Technical assistance is also available to captains and crew on best fishing practices for setting from the new position, including timing for clipping branch lines to the main line and practices for throwing baited hooks.

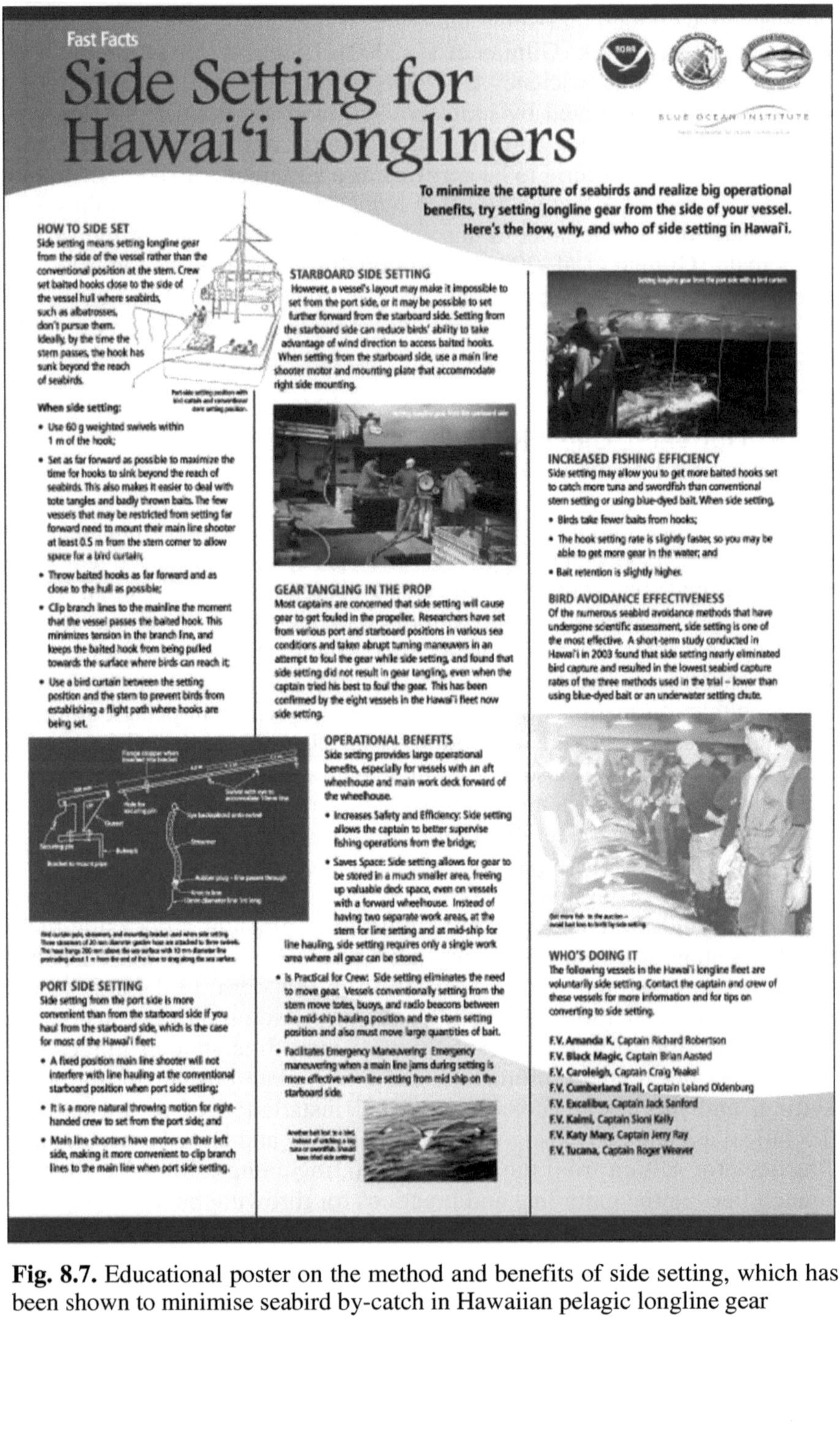

Fig. 8.7. Educational poster on the method and benefits of side setting, which has been shown to minimise seabird by-catch in Hawaiian pelagic longline gear

These education and outreach programs are an investment to bring about changes in behaviour and attitudes by having an industry that is better informed of prescribed fisheries by-catch avoidance methods, and, in some cases, operational benefits from employing these techniques. Showcasing the results of industry-led research to abate fisheries by-catch also has the benefit of broadly disseminating the results so that the effective methodology can be replicated in other fleets worldwide and ineffective components can be improved.

8.7.6 One Fleet Pilot Project

The Hawaii Longline Association worked with an environmental organisation and fishery management authorities to examine the state of knowledge of employing fleet communication programs to reduce fisheries by-catch, and is now planning to institute a pilot program to reduce by-catch of sea turtles and albatrosses. Instituting a fleet communication system to report near real-time observations of by-catch hotspots enables a commercial fishery to operate as a coordinated 'One Fleet' to substantially reduce fleet-wide capture of protected by-catch species, including fish, seabirds, sea turtles and marine mammals. This benefits the by-catch species, reduces waste, can provide economic benefits to industry by reducing the risk of exceeding government-established seasonal by-catch thresholds, and can avoid possible future declines in target species catch resulting from by-catch of juvenile and undersized individuals. We analysed case-studies of fleet communication programs in three United States fisheries; the North Atlantic longline swordfish fishery; the North Pacific and Alaska trawl fishery; and the Alaska demersal longline fishery. Available information from these case-studies supports the inference that they have substantially reduced fisheries by-catch and provided large economic benefits that outweigh relatively nominal operational costs.

It is not yet known how likely it is that the Hawaii longline swordfish and tuna fleets will annually exceed seasonal sea turtle by-catch limits. This makes it difficult to assess if economic benefits from instituting a 'One Fleet' protocol, resulting from enabling the fleet to operate for a longer time period, will outweigh the economic costs from managing the fleet communication program. Furthermore, it may not be possible to determine definitively the effect of instituting the fleet communication program on sea turtle and seabird by-catch rates, due to the lack of a suitable control for comparison. Historical by-catch rates would not provide a suitable comparison because the fleet is now using different methods designed to minimise seabird and sea turtle by-catch. Furthermore, comparison of by-catch rates from different time periods can be confounded by numerous variables, including weather, seabird and turtle behaviour, fishing practices, location of fishing grounds and consistency in observer methods. However, if some of the Hawaii longline vessels opted not to participate in the fleet communication program, a comparison of by-catch rates of participating and non-participating vessels could provide an understanding of the effect on by-catch rates from this single factor, assuming that there are no other substantial differences between the two categories of vessels. This was possible for the Alaska demersal longline fisheries fleet communication program. Non-monetary benefits to the Hawaii longline industry from instituting a 'One Fleet' program to reduce turtle and bird by-catch could be substantial, such as from positive media coverage and other values not described by established financial indicators.

8.7.7 Conclusions

In Hawaii and elsewhere, we have seen that fishers are some of the most qualified people to develop and improve by-catch avoidance strategies. Fishers have a large repository of knowledge and information related to by-catch, which can be tapped to contribute to finding effective and practical solutions. Mitigation methods that effectively avoid by-catch, do not reduce fishing efficiency, or better yet, increase fishing efficiency and provide operational benefits, have the greatest chance of being accepted by industry. Fishers and fishery associations need to become active participants to address by-catch problems by being involved in research and commercial demonstrations, implementing best practices, and supporting adoption of regulations based on the best available science before restrictions, embargos and possible closures are imposed on them.

Most countries have a low degree of political will to address fisheries by-catch problems and, as is the case in Hawaii, have scarce resources for enforcement of by-catch management measures. Few national fishery management authorities have frameworks to manage interactions between sensitive by-catch species and fishing vessels and many do not require employment of effective by-catch avoidance methods. A bottom-up approach that fosters a sense of industry ownership for effective by-catch mitigation methods, and concomitant compliance with requirements for using by-catch avoidance methods are needed in these countries.

While the effectiveness of this approach to address fisheries by-catch is broadly recognised, there has been far too little funding allocated for cooperative research and commercial demonstrations to find solutions to sea turtle, seabird and other by-catch problems in longline gear. In the United States, this may be a result of the government's fear of being sued if they propose to conduct or fund experiments in United States fisheries that result in injury to protected resources, even though these experiments will potentially result in substantial reductions in mortality of these species when best practices are identified and spread to multiple fisheries. Some United States fishery management authorities are funding experiments to test technical measures to reduce sea turtle by-catch in longline fisheries abroad, in part, to avoid problems with trying to receive permits and risk being sued by conducting the research in domestic fisheries. But too little research is being supported, there is insufficient coordination resulting in duplicative efforts and solutions found abroad may not be relevant to domestic fisheries. The amount of research being conducted is too small, research needs to occur in individual fisheries to find solutions that we can have confidence will work in our fisheries, and the agencies designing the experiments need to do more to tap into fishers' knowledge to identify new promising strategies.

8.7.8 Acknowledgements

The impetus for preparing this case-study came from insights derived during three years of research on methods to reduce seabird by-catch in Hawaii pelagic longline fisheries. We are grateful for input from Jerry Ray, Barry Woods, George Ching, Kelly Malakai, and Beverly Ray, captain and crew of the F.V. *Katy Mary*, Nigel Brothers, a consultant from Australia, Dr. Chris Boggs and Donald Kobayashi of NOAA Fisheries, and Kitty Simonds and Paul Dalzell of the Western Pacific Regional Fishery Management Council.

8.8 Case Study 7 – Seabird By-catch Mitigation: The Southern Ocean (CCAMLR) Experience

(by J.P. Croxall, K. Rivera and C.A. Moreno)

8.8.1 The Problem

Decreases in albatross populations at sub-Antarctic islands became evident in the mid-1980s, particularly at South Georgia and Iles Crozet where the longest sets of annual population counts were derived (Croxall et al. 1990, Jouventin and Weimerskirch 1990, Prince et al. 1994) (Fig. 8.8). Three sets of observations and data linked these population declines to incidental mortality associated with longline fisheries and thus brought the issue to widespread attention, including that of fishery management organisations:

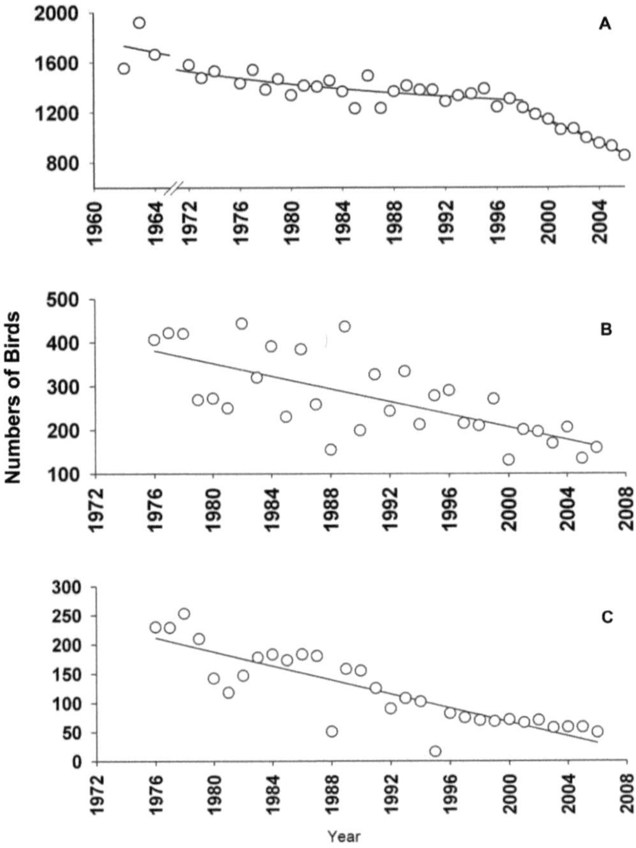

Fig. 8.8. Changes in population size of albatrosses in study colonies at Bird Island, South Georgia (BAS unpublished data), (A) wandering albatross (whole island counts), (B) grey-headed albatross (Colony E) and (C) black-browed albatross (Colony H)

1. Analysis (in 1989) of the 81 recoveries (from 20,000 banded) of wandering albatrosses (*Diomedea exulans*) from South Georgia, indicated that fisheries, particularly those using longline gear, were the main cause of this mortality (Croxall and Prince 1990).
2. Direct estimates (in 1988) of albatross by-catch rates on vessels using longlines to catch southern bluefin tuna (*Thunnus maccoyi*) in the Tasman Sea (Brothers 1991), indicated that, even with rates of < 0.5 birds per thousand hooks, the total annual albatross by-catch from tuna longline fishing could easily exceed 40,000.

3. Direct observations (1991) of albatross and petrel by-catch on vessels fishing for Patagonian toothfish (*Dissostichus eleginoides*) using longlines around South Georgia, suggested that over 3500 petrels (including more than 1000 albatrosses) could be killed annually in this fishery in this region (Dalziell and de Poorter 1993). This longline fishery started in 1989.

In 1991 the above observations were brought to the attention of the Convention for the Conservation of Antarctic Marine Living Resources (CCAMLR), the management authority responsible for regulating fishing in the Southern Ocean and within whose boundaries many of the most affected albatross populations breed (e.g., at South Georgia, Iles Crozet, Kerguelen and Prince Edward Islands).

8.8.2 The Context

The commercial harvesting of Antarctic marine living resources had followed a familiar pattern of prospecting, exploitation and over-exploitation. By the late 1970s, just two centuries after the discovery of the region's resources, most, if not all, populations of Antarctic fur seal (*Arctocephalus gazelle*), several species of great whale and marbled rock cod *Notothenia rossii* were commercially unviable and nearly biologically extinct. Fisheries were switching to Antarctic icefish (*Champsocephalus gunnari*) (already over-exploited by 1980) and Antarctic krill (*Euphausia superba*). There was an overriding fear that not only would recently protected whale populations fail to recover, but that other species dependent on krill and its associated food chain would be affected by its harvesting.

Therefore, in 1977, the contracting parties to the Antarctic Treaty, who had been successful in depoliticising governance and promoting scientific collaboration in respect of the Antarctic Continent, started to negotiate an international convention, primarily to prevent over-exploitation of marine resources, especially Antarctic krill. The resulting CCAMLR Convention, signed in 1980 and in force since 1982, applies to the whole Southern Ocean south of the Antarctic Polar Front – an area of 32 million km^2 (see Fig. 8.9). The marine living resources involved in the Convention include all species in the Convention Area other than whales and seals, for which there were existing Conventions. The CCAMLR Convention was the first in the marine environment to try to combine the requirements of sustainable harvesting with adequate protection for non-target species potentially affected by harvesting. In fact, in three of its fundamental principles, it was foreshadowing, by at least a decade, the widespread adoption of the

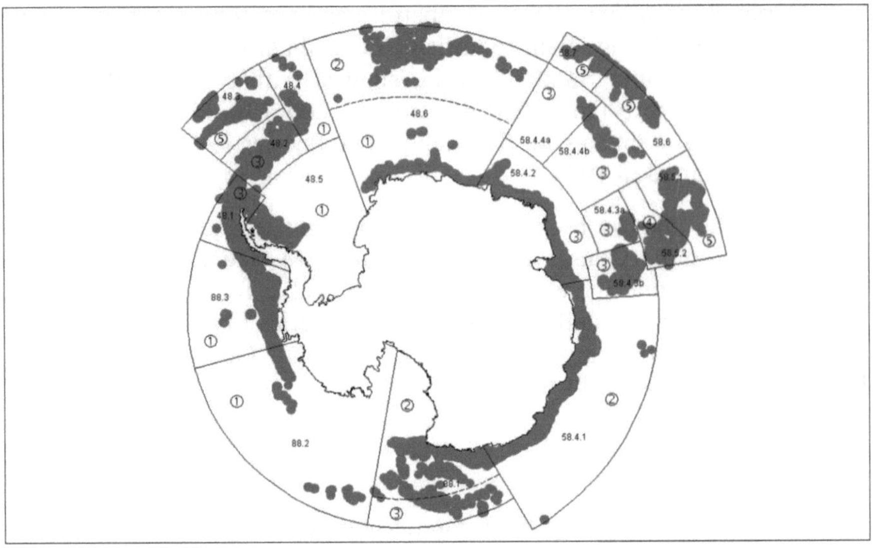

Fig. 8.9. Map of the CCAMLR Convention Area, showing its subdivision into statistical areas and the assessment of each of these for potential risk of interaction between seabirds, especially albatrosses, and longline fisheries. Key: 1: low; 2: medium to low; 3: average; 4: average to low; 5: high. Shaded areas represent seabed areas of depth between 500 and 1800 m, the principal fishing grounds for toothfish (source: CCAMLR 2004, Fig. 7.3)

precautionary principle and the need for ecosystem-based approaches to the management of marine systems. Thus, Article II of the CCAMLR Convention contains the requirements:

1. to balance the needs of sustainable harvesting with those of conservation; and
2. to provide protection for dependent and related species, coupled with the restoration of depleted stocks and populations;
3. to avoid changes that are potentially irreversible within two to three decades.

In 1991 the situation for CCAMLR was that, given the population dynamics of albatrosses, their population decreases were of a magnitude potentially irreversible within two to three decades. However, the main cause of these changes likely reflected events in adjacent waters that were not under the jurisdiction of CCAMLR. Nevertheless, by allowing longline fishing in the knowledge that potentially high levels of albatross by-catch were likely, CCAMLR was clearly not acting in the precautionary manner prescribed under its Convention.

8.8.3 Tackling the Problem

As a result of the above situation, CCAMLR, through the representatives of its 24 member states, at meetings of its Working Group on Fish Stock Assessment, Scientific Committee and Commission started the process of developing mechanisms for regulating by-catch in longline fisheries (including initially acquiring data and information to enable it to do this). A timetable indicating the evolution and development of this process is set out in Table 8.2.

Lest progress be thought to be exceptionally slow, it should be noted, first, that measures legally binding on all members of CCAMLR (e.g., as conservation measures) must be adopted by consensus. Second, once sufficient data

Table 8.2 Milestones in the development of effective mitigation measures to prevent seabird by-catch in longline fisheries in the CCAMLR Convention Area

1.	1982	CCAMLR Convention comes into force.
2.	1986	Reports of incidental mortality required.
3.	1989	Longline fishing for Patagonian toothfish starts (around South Georgia); incidental mortality becomes a CCAMLR agenda item.
4.	1990	First unofficial report of seabird by-catch; reporting forms on incidental mortality data and formats agreed as part of a Conservation Measure.
5.	1991	First direct observations of seabird by-catch; first Conservation Measure on mitigation of incidental mortality of seabirds.
6.	1993	Working Group on Incidental Mortality Associated with Longline Fishing established (first meeting in 1994).
7.	1993	International scientific observers required on all (four) vessels longline fishing in the South Georgia area.
8.	1994	First outreach materials to fishers and fishery managers and approaches to other RFMOs.
9.	1995	Closed season for longline fishing for toothfish (1 August to end February) to assist reducing incidental mortality of seabirds.
10.	1996	First adequate (though incomplete) scientific data (from international scientific observers) on seabird by-catch (from 3 of 16 vessels fishing).
11.	1997	Highest estimated seabird by-catch (> 6000 birds) in regulated fishery.
12.	1997	First comprehensive seabird by-catch risk assessment for different parts of the Convention Area.
13.	1997	Closed fishing season extended by 1 month (to 1 April), to protect seabirds until improved compliance with Conservation Measures.
14.	1998	Closed fishing season extended by two weeks (to 15 April).
15.	1999	First (autoline) vessels achieve full compliance with all mitigation measures for seabird by-catch (Conservation Measure 29).

Table 8.2 (cont.)

16.	1999	Closed fishing season extended to 1 May to protect seabirds until full compliance with relevant Conservation Measures.
17.	2001	First Spanish-system longline vessel achieves full compliance with all relevant mitigation measures (Conservation Measure 29).
18.	2002	First exemptions to night setting requirements (subject to seabird by-catch limit) for areas of lower risk for seabird by-catch.
19.	2003	Seabird by-catch in regulated fishery (except French EEZ) at record low (15 birds).
20.	2003	Half of vessels longline fishing comply with relevant mitigation measures for seabird by-catch (Conservation Measure 25-02 which replaced CM 29).
21.	2003	Important revision of mitigation measure requirements (Conservation Measure 25-02), incorporating use of integrated weight longline; additional exemptions agreed, subject to seabird by-catch level limits.
22.	2004	Seabird by-catch levels in French EEZ reduced by 75% following implementation of CCAMLR recommendations.
23.	2004	Unified system (for whole Convention Area) of mitigation requirements in relation to seabird by-catch risk.

were obtained to assess the magnitude of the problem, steady progress was made. Some of the main positive outcomes of the process set out in Table 8.2 are summarised below.

1. *By-catch reduced:* Once a full range of mitigation measures (Table 8.3), including a closed season, were imposed and monitored effectively, seabird by-catch numbers and rates at South Georgia (statistical subarea 48.3; see Fig. 8.8) were reduced ten-fold within a single year (Table 8.4).

Table 8.3 Principle types of mitigation measures implemented by CCAMLR.

Action	Rationale
No offal discharge	Avoid attracting birds
Streamer lines	Keep birds away from sinking longline
Weighted lines	Sink lines too fast for birds to access
Night setting	Albatrosses are diurnal
Closed seasons	Protect birds when breeding
Scientific observers on every vessel	

Table 8.4 Total estimated seabird by-catch and by-catch rate (birds per thousand hooks) in longline fisheries for toothfish *Dissostichus* spp. in the CCAMLR Convention Area (source: CCAMLR 2004).

Subarea	Year							
	1997	1998	1999	2000	2001	2002	2003	2004
South Georgia (Subarea 48.3)								
Estimated by-catch	5755	640	210	21	30	27	8	18
By-catch rate	0.23	0.032	0.013	0.002	0.002	0.0015	0.00003	0.001
Indian Ocean (Subarea 58.6, 58.7)								
Estimated by-catch	834	528	156	516	199	0	7	39
By-catch rate	0.52	0.194	0.034	0.046	0.018	0	0.003	0.025
Ross Sea (Subarea 88.1, 88.2)								
Estimated by-catch	-	0	0	0	0	0	0	0
By-catch rate	-	0	0	0	0	0	0	0.0001

1. *Improvements to mitigation:* Improved ability at using and managing the technical mitigation methods (streamer lines, line weighting, offal discharge) soon produced further by-catch reduction at South Georgia by ten-fold again over the next 2 years – with rates stabilising thereafter.

   Reductions were slower to achieve in the Indian Ocean (where closed seasons were not implemented) but, ultimately, similar proportionate reductions were achieved in the areas around the Prince Edward Islands (part of statistical subareas 58.6 and 58.7; see Fig. 8.9). Years of minor increases in by-catches (e.g., 2004) could be clearly associated with a drop in the standard of implementation of the technical mitigation measures (CCAMLR 2004).

   Even the recent massive by-catches of white-chinned petrels *Procellaria aequinoctialis* (about 25,000 birds over years 2002 and 2003 combined) in French-managed fisheries in the Indian Ocean proved susceptible to implementation of the technical mitigation measures used elsewhere, reducing by-catch by 75% in one season (2004) (Fig. 8.10). Clearly streamer lines, line weighting and associated best practice with discharge of offal can produce major improvements in by-catch quite independently of those achieved by closing areas to fishing during the breeding season of seabirds.

2. *Use of the precautionary approach – Seabird by-catch limits*: Management of seabird by-catch in CCAMLR's new and exploratory fisheries (i.e., starting longline fishing in a new statistical subarea or division) has been exemplary in terms of adopting a precautionary approach, particularly in defining by-catch risk levels and attendant

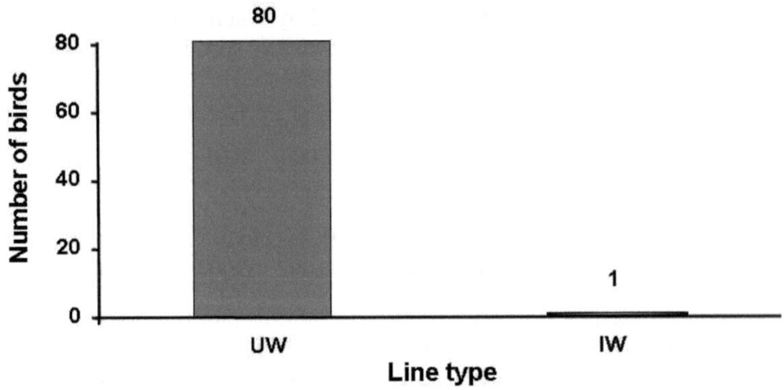

Fig. 8.10. Incidental mortality of white-chinned petrels in controlled experiments using unweighted (UW) and integrated weight longlines (IW) (source: Robertson et al. 2006)

area-specific mitigation requirements and management actions (see CCAMLR 2004, Table 7.17). So far all regulations have been strictly observed with no, or almost no, seabird by-catch whatsoever. Furthermore, regulated relaxation of mitigation requirements (subject to seabird by-catch limits) have also been entirely successful at avoiding by-catch.

3. *Adaptive management:* Mechanisms for the stepwise removal of some mitigation requirements (e.g., closed seasons), consequent on complete compliance with the necessary mitigation measures have been agreed and implementation has either commenced and/or the preconditions met. However, greater relaxation of these regulations (e.g., allowing longline fisheries to operate with technical measures alone in the highest risk by-catch areas during the main seabird breeding season), may prove to be quite challenging, especially for avoiding by-catch of white-chinned petrels and for operations involving the Spanish system of longline fishing.
4. *Easier methods for fishers:* Development of new methods which are easier and more effective for fishers to use (e.g., longlines with integrated weight) are enabling autoline vessels to fish with greater freedom and efficiency than hitherto.

8.8.4 Drivers and Obstacles

Here we summarise those factors which, in our opinion, had the greatest positive or negative effects on the speed of progress and success of outcomes in this case study. Several of them may still be powerful influences on future developments.

8.8.4.1 Positive Influences

1. Placement of independent scientific observers on vessels.
2. Creation of a formal working group which comprised all stakeholder constituencies – fishers, fishery managers, fishery scientists, technical experts, seabird biologists – to analyse and assess data and to provide advice. In CCAMLR, this was the working group on Incidental Mortality Associated with Fishing (IMAF)).
3. Collaborative research into practical solutions involving fishing companies and scientists and supported by governments.
4. High value of fishery so that the initial introduction of mitigation measures were neither disproportionately costly nor powerful disincentives to continue to participate in the fishery.
5. Relative geographical restriction of the toothfish fishery, which simplified management especially by coastal states around sub-Antarctic islands.
6. Vessel compliance with by-catch reduction measures and reporting requirements are fishery permit conditions.
7. Increasing recognition of the CCAMLR process and recommendations as 'role models' leading to the uptake of CCAMLR-style seabird avoidance measures in other parts of the world.

8.8.4.2 Negative Influences

1. Traditional commercial and operational secrecy at the start of a new fishery.
2. Remoteness of the region and resulting difficulty of policing in respect of Illegal, Unregulated and Unreported (IUU) fishing.
3. The Spanish system of longlining made simplifying mitigation measures (such as integrated weighting for autoliners) very difficult.
4. Lack of ability to test scientifically the contribution that each of the different mitigation measures makes to overall by-catch reduction. Consequent difficulty in proposing best practice combinations for new areas, circumstances, vessels, etc.
5. Closed seasons, although effective at reducing local by-catch rates, risk displacing fishing to other areas where management and mitigation may be much less effective.

8.8.5 Next Steps

The main challenges for CCAMLR within its Convention Area relate to: (i) further reducing seabird by-catch in the French Economic Exclusion Zone, and (ii) eliminating IUU fishing and its attendant by-catch.

However, now most by-catch of Convention Area seabirds occurs in adjacent regions. In this regard, CCAMLR needs to:

1. Collaborate with adjacent Regional Fishery Management Organisations (RFMOs), especially IOTC, ICCAT, CCSBT and the new Indian Ocean RFMOs, to ensure that seabird by-catch (especially of birds breeding in the Convention Area) is eliminated or minimised by the use of a suite of measures similar to those employed by CCAMLR.
2. Assist the development by such RFMOs of expert groups to advise on collection and analysis of by-catch data and on potential practical solutions to by-catch problems. Obtaining advice from, or participation of, experts with experience of the CCAMLR IMAF group could assist these RFMOs share and exchange information and assist with the transfer and uptake of the effective ways that CCAMLR has reduced seabird by-catch.
3. Work with relevant CCAMLR members to ensure that their vessels operating in high seas areas adjacent to the Convention Area are employing mitigation measures as effective as those required within the Convention Area.
4. Promote and assist the development of mitigation methods that operate effectively without comprehensive reporting, monitoring and compliance, such as further development and implementation of underwater setting devices and integrated line weighting.
5. For states into whose waters CCAMLR seabirds migrate (especially Argentina, Australia, Chile, New Zealand and South Africa), ensure that domestic legislation with respect to mitigation is as effective as that required by CCAMLR.
6. Work with relevant CCAMLR members to ensure that successful mitigation by their vessels of seabird by-catch in the CCAMLR area is complemented by equally successful mitigation by these and other vessels in their domestic fisheries (and, indeed, wherever their vessels participate in fisheries where there are risks of seabird by-catch).
7. Develop improvements to the Spanish system of longline fishing, particularly to enable simplification of the implementation of mitigation measures.
8. Reduce by-catch in appropriate parts of the French EEZ to levels comparable to the rest of the Convention Area.
9. Continue close monitoring of CCAMLR fisheries to ensure full compliance with conservation measures and prevent increases in by-catch (as seen in 2004).
10. Support and promote initiatives by industry, governments, and RFMOs to combat IUU.

8.8.6 Acknowledgements

We thank all our colleagues in the CCAMLR Working Group on Incidental Mortality Associated with Fishing for their exemplary collaboration over the last decade. We thank the many other individuals who have facilitated scientific and management research and action on this topic, particularly in the Southern Ocean.

8.9 Summary and Conclusions

(by Stephen J. Hall)

The stories told in the preceding case-studies in this chapter describe how changes in the behaviour of individuals and institutions have occurred in the face of various by-catch issues. It also describes how these changes have delivered conservation benefits that contributed to the long-term sustainability of fisheries. Each story is unique, and shows with varying degrees of emphasis some of the key factors that have led to successful outcomes. Behind this uniqueness, however, there are common threads that point to general lessons about how to get fishers to change their behaviour. The purpose of this section of this chapter is to draw those threads together.

Before drawing lessons about how change occurs among fishers and how to support it, it is worth considering how fishing differs from other industries. Its distinctiveness comes, not only from the technical peculiarities of fishing, but also from the socio-economic contexts in which fishing occurs. The public's empathy with some by-catch species, property and access rights regimes and the cultural perspectives associated with fishing are all important. These and other issues combine to make fishing different from other industries. While admitting these differences, however, there is a strong case for arguing that, when thinking about how change occurs, it is the similarities between these stories and those elsewhere that are most important. It is from those similarities that we can draw general lessons.

In drawing general lessons about change among fishers, however, we must also recognise the huge volume of literature on change management that fills the shelves of business school libraries and book shops. This literature provides a plethora of frameworks and models for describing and understanding change, each of which has strengths and weaknesses. To these I add my own, synthesised from multiple sources, which treats change from the perspective of those whose behaviour one is seeking to influence. Given this huge literature on change management, one could view anything newly written as merely packaging 'old wine in new bottles'.

I would argue, however, that context matters. Given the peculiarities of the fishing sector noted above, I hope that placing the key features of the case-studies in this chapter into a common framework will be informative and useful.

8.9.1 A Change Model

Drawing on the work of Kotter (1990), Senge (1990) and many others, the model for change developed here is built around three concepts. (i) where individuals, communities or institutions are on a continuum that reflects readiness for change; (ii) the leadership actions needed to cause change at each point on the continuum; and (iii) the drivers that deliver those leadership actions within the context of fisheries (Fig. 8.11).

In essence, the model argues that to get fishers to move from denial to commitment, one must first apply pressure for change by providing information. One must then find champions for change among fishers to help move the industry through the phase of resistance to one of exploration in which innovation and learning among fishers predominates. Accepted solutions developed during the innovation and learning period then become widely adopted through peer pressure mechanisms. These then take individuals beyond the exploration phase to commitment so that improved approaches become accepted practice.

Below I examine the validity of this model using the stories presented in this chapter.

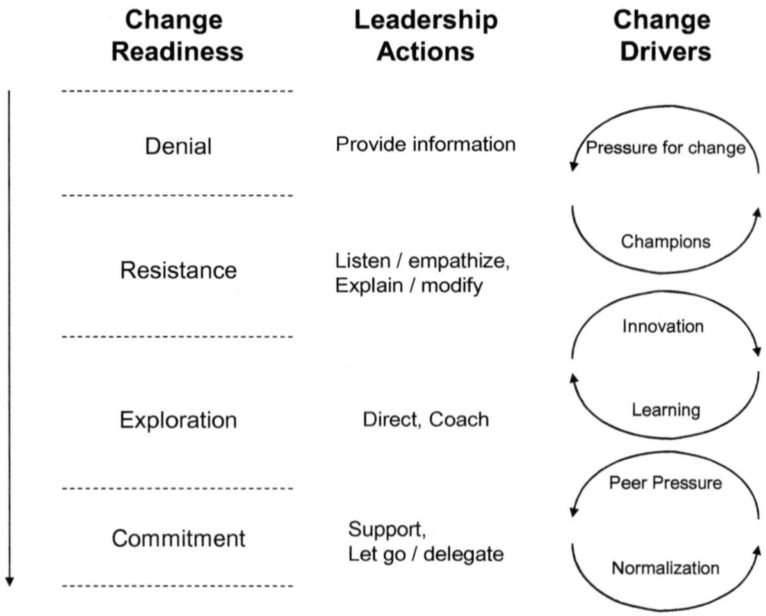

Fig. 8.11. A change model

8.9.1.1 Applying Pressure and Finding Champions

Pressure for change can, of course, come from various sources, but a common thread in many of the stories told here is how public opinion, framed by media attention, has catalysed action by fishers. The leadership action needed during the early phase of change is to provide information that will alter opinions – and no agent does this better than the popular press. Adverse media attention and the consequent socio-political pressure have been powerful stimuli for action. The gory videos of dolphin deaths in the tuna seines (Case-study 1) and the media attention to the death of 312 white chinned petrels by an auto-liner in New Zealand (Case-study 4) are good examples. Before fishers will start acting, someone usually needs to make a fuss.

But, although the press have proved important, it would be wrong to suggest that information provided by others does not also play a role. This role can be especially important when fishers' interpretation of personal experiences run contrary to the messages they are hearing elsewhere. An excellent example of this comes from Peckham et al. (Case-study 5) where fishers' perceptions from high turtle catches worked against the message that they were endangering populations. Here patient explanation about how turtles aggregate in fishing areas and information about the bigger picture had an important role to play in persuading fishers of the need for change. Combining this with information on individual turtle movements from satellite tracking, which no doubt created an emotional connection between fishers and the turtles, was especially powerful. But even when media attention is high, or when other information channels are effective, it almost always needs the initiative of key individuals to get fishers moving.

Champions usually need to emerge early in the story, even when threats to the fishery from litigation or market forces are clear. One important role that these individuals seem to play is in helping others understand that 'perception is often reality'. It often takes an insider to persuade others that, even if the media has distorted an issue, it is the public and government's perception of the truth that will affect their business. Although the most obvious examples of such champions in the case-studies tend to be men, one should not forget the role played by women. In their role as marketers of fish, they are often more aware of how markets can change with external pressure and can play a key influencing role (see Case-study 2).

As our model implies, finding champions is not only important for persuading others about the need for change. We also need them to lead the way in finding solutions. Thomas and Molloy (Case-study 4) are most clear on this matter: 'Picking respected and committed fishers as role-models to champion behaviour change is a cornerstone of the Southern

Seabird Solutions approach'. With such champions on-board the shift to a learning-and-innovation-cycle can take place.

8.9.1.2 Fostering Innovation and Learning

A fishing master developed tori poles to reduce seabird by-catch in the Japanese long-line fishery, as early as 1988. Tuna fishers devised new manoeuvres to avoid catching dolphins. Hawaiian long-liners helped refine hook technologies to reduce turtle capture. If there is one lesson above all others that comes from the chapters presented here, it is that we need to involve fishers in solving the technical problems of reducing by-catch. This is important for two reasons: first, technical solutions will get better faster; and second, engaging fishers in testing and refining innovations helps these innovations gain acceptance.

While accepting the value of developing a learning and innovation cycle within the fishery is non-controversial (see Hall and Mainprize 2004, for review), the best way to achieve it is unclear. In many respects, the best approach will depend on the particular setting. However, one general lesson is that researchers and extension workers often have a key catalytic role to play by promoting knowledge-sharing and stimulating learning and innovation. Engaging fishers in this cycle is especially important in remote areas where there is limited enforcement capacity and adopting new approaches rests solely in the hands of fishers (Peckham et al., Case-study 5).

One good example of the catalytic role played by extension agents is the 'skipper exchange program' described by Thomas and Molloy (Case-study 4). This approach provides an excellent vehicle, not only for sharing ideas and best-practice solutions, but also for recognising and rewarding champions. A testament to its power is that one visit from a New Zealand skipper led to the wholesale adoption of new weighted longlines by the Reunion fleet in Chile. Other models for sharing knowledge and stimulating innovation abound.

While the virtues of the learning and innovation cycle are clear, it is also important to recognise the costs of its absence. The crewing of Japanese vessels with foreign nationals illustrate this point (Nakano and Clarke, Case-study 3). Because foreigners are not seen as apprentices, the Japanese fishers do not pass on skills in by-catch mitigation. As a result, a culture of innovation does not develop. Strong social hierarchies within Japanese fishing communities also appear to inhibit both acceptance of the need for change and the learning and innovation needed to find solutions.

8.9.1.3 From Peer Pressure to Accepted Practice

As a cycle of learning and innovation develops and the number of fishers involved increases, it is reasonable to suppose that a sense of mutual accountability for improving should also develop. Recognition of the need for such accountability often seems to form early in the change process with the realisation that the poor performance of a few boats could affect everyone. Accepting such accountability usually appears much later, however, when there is general buy-in by the majority, and instruments are in place to monitor and report individual performance.

Such monitoring stimulates two change drivers, both of which amount to a form of peer pressure. The first is that it helps create internal pressures in competitive individuals who want to improve to be better than their peers. The key to tapping into that competitive streak is collecting data on performance and sharing those data in suitable ways. Experiences in the tuna fleets of the eastern Pacific point the way to best-practice in this regard. Fleet performance statistics are shared and there is private feedback on individual performance, combined with discussion and coaching on how to improve.

The second change driver comes from the external pressures imposed by peers who expect others to 'pull their weight'. This pressure can come in subtle (and possibly less subtle!) social interactions. It may then be a short step from here to creating legislative or management instruments such as individual or fleet by-catch quotas, or by-catch hot-spot reporting systems, that essentially serve to normalise procedures and make them recognised practice.

8.9.2 Conclusions

I hope the model described here provides a useful way to think about these case-studies and helps give further insights into how change in fisheries occurs. I also hope that this model, or something like it, will be used in discussions with the players involved in a fishery to help promote change. I base this hope on the premise the more fishers, NGOs, legislators, and others understand the bigger picture of what is happening, the greater their engagement and the more informed are their decisions.

One example of where discussion of the change model itself could be useful is for gaining agreement on where a particular fishery is on its journey and on the leadership actions that are most suitable for moving it forward. Getting agreement on this could, for example, avoid continued litigation or harping on the need for change when such actions could be counter-productive and slow or even halt progress. The situation faced by Seabird Solutions in New Zealand seems to be an example of this problem (Thomas and Molloy, Case-study 4). The success of such discussions depends of course on the key players having negotiable positions, something that is by no means guaranteed. This appears to be the problem in Hawaii, for example, where continued litigation to close the fishery completely is stalling the introduction of further improvements (Gilman et al., Case-study 6).

The stories presented in this chapter are highly varied and describe fisheries at different stages along a continuum of change. While it is tempting to highlight deficiencies, it is important to remember that the common theme running through all of them is success. To greater or lesser degrees, they describe initiatives that are reducing by-catch and making fisheries more ecologically sustainable. What is especially encouraging is some fisheries seem to have reached the normative stage, and have fully institutionalised a set of improved practices. For fisheries that have reached this stage, the time may well be right to loop back to start another round of improvement.

References

Barabasi A, (2002) The new science of networks. Perseus Books Group, United Kingdom, 280 pp

BirdLife International (2003) FAO's positive efforts to cut seabird deaths undermined by 14 'longline laggards'. BirdLife International, Cambridge, United Kingdom, 3 pp

Brothers N (1991) Albatross mortality and associated bait loss in the Japanese Longline Fishery in the Southern Ocean. Biol Conserv 55:255–268

Brothers NP, Cooper J, Lokkeborg S (1999a) The incidental catch of seabirds by longline fisheries: Worldwide Review and Technical Guidelines for Mitigation. FAO Fisheries Circular no. 937. Food and Agriculture Organization of the United Nations, Rome

CCAMLR (2004) Report of the twenty-third meeting of the Scientific Committee. CCAMLR, Hobart, Tasmania, Australia, 680 pp

Crowder LB, Crouse DT, Heppell SS, Martin TH (1994) Predicting the Impact of Turtle Excluder Devices on Loggerhead Sea-Turtle Populations. Ecol Appl 4:437–445

Croxall JP, Prince PA (1990) Recoveries of wandering albatrosses *Diomedea exulans* ringed at South Georgia 1958 – 1986. Ringing Migr 11:43–51

Croxall JP, Rothery P, Pickering SPC, Prince PA (1990) Reproductive performance, recruitment and survival of wandering albatrosses *Diomedea exulans* at Bird Island, South Georgia. J Anim Ecol 59:775–796

Dalziell J, de Poorter M (1993) Seabird mortality in longline fisheries around South Georgia. Polar Rec 29:143–145

Delgado SG (2005) Local perceptions and ocean conservation: A study of human consumption, exploitation, and conservation of endangered sea turtles in Baja

California Sur, México. Life Sciences Communication. University of Wisconsin at Madison, USA, 137 pp

FAO (2003) Implementation of the NPOA-Seabirds. Committee on Fisheries Progress Report from 24th Session on Implementation of the International Plan of Action for Reducing Incidental Catch of Seabirds in Longline Fisheries. United Nations Food and Agriculture Organization Committee on Fisheries, Rome, 3 pp

Gilman, E (2001) Integrated management approach to address incidental mortality of seabirds in longline fisheries. Aquat Conserv 11:391–414

Gilman E, Freifeld H (2003) Seabird mortality in north Pacific longline fisheries. Endangered Species Update 20:35–46

Gilman E, Carboneras C, Gandini P, Balogh G, Leadbitter D, Neves T (2002) Incentive instruments for sustainable bycatch of sensitive species in longline fisheries. In: Proceedings of the Incentives Session, Second International Fishers' Forum, 19–22 November 2002, Honolulu, Hawaii, USA. U.S. Western Pacific Regional Fishery Management Council, Honolulu, USA, 124 pp

Hall SJ, Mainprize B (2004) Towards ecosystem-based fisheries management. Fish Fish 5:1–20

Hall MA, Alverson DL, Metuzals KI (2000) Bycatch: Problems and solutions. Mar Pollut Bull 41:204–219

Jacobson SK (1999) Communication skills for conservation professionals. Island Press, Washington DC, USA, 351 pp

Jouventin P, Weimerskirch H (1990) Long-term changes in seabird and seal populations in the Southern Ocean. In: Kerry KR, Hempel G (ed) Antarctic ecosystems: Ecological change and conservation. Springer-Verlag, Berlin, pp 208–213

Kamezaki N, Matsuzawa Y, Abe O (2003) Loggerhead turtles nesting in Japan. In: Bolten AB, Witherington B (eds) Loggerhead Sea Turtles. Smithsonian Books, Washington DC, USA, pp 210–218

Kotter JP (1990) A force for change: How leadership differed from management. The Free Press, New York

MacKenzie-Mohr J, Smith W (1999) Fostering Sustainable Behavior. New Scoiety Publishers, Cabriola Island, British Columbia, Canada, 176 pp

Nichols WJ (2003) Biology and conservation of sea turtles in Baja California, Mexico. Doctoral dissertation. University of Arizona, Tucson, USA, 455 pp

Peckham SH, Nichols WJ, Dutton P, de la Toba V, Caballero-Aspe E, Salazar-Oropeza O (2004) Reducing bycatch of loggerhead turtles in coastal fisheries of the Baja California peninsula, Mexico. Twenty-Fourth Annual Symposium on Sea Turtle Biology and Conservation, San Jose, Costa Rica.

Pesenti C, Nichols WJ, Rangel-Acevedo R, Laudino-Santillan J, Montano-Medrano B, Lopez Castro MC, Peckham SH (2005) The Grupo Tortugero: Open networks as models for conservation. Poster presentation at the Twenty-Fifth Annual Symposium on Sea Turtle Biology and Conservation. Savannah, GA, USA. Available on-line at http://www.grupotortuguero.org/files/file/138Pesenti_etal_2005_ISTS.pdf

Polovina JJ, Howell E, Parker DM, Balazs GH (2003) Dive-depth distribution of loggerhead (*Carretta carretta*) and olive ridley (*Lepidochelys olivacea*) sea turtles in the central North Pacific: Might deep longline sets catch fewer turtles? Fish Bull 101:189–193

Prince PA, Rothery P, Croxall JP, Wood AG (1994) Population dynamics of black-browed and grey-headed albatrosses at Bird Island, South Georgia. Ibis 136:50–75

Robertson G, McNeill M, Smith N, Wienecke B, Candy S, Olivier F (2006) Fast sinking (integrated weight) longlines reduce mortality of white-chinned petrels (*Procellaria aequinoctialis*) and sooty shearwaters (*Puffinus griseus*) in demersal longline fisheries. Biol Conserv 132:458–471

Santora C (2003) Management of turtle bycatch: Can endangered species be protected while minimizing socioeconomic impacts? Coast Manag 31:423–434

Senge PM (1990) The Fifth Discipline: The art and practice of the learning organization. Random House, New York

Reviews: Methods and Technologies in Fish Biology and Fisheries

1. J.R. Sibert and J.L. Nielsen (eds.): *Electronic Tagging and Tracking in Marine Fisheries.* 2001 ISBN 1-4020-0125-8
2. D. Symes and J. Phillipson (eds.): *Inshore Fisheries Management.* 2001 ISBN 1-4020-0128-2
3. C.O. Cunningham (ed.): *Molecular Diagnosis of Salmonid Diseases.* 2002 ISBN 1-4020-0506-7
4. T.S. Gray (ed.): *Participation in Fisheries Governance.* 2005 ISBN 1-4020-3777-5
5. R.J. Baird: *Aspects of Illegal, Unreported and Unregulated Fishing in the Southern Ocean.* 2006 ISBN 1-4020-5338-X
6. T.M. Bert (ed.): *Ecological and Genetic Implications of Aquaculture Activities.* 2007 ISBN 978-1-4020-0884-9
7. S.J. Kennelly: *By-catch Reduction in the World's Fisheries.* 2007 ISBN 978-1-4020-6077-9

MIX
Papier aus verantwortungsvollen Quellen
Paper from responsible sources
FSC® C105338

If you have any concerns about our products,
you can contact us on
ProductSafety@springernature.com

In case Publisher is established outside the EU,
the EU authorized representative is:
Springer Nature Customer Service Center GmbH
Europaplatz 3, 69115 Heidelberg, Germany

Printed by Libri Plureos GmbH
in Hamburg, Germany